RISK-BASED DECISION MAKING IN WATER RESOURCES VI

Proceedings of the sixth conference
sponsored by the
Engineering Foundation

and co-sponsored by the
Universities Council on Water Resources
Task Committee on Risk Analysis and Management of the
Committee on Water Resources Planning of the
ASCE Water Resources Planning and Management Division

Supported by the
National Science Foundation
U.S. Army Corps of Engineers, Institute for Water Resources

Approved for publication by the
Water Resources Planning and Management Division
of the American Society of Civil Engineers

Santa Barbara, California
October 31 - November 5, 1993

Edited by Yacov Y. Haimes, David A. Moser, and Eugene Z. Stakhiv

Technical editing by Gail Hyder Wiley

Published by the
American Society of Civil Enginee
345 East 47th Street
New York, New York 10017-2398

ABSTRACT

This proceedings, *Risk-Based Decision Making in Water Resources VI,* contains edited papers presented at the sixth Engineering Foundation Conference held in Santa Barbara, California, October 31 - November 5, 1993. During this conference, the sense of the need for a stronger theoretical and methodological foundation to deal with the risk of extreme and catastrophic events remained high on the agenda of the participants. Many of the issues that emerged during the fifth conference, such as the possibility of global climate change, the safety of dams under extreme hydrologic conditions, the protection of lives and properties from catastrophic natural hazards, and the nationwide state of deterioration of our water resources, were addressed in diverse ways by conference participants, but much more work remains.

ISBN 0-7844-0032-6
ISSN: 1063-5076
Manufactured in the United States of America.

PREFACE

These proceedings of the sixth Engineering Foundation Conference on risk-based decisionmaking represent the natural growth of the field and the culmination of the theory, methodology, and applications in risk assessment and management of water resources and natural and man-made hazards. The most striking observation is that the objectives and goals set forth for the first conference in September 1980 in Asilomar, California, and the issues raised in the preface of the 1980 proceedings are as relevant and as timely today as they were then. The following is a sample of what appeared in the preface of the above proceedings.

The conference's goals and objectives were to:

1) Familiarize the participants with the state-of-the-art in risk/benefit analysis.
2) Explore the feasibility of using risk/benefit analysis in water resources planning and management.
3) Provide a medium conducive to the exchange of information on the conference theme among educations, analysts, managers, and policymakers.
4) Identify and articulate future desired actions designed to alleviate some of the present problems we face in risk/benefit analysis and risk assessment in general.

These proceedings attempt to provide some answers to myriad questions of risks and uncertainties that arise in the formation of public policy in general, and in water resources planning and management in particular. A sample of some generic questions is listed below:

a) What is the efficacy of models on risk and analysis and decision making?
b) To what extent are these models and methodologies credible?
c) To what degree does the value of information increase with more models on risk?

d) How important are the methodologies for the quantification of risk as part of the risk assessment process, and how critical is the understanding of the process itself?
e) Who should decide on acceptability of what risks, for whom, and in what terms, and why? (W. W. Lowrance)

The unescapable message here is that these conferences have been addressing epistemological issues that are in many respects timeless. In spite of the incredible advancement of our knowledge in risk-based decisionmaking in the decade since the Asilomar Conference, the professional community has continued to be as challenged today by these questions and investigations as it has been in the past.

In the preface to the proceedings of the second conference, which was held in November 1985 in Santa Barbara, California, we expanded on the theme of the 1980 conference. We wrote:

> The conference was geared toward balancing issues of practical concern as well as understanding some of the philosophical underpinnings and theoretical premises of risk analysis. ... The themes of concern to practicing engineers appeared to focus on:
> - legal liability issues, which are further dependent on
> - standards-based design approaches versus generic risk analysis methods
> - the role of individual versus collective choice and responsibility
> - the incorporation of subjective, professional judgments into highly deterministic analytical techniques
> - the usefulness of risk analysis as an aid to decisionmaking under uncertainty

In the preface to the third conference, which was held in November 1987 in Santa Barbara, California, several changes in the profession were noted, as follows:

> We are realizing and appreciating both the efficacy and also the limitations of mathematical tools and systemic analysis.
>
> We are able to acknowledge the fact that the ultimate utility of decision analysis, including risk-based decisionmaking, is not necessarily to articulate the *best* policy option, but rather to avoid the worst, the disastrous

policies—those actions in which the cure is worse than the disease.

We are realizing that viable risk management must be done within a multiobjective framework, where trade-offs can be explicitly evaluated to reflect social risk preferences, professional or expert opinion, and of course, the economic consequences of alternative courses of action.

In the preface to the fourth conference, which was held in October 1989 in Santa Barbara, California, we reflected on the increasingly growing popularity of risk-based decisionmaking. Societal and technolgical factors, we wrote, contribute to the increasing popularity of risk analysis. The reasoning follows:

> As a society, we are becoming increasingly more conscious of health and safety; we are becoming more protective of our environment; and we are becoming an increasingly litigious society. Our concern in assessing and managing risk has emerged as a natural by-product of these trends.
>
> Technology forms the basis for the second factor favoring the growth of risk analysis. The emergence of innovative and complex technology necessitates its better use, management, and control. We are realizing the imperative nature of the risk management of technological systems. Such risk management is and must be an integral part of technology management, just as it should be an integral part of good management in general.
>
> We believe that these two basic factors—the social and the technological—have forced us to realize that accounting for uncertainties and managing risks play dominant roles in every managerial decision. It is their realization that brings us to these Engineering Foundation Conferences—bringing professionals from many disciplines with diverse experiences together under one roof.

The preface to the proceedings of the fifth conference, held in November 1991 in Santa Barbara, California, reviewed the previous conference themes and stressed the growing need for a "stronger theoretical and methodological foundation to deal with risk of extreme and catastrophic events." Some of the issues that seemed to call for more concerted effort included the possibility of global climate change, the safety of dams under extreme hydrologic conditions, the protection of lives and properties from catastrophic natural hazards, the protection of ground water and surface water resources from major contamination, and the nationwide state of deterioration of our water resources and public works infrastructure.

During the sixth conference, in October–November 1993 in Santa Barbara, California, the sense of the need for a stronger theoretical and methodological foundation to deal with the risk of extreme and catastrophic events remained high on the agenda of the participants. Many of the issues, enumerated above, that emerged during the fifth conference were addressed in diverse ways by conference participants, but much more work remains.

During the opening of the conference, we again asked ourselves and the participants to use the five-day gathering for personal and professional growth. Jonathan Bulkley's notes from the first session articulate the conference's goals and objectives.

We hope that the interest in these Engineering Foundation Conferences remains as high as it has been in the past and we look forward to the seventh conference in 1995.

Several organizations and individuals were instrumental in making this conference possible. Dr. Eleonora Sabadell, Program Manager of the National Science Foundation Research Program, "Natural and Man-Made Hazards," for her essential support of this conference and for her encouragement. Kyle E. Schilling, Director of the U.S. Army Institute for Water Resources, was equally generous in providing financial support for the conference, as was Dr. Charles Freiman of the Engineering Foundation.

We thank, also, the ASCE Task Committee on Risk-Based Decision Making and the Universities Council on Water Resources for co-sponsoring this conference and providing the valuable resource base for a large number of eminent speakers and conference participants. Ultimately, the value of a conference such as this resides in the intangible nature of the ideas that were presented and debated and the influence that the discourse had on the perspectives of each of the participants. We hope that these perspectives are reflected in the proceedings of this conference.

All papers have been reviewed, edited, and accepted for publication in these proceedings by the editors. The papers are eligible for discussion in the *Journal of Water Resources Planning and Management* and also eligible for ASCE awards.

We finally acknowledge the invaluable editorial work provided by Gail Hyder Wiley; the administrative assistance provided by Sharon Gingras, Manager, Center for Risk Management of Engineering Systems, University of Virginia; the staff of the Engineering Foundation; and Shiela Menaker, Manager, Book Production, of the ASCE for her hard work in bring these proceedings to their final printed form.

Yacov Y. Haimes
Charlottesville, Virginia

David A. Moser and
Eugene Z. Stakhiv
Fort Belvoir, Virginia

CONTENTS

Federal Risk Management Policy: Where Are the Problems?

Anthony J. Thompson[1]

Abstract

Federal risk management policy involves both risk assessment and risk management elements. Risk assessment consists of hazard identification, dose-response assessment, exposure assessment, and risk characterization. Risk management, on the other hand, involves making policy decisions based on risk assessment results and social, economic, and political factors.

These remarks summarize the role of risk assessment and risk management in the regulatory, judicial, and legislative arenas. They then focus on the new direction for risk assessment policy as the government and industry grapple with risk management issues.

Introduction

Federal risk management policy involves both risk assessment and risk management elements. Risk assessment consists, to the extent possible, of an objective scientific assessment of the expected adverse health effects of exposure to potentially hazardous substances. There are four main components to risk assessment: (1) hazard identification, (2) dose-response assessment, (3) exposure assessment, and (4) risk characterization. Risk assessments are "not designed for making judgments but to illuminate them." (Wilson and Crouch 1982).

Risk management, as the National Research Council notes, "is the public process of deciding what to do where risk has been determined to exist." (National Research Council 1983). Risk management involves

---

[1]These remarks were given on November 1, 1993. On February 22, 1994, Mr. Thompson became a partner in the Environmental Law Group at Shaw, Pittman, Potts, and Trowbridge; 2300 N Street, NW; Washington, DC 20037

making policy decisions based on risk assessment numbers and social, economic, and political factors. These decisions directly affect the promulgation of federal and state regulations that apply to a wide range of facilities. For a risk manager to regulate effectively, there must be an understanding of what the risk is and how big it is. Risk management "is informed, but not determined by science" (Nichols and Zeckhauser 1986).

Regulatory Arena

Statutes

Each environmental statute has a slightly different "formula" that risk assessment must address. The following are several examples of these statutory risk assessment formulas.

- EPA/Clean Air Act (CCA) Pre-1990 "air toxics": "protect the public health" + "ample margin of safety" = emissions standards.

- EPA/Clean Water Act (CWA): "reasonably be anticipated to pose an unacceptable risk to human health or the environment" or "present an imminent and substantial danger to the public health or welfare" = levels at which to prevent discharge of pollutants.

- EPA/Safe Drinking Water Act (SDWA): prevent "known or anticipated adverse effects" + "adequate margin of safety" = Maximum Contaminant Level Goals (MCLGs); "as close as feasible" to MCLGs = Maximum Contaminant Level (MCL).

- EPA/Federal Insecticide, Fungicide and Rodenticide Act (FIFRA): "any unreasonable risk to human health or the environment" + economic, social, and environmental costs and benefits of the pesticide = prevent use of these pesticides.

- Occupational Safety and Health Act (OSHA)/ Mine Safety and Health Act (MSHA): "extent (technically and economically) feasible" + "best available evidence" + "no material impairment of health or functional capacity for working lifetime" = exposure standard for toxic materials.

Other regulatory activity

In recent years, there has been a good deal of activity at the federal level evaluating how risk assessments are used and viewed in the regulatory context. For example, in 1986, EPA issued carcinogen assessment guidelines that, while they did not call for realistic quantified risk assessments, did emphasize that "risk assessments will be conducted on a case-by-case basis, giving full consideration to all scientific information."

Several years later, in November 1992, EPA issued a "Draft Working Paper on Possible Revisions to Agency Guidelines for Making Cancer Risk Assessments." One of the major changes in the draft guidelines from the 1986 guidelines is an "increased emphasis on providing characterization discussions for each part of a risk assessment (hazard identification, dose-response, exposure, and risk assessments). These serve to summarize these assessment components, with emphasis on explaining the extent and weight of evidence, major points of interpretation and rationale, strengths and weaknesses of the evidence and analysis, and alternative conclusions that deserve serious consideration."

The Draft guidelines address the following four questions:

1. "Can the agent present a carcinogenic hazard to humans?

2. At what levels of exposure?

3. What are the conditions of human exposure?

4. What is the overall character of the risk, and how well do data support conclusions about the nature and extent of the risk?"

One of the most vocal of government entities on risk assessment issues has been EPA's Science Advisory Board (SAB). In its report on EPA's risk assessment methodology for radionuclides, the SAB states that "the general field of risk assessment is rapidly moving away from the practice of giving single, perhaps worst-case estimates, to that of providing best estimates along with a statement of the uncertainty of that best estimate" (SAB-PAC-93-014, July 1993). Recently, the SAB emphasized that "[q]uantitative uncertainty analysis should be an integral part of performing human health and ecological risk assessments for toxic chemicals, radionuclides, physical stressors, and biotic stressors. Uncertainties associated with both exposure and effects must be accounted for in risk assessments and subsequent risk management deci sions and communications." (SAB, Letter to Carol Browner, EPA Administrator, SAB-RAC-COM-93-006, July 23, 1993)

The SAB also has taken an active role in reviewing and commenting on the need for risk assessment in setting priorities. The SAB recommends that EPA "target its environmental protection on the basis of opportunities for the greatest risk reduction," and "reflect risk-based priorities in its budget process" (SAB-EC-90-021, September 1990).

Similarly, in its 1990–91 report titled "Regulatory Programs," the Office of Management and Budget (OMB) critiqued in detail federal risk assessment policies. The report notes that "conservatism in risk assessment distorts the regulatory priorities of the Federal Government, directing societal resources to reduce what are often trivial carcinogenic risks while failing to address more substantial threats to life and health." OMB recommends that risk assessments include scientifically credible value estimates, weight-of-evidence determinations, full disclosure, and the avoidance of perverse outcomes.

Several other agencies are also actively reviewing risk assessment issues. The Office of Science and Technology Policy has created two interagency committees under the Federal Coordinating Council on Science, Engineering and Technology to address the need to resolve different (and often contrary) risk assessment approaches. The National Academy of Sciences has created a Committee on Risk Assessment Methodology to review interagency risk assessment issues. EPA, the Department of Health and Human Services, and the Consumer Product Safety Commission (CPSC) are currently reviewing and revising their risk assessment guidelines. EPA, for example, has issued draft guidelines for neurotoxicity risk assessment.

Leaders within these agencies have become more aggressive in trying to delineate a consistent risk assessment policy. In February 1992, Henry Habicht, then the EPA deputy administrator, wrote a memorandum to all EPA assistant and regional administrators, providing guidance on risk characterization for risk managers and risk assessors. The memo emphasizes that (1) risk needs to be fully characterized, including statements of confidence about data and methods; (2) a basis needs to be given for greater consistency and comparability in risk assessments; and (3) professional scientific judgment needs to be recognized as playing an important role in the overall statement of risk.

In February 1993, William Farland, director of EPA's Office of Health and Environmental Assessment, wrote to the director of the

National Toxicology Program (NTP) regarding its hazard identification procedures, stating:

> Although it is reasonable to presume that animal carcinogens are human carcinogens, the position is rebuttable. Mechanistic and exposure information may modulate significantly the case for human carcinogencity. ... Without developing some appreciation of the uncertainties for human cancer potential for each chemical in the Report [NTP Annual Report on Carcinogens] the naive reader is forced to conclude that all listed substances have the same hazard potential. Certainly that is not the case ... *This Agency has learned over the years the importance of going beyond animal cancer studies in making hazard determinations, and the International Agency for Research on Cancer is beginning to recognize the significance of mechanistic information on their classification decisions. We urge the NTP to do likewise.* (Emphasis added.)

This concept is important because presently two positive animal studies can lead to a substance being listed as a "probable" human carcinogen by NTP, which in turn has regulatory consequences for that substance (i.e., kicks in OSHA "HAZCOM" requirements and Proposition 65 in California).

Judicial Arena—Court Decisions

The regulatory formulas set forth above have been affected by judicial decisions on various agency regulations. Most of the decisions are made by nonscientists (judges) who are trying to look at complex scientific and technical issues and apply common sense to their potential impact in the "real world." The following briefly discusses several of the key decisions.

Industrial Union Dept., AFL-CIO v. American Petroleum Institute, *448 U.S. 607 (1980) (the "Benzene" decision)*

In the Benzene decision, the Supreme Court rejected OSHA's generic carcinogen policy. The Court reasoned that it does not make sense to regulate a substance to the lowest possible level merely because "someone, somewhere might get cancer" without demonstrating that the chosen level would prevent a significant risk of cancer. The Court found that "'[s]afe' is not the equivalent of 'risk-free,'" and that "a workplace can hardly be considered 'unsafe' unless it threatens the workers with a significant risk of harm." Absolute scientific certainty, however, is not required. The Supreme Court held that "[s]o long as they are supported by a body of reputable scientific thought, the Agency is free to use conservative assumptions in interpreting the data with respect to carcinogens, risking error on the side of overprotection rather than underprotection."

Gulf South Insulation v. CPSC, *701 F.2d 1137 (5th Cir., 1983) (the* "Gulf South" *decision)*

The use of risk assessment in the regulatory context arose a few years later in the context of the CPSC in the Fifth Circuit *Gulf South* decision overturning the ban on the use of urea-formaldehyde foam insulation (UFFI) in schools and homes. The Fifth Circuit noted that "[t]he failure to quantify the risk at the exposure levels actually associated with UFFI is the finding's Achilles heel. Predicting how likely an injury is to occur, at least in general terms, is essential to a determination of whether the risk of that injury is *unreasonable*." (Emphasis added.) Among the problems with CPSC's analysis were "(1) reliable data were not used and insufficient testing had been conducted; (2) the methodology used by the CPSC failed to indicate the likelihood that adverse health effects would occur; and (3) the estimated threshold exposure level of harm had no relationship to the real world characteristics of an average home."

Natural Resources Defense Council v. EPA, *824 F.2d 1146 (D.C. Cir., 1987) (the "Vinyl Chloride" decision)*

The risk issue appeared again later in the context of the Clean Air Act Section 112 "air toxics" provisions in the so-called Vinyl Chloride decision, the D.C. Circuit found that "Congress' use of the word 'safety,' moreover, is significant evidence that it did not intend to require the Administrator to prohibit all emissions of non-threshold pollutants." The court held that "the Administrator's decision does not require a finding that 'safe' means 'risk-free,' or a finding that the determination is free from uncertainty. Instead, we find only that the Administrator's decision must be based upon an expert judgment with regard to the level of emission that will result in an 'acceptable' risk to health. In this regard, the Administrator must determine what inferences should be drawn from available scientific data and decide *what risks are acceptable in the world in which we live.*"—i.e., a "reality" check. (Emphasis added.)

AFL-CIO v. OSHA, *965 F.2d 962 (11th Cir., 1992) (the "Air Contaminants" decision)*

More recently, these ideas were reemphasized in the Eleventh Circuit's Air Contaminants decision. The Eleventh Circuit, in vacating the threshold limits for more than four hundred chemical substances, held that "the agency 'has no duty to calculate the exact probability of harm' or 'to support its finding that a significant risk exists with anything approaching scientific certainty.' However, OSHA must provide at least an estimate of the actual risk associated with a particular toxic substance and explain in an understandable way why that risk is significant."

The court noted that OSHA "requires some assessment of the level at which significant risk of harm is eliminated or substantially reduced." The court further advised that OSHA may not "base a finding of significant risk at lower levels of exposure on unsupported assumptions using evidence of health impairments at significantly higher levels of exposure."

Legislative Arena

Risk assessment principles and policies are also beginning to emerge in legislation. The following looks at the Clean Air Act Amendments of 1990 and several bills pending before Congress.

Clean Air Act—1990 Amendments (Public Law 101-541, 104 Stat. 2399, November 15, 1990)

Numerous sections of the Clean Air Act 1990 Amendments involve risk assessment issues. For example, under Section 112(o), Congress directed EPA and the National Academy of Sciences to conduct a review of EPA's risk assessment methodology for determining carcinogenic risks from hazardous air pollutants.

Under Section 303 of the 1990 Amendments, Congress created a Risk Assessment and Management Commission. The commission's charge is to "make a full investigation of the policy implications and appropriate uses of risk assessment and risk management in regulatory programs under various federal laws to prevent cancer and other chronic human health effects." Although the commission was created under the Clean Air Act, its charter is not limited to either Clean Air Act issues or to statutes falling within EPA's jurisdiction. Rather, it encom passes all federal statutes that protect against the risks of cancer and other chronic effects from hazardous substances.

In addition, a number of different areas under the Clean Air Act will require some form of risk analysis. To illustrate:

1. Under Section 112(c)(9)(B)(i), EPA may only delete a pollutant from the list of hazardous air pollutants if it determines that the substance poses a cancer risk of less than one in one million to the maximally exposed individual.

2. Under Section 112(f), EPA is to report to Congress on methods for assessing the risk to the public that remains after the application of maximum available control technology (MACT). EPA is then to make recommendations regarding legislation to reduce those risks.

3. Under Section 112(k), EPA is required to identify at least thirty hazardous air pollutants emitted from area sources that present the greatest threat to public health in urban areas.

The Chafee-Lautenberg Amendment [to EPA's FY 1993 Appropriation Bill, Pub. L. No. 102-398, 106 STAT 1618 (1992)]

This amendment requires EPA to "conduct a risk assessment of *radon* considering: (a) the risk of adverse human health effects associated with exposure to various pathways of radon; (b) the costs of controlling or mitigating exposure to radon; and (c) the costs for radon control or mitigation experienced by households and communities, including the costs experienced by small communities as the result of such regulations. Such an evaluation shall consider the risks posed by the treatment or disposal of any water produced by water treatment. *The Science Advisory Board shall review the Agency's study* and submit a recommendation to the Administrator on its findings."

The SAB, in reviewing EPA's study on radon, stated:

> The SAB strongly supports the use of a relative risk reduction orientation as an important consideration in making risk reduction decisions on all sources of risk, including those attributable to radon ... [T]he relative risk approach calls for giving the highest priority to mitigating the largest sources of risk first, especially when the cost-effectiveness of risk reduction of such sources is high. The SAB recognizes

> that the large number of laws under which EPA operates makes it difficult to implement a relative risk reduction strategy uniformly across the Agency. Radon is an excellent example of the problem with radon in drinking water governed under one statute (Safe Drinking Water Act) while radon in indoor air is not currently subject to regulation under a specific statute. *The SAB strongly encourages the Agency and the Congress to work together to consider changes in existing statutes that would permit implementation of relative risk reduction strategies in a more efficient and effective manner* (SAB, Letter to Carol Browner, EPA Administrator, EPA-SAB-EC-LTR-93-010, July 30, 1993). (Emphasis added.)

The Environmental Risk Reduction Act (S. 110)

This bill, sponsored by Senator Daniel Patrick Moynihan (D-N.Y.), emphasizes that scientific research, and not political pressure, should be the basis for government risk assessments. The bill provides that the ability to reduce risks requires

(a) accurate, quantitative estimates of the exposure to humans and ecosystems to all important risk factors;
(b) accurate techniques for predicting the effects of such exposures;
(c) an adequate understanding of technical, economic, social, and legal alternatives to reduce exposure to risk factors; and
(d) accurate estimates of the costs and benefits of alternatives for reducing risks.

The bill provides for the creation of two expert advisory committees. The first committee, Committee on Relative Risks, would provide advice on ranking of relative risks. The second committee, Committee on Environmental Benefits, would provide advice on estimating quantitative benefits of reducing risk. The bill also provides for risk assessment research, an interagency panel on risk assessment and reduction, and reports to Congress.

The bill also would require the EPA administrator to develop risk assessment guidelines to ensure consistency in risk assessments by setting forth minimum standards. The bill would require that at a minimum the guidelines include

- hazard identification—is exposure causally linked to effect?

- an assessment that measures or estimates exposure of well-defined populations
- an assessment determining the magnitude of response of affected populations to different exposure levels to a stressor *under representative environmental conditions*
- risk characterization—description of and probable effects from alternative risk management options (including no action) and a quantitative estimate of the uncertainties

EPA Cabinet Bill (S.171)

Senator Bennett Johnston (D-La.) has sponsored an amendment to S.171—the bill elevating EPA to a cabinet position—that would require risk assessment and cost-benefit analysis for EPA regulations. The Johnston amendment provides that in promulgating final rules, EPA must publish in the *Federal Register:*

> 1. an estimate, performed with as much specificity as practicable, of the risk to the health and safety of individual members of the public addressed by the regulation and its affect [sic] on human health or the environment and the costs associated with implementation of, and compliance with, the regulation;
>
> 2. a comparative analysis of the risk addressed by the regulation relative to other risks to which the public is exposed.

In addition, the secretary must certify that the regulation will substantially protect human health and the environment. If the secretary cannot make the required certification, the secretary must explain the reasons in the final regulation. Representative Henry Waxman (D-Calif.), as well as several environmental groups, oppose the Johnston amendment.

The Risk Communication Act of 1993—(H.R. 2910)

This bill requires federal agencies to disclose the assumptions on risks and techniques used in making regulatory assessments. It is intended to encourage "credible science." The bill sets forth principles for risk assessment and a framework for presenting risk assessment information. For example, the bill requires the EPA administrator to "explain the range of exposure scenarios used in any risk assessment, and, to the extent feasible, provide a statement of the size of the corresponding pop ulation at risk and the likelihood of such exposure scenarios," and to "provide appropriate comparisons with estimates of other risks, including those that are familiar to and routinely encountered by the public."

Toward a New Direction

The early 1970s

In the 1970s, many people believed that there would be a cancer "balloon" from environmental exposure to hazardous substances generated by industrialization and, therefore, that control of environmental carcinogens would reduce cancers in the U.S. by 60% to 70%. Regulatory attention, therefore, turned to control of exposures to carcinogens. For example, EPA established the Carcinogen Assessment Group and an Interagency Regulatory Liaison Group, comprising EPA, FDA, OSHA, CPSC, and DOA, was formed to write a report on an interagency consensus of carcinogenic risk assessment. In essence, the government was starting to develop its policy on potential carcinogens from scratch.

As a result, in the early years, EPA relied on a series of the most conservative assumptions at every juncture of the "risk" analysis. EPA adopted the linear nonthreshold model (from the Atomic Energy Commission). All humans were assumed to be equally susceptible to cancer, despite demonstrable variability, and animals were assumed to be surrogates for humans. All tumors (even benign ones) were presumed to be malignant. Estimated cancers were assumed to be real increases in cancer. Carcinogens were assumed to be independent actors, even though there were apparent interactions among various substances. Thus, regulatory risk "assessment" was essentially a process marked by a series of default assumptions. These assumptions produced extremely conservative numbers greatly inflated in terms of actual risk—but perhaps not unreasonable in the existing context.

This default assumption approach in turn was reflected in major regulatory policies of several federal agencies. For example:

1. OSHA's Carcinogen Policy (1980): Wherever a carcinogen is involved, there is no safe level of exposure absent clear proof to the contrary and, therefore, controls can be imposed to levels as low as feasible.

2. EPA's proposed Airborne Carcinogen Policy (1979): Based on the same nonthreshold assumption of causation that OSHA used: There is no safe level of risk for a carcinogen, so if there

is public exposure, the limit on public exposure must be taken as low as it can go. (EPA never published a final policy.)

These "generic" approaches were designed to avoid the potential for protracted regulatory proceedings addressing potential carcinogenic impacts.

Change

Changes in risk assessment/risk management approaches are necessary. Estimates of the probability and severity of risk need to be directed at "real world" conditions, not, as has often been done, exclusively on the basis of hypothetical isolated situations that bear little or no resemblance to actual conditions.

Specifically, there should be increased attention on non-point sources of pollution. The source of an estimated 85% of aggregate health effects from hazardous air pollutants is small area sources, including emissions from small commercial establishments, residences, and motor vehicles.

"Cancer phobia" and other misperceptions, moreover, must be addressed if we are to prioritize societal efforts adequately. The cancer "balloon" never happened. Doll and Peto estimate "environmentally caused cancers" to be about 2%. It is now understood that the relationship between pollution and cancer is much weaker than originally perceived, and that, given existing regulatory controls, the relationship between pollution from point sources and cancer probably is even less significant now.

Procedures that have served to evaluate carcinogens no longer provide, in and of themselves, an adequate framework for evaluating the wide range of biological information that is available; nor do they provide regulators and the public with the best available information. For example, University of California at Berkeley biologist Bruce Ames notes the public is generally unaware that "Americans eat an estimated 1,500 milligrams of *natural* pesticides per day which is about 10,000 times more than they consume of synthetic pesticide residues." The difficulty in bringing about change, however, is that existing programs have developed lives of their own.

Indeed, risk assessment tools have moved way beyond the early conservative default assumption approach of the 1970s. There are a variety of tools available for effective risk assessment and management.

Tools to evaluate risk factors include epidemiology, animal studies, and *in vitro* studies. Modern source term assessment provides more sophisticated measurement and modeling techniques. Realistic population exposure scenarios and validated model data or actual measurements regarding transport of contaminants from source to receptor can be used to provide a more accurate assessment of the real risk *"in the world in which we live."* (Emphasis added.) For example, it is not reasonable to rely upon the hypothetical man living at the point of highest exposure for 24 hours per day, 365 days per year for 70 years because, as EPA admits, it would be "unusual for a person to live a lifetime at the same location," and it is unlikely a facility will operate continuously at the same level for that period of time. Therefore, the probability of a substantial number of individuals actually incurring these "maximum calculated risks is small" (EPA, FEIS 1983). Indeed, the probability of *any* person incurring these risks is virtually nonexistent.

There is also a need today to understand the nature of the contaminant and likely pathology of exposure: Is it from airborne? dermal? ingestion? all of those? What is the dose distribution? We need to assess dose and dose response, the intake and uptake rates, and pharmacokinetic assessments. Risk assessments should also use best estimates of risk with upper and lower bounds of uncertainty to avoid a "deterministic" approach to risk assessment (i.e., the use of single numbers or point estimates as input to risk assessments) and move to "probability distribution analysis."

Policymakers also need to consider the proper role of negative data. For example, the SAB, in commenting on the risk of radon in drinking water recently stated: "[E]ven at the upper bound of the uncertainty analysis for ingested radon, for most situations the risk from radon ingested in drinking water is still much lower than the risk from airborne radon entering the house directly from the soil. Indeed, for many homes the risk from the radon in water is even lower than that from radon in the outdoor air" (SAB, Letter to Carol Browner, EPA Administrator, EPA-SAB-EC-LTR-93-010, July 30, 1993).

The quality of underlying data also should be evaluated and there should be a periodic review of the scientific basis for assumptions. Risk assessment should be an "iterative" process.

Risk assessment cannot afford "intellectual" exercises by the expert scientific community. The expert scientific community must recognize the potential impacts of their actions. For example, the National Research Council 1988 study by the Committee on Biological Effects of Ionizing Radiations (BEIR IV) failed to look at the quality of the informa tion from underlying cohorts, several of which were seriously flawed. EPA, in turn, relied on BEIR IV in setting the radionuclides in drinking water standard, thereby proposing a standard based on faulty assumptions. Similarly, NTP's hazard identification listing which can be based on the "strength of the evidence" of two animal studies has a regulatory impact (OSHA's hazard communication standard, Prop 65), as well as potential competitive and "toxic tort" litigation implications. (See Current hazard identification programs: Potential societal and regulatory consequences. *Toxics Law Reporter*. September 8, 1993, pages 417–20.)

It is worth noting that in toxic tort litigation, the legal "formula" that risk assessment must address is different than in the regulatory context. In litigation, the test the court and jury apply is whether the specific substance at issue was more likely than not to have caused the plaintiff's particular cancer. Listings by NTP, International Agency for Research on Cancer (IARC), and other regulatory determinations at most demonstrate some more or less reliable "statistical" basis for estimating increased risk. These listings and regulations, however, often are used to provide a "colorable indicia" of a cause and effect relationship and, thereby, to avoid dismissal in a toxic tort suit, (i.e., to get to the jury solely on the basis of such findings).

The Supreme Court recently held in *Daubert v. Merrell Dow Pharmaceuticals, Inc.*, 113 S. Ct. 2786 (1993) that scientific testimony is admissible under the Federal Rules of Evidence based on the judge's "preliminary assessment of whether the reasoning or methodology underlying the testimony is scientifically valid and of whether that reasoning or methodology properly can be applied to the facts in issue." In making this assessment, the Court advises judges to consider whether the scientific theory or technique can be (and has been) tested, whether it has been subjected to peer review and publication, the known or potential rate of error, and whether it has been generally accepted in the scientific community. While none of these factors are dispositive, they play a role in determining the admissibility of scientific evidence at trial. The Court's decision is likely to have an impact on the regulatory process as well.

Change is needed to address regulatory priorities better. Poor allocation of resources results in trading lives for lives. As former EPA Administrator William Reilly has stated: "We need to communicate more clearly about risk and about choice, about consequences and about costs. ... As we move forward, at a time when new money is hard to find, the premium will be on clear thinking and frank communication about difficult trade-offs." Similarly, former EPA Administrator William

Ruckelshaus has stated that the preferred risk assessment approach is to adopt "as policy the full disclosure of the risks involved in regulatory decisions."

Recently, the American Industrial Health Council (AIHC) has asked the Clinton Administration to coordinate risk assessment at the federal level by appointing one senior government official to oversee all risk assessment activities. Ronald Lang, president of AIHC, notes that "[b]oth Congress and the Administration will be making critical decisions over the next year or two, and the danger is that these decisions will be based upon 20-year-old risk assessment methods, with the result that the government's attention and its funds will be terribly misdirected." AIHC emphasizes the need to base regulatory decisions on laboratory, epidemiological data, and real-world exposures.

Communicating risk assessment

It is imperative that risk assessments be as objective as possible and clearly presented to the regulators, employees, and the public in terms that are *understandable* and therefore *meaningful*. Communication to employees and the public must be presented in terms a nonscientist can understand. For example, "1×10^{-6}" does not mean much to the average citizen, whereas the statement that "nobody is likely to be harmed" does. Information must be related to other information the public may already know and understand or by analogies that are understandable.

Comparative risk information can be used. For example, (a) the risk of being struck by lightning is 5×10^{-7} and that means the risk of cancer from a certain substance is x% smaller than the risk of death from a lightning strike; or, (b) one part per trillion is like one drop of vermouth in a martini made with twenty-five million gallons of gin.

In addition, scientists need to demand that their work not be misused and to explain *what* scientific tools they are using *and why*—(e.g., explain statistical risk estimates, the meaning of confidence levels, and default assumptions). This means coming down from the "ivory tower." If information is misused by special interests, (e.g., the alar apple panic—an environmental representative called it the "most potent carcinogen of all pesticides") or by regulators (e.g., misuse of NTP reports by regulatory agencies), the scientific community needs to step up and help to clarify matters. Better information needs to go into the regulatory process so that better decisions come out. In the final analysis, scientists need to get involved in the regulatory arena *to understand the context in which their work is being used.*

Kenneth Olden, director of National Institute of Environmental Health Services and NTP, in a speech before the 1992 Annual Meeting of the Society of Risk Analysis, noted that

> [t]o improve the risk assessment process, there needs to be better and more substantial interactions between the National Toxicology Program, the regulatory agencies and private industry. NTP, EPA, FDA, and industrial scientists involved in toxicological testing and research must collaborate in generating the mechanistic information necessary to improve risk assessment. Regulatory and advisory standards must be communicated as interim decisions based on the best data available at the time. *We should tell the Public exactly what the uncertainties are, and signal our intent to reconsider the decision as the uncertainties are eliminated through further research.* (Emphasis added.)

The SAB recently has also stated that "[r]egulatory action must be based on realistic estimates of risk and these require a full disclosure of uncertainty. The disclosure of uncertainty enables the scientific reviewer, as well as the decision-maker, to evaluate the degree of confidence that one should have in the risk assessment" (SAB-PAC-93-014, July 1993).

Some within the environmental community and perhaps organized labor will resist more state-of-the-art and, often, less conservative risk assessment methodologies since these methodologies could raise questions about the validity of some of their cherished regulatory programs and goals. Some have argued the following points.

1. Qualitative Risk Assessment (QRA), rather than being a scientific tool for objective and sound health policy, has been a "political solution to conservative arguments against environmental regulation ... it gives the appearance of 'scientific objectivity.'"

2. It is, however, nothing more than a sophisticated form of the "dilution" solution to pollution.

3. As such, QRA sets limits for discharges rather than determining if *any* discharge should be allowed (Ginsburg 1992).

Others within the environmental community have taken positions with respect to risk assessments that are so extreme that even state-of-the-art "best estimates" with upper and lower bounds of uncertainty could expose a source of potentially hazardous contamination to regulatory shutdown. For example, the Natural Resources Defense Council (NRDC) has suggested (1984) that government should respond to even a proposed single estimated "statistical death" with the same urgency the police would exhibit to a "tip that a dangerous person has threatened to shoot randomly into a Times Square crowd until he kills one person" (NRDC Comments 1984).

Conclusion

A conservative approach to risk assessment might have been justified in the past, when the scientific community feared a major impact of toxic pollutants on the cancer rate, and when tools to assess the precise extent of the potential danger were by today's standards underdeveloped. Under such circumstances it might have been prudent to err routinely on the side of over-regulation.

Such an approach can no longer be defended. It is clear that the potential carcinogenic health impact of toxic pollutants is smaller than initially predicted. Beyond that, the *science* of risk assessment has improved dramatically in recent years, and there is a growing consensus about its capabilities and how they can be *prudently used* to look at potential carcinogenic and noncarcinogenic health threats. The net result of these developments is that risk assessment today must take full advantage of the best data and methodologies to produce the *most accurate, and not just the most conservative,* explanation of public health threats posed by a particular substance, so that risk management decisions can be properly informed.

References

Doll, R., and R. Peto. 1981. *The causes of cancer*. New York: Oxford University Press.

Environmental Protection Agency. 1979. Proposed airborne carcinogen policy. 44 *Federal Register* 58660, 10 October.

Environmental Protection Agency. 1983. *Final environmental impact statement for standards for the control of byproduct materials from uranium ore processing (FEIS)*. Vol. I and II, September.

Environmental Protection Agency. 1992. Draft working paper on possible revisions to agency guidelines for making cancer risk assessments. EPA/600/AP-92/003, November.

Farland, W., EPA Director, Office of Health and Environment. 1993. Letter to the director of the National Toxicology Program, February.

Ginsburg, R. 1992. Quantitative risk assessment: The illusion of safety. *Facets of Groundwater*, September.

Habicht, H., EPA Deputy Administrator. 1992. Memorandum to all EPA assistant and regional administration on risk characterization, February.

National Research Council. 1983. *Risk assessment in the federal government: Managing the process*. Washington, D.C.: National Academy Press.

National Research Council. 1988. *Health risks of radon and other internally deposited alpha-emitters, BEIR IV*. Washington, D.C.: National Academy Press.

Natural Resource Defense Council. 1984. Comments re: Proposed withdrawal and proposed standards for benzene emissions.

Nichols, A., and R. Zeckhauser. 1986. The perils of prudence: How conservative risk assessments distort regulation. *Regulation*, November/December.

Office of Management and Budget. *1990-91 regulatory programs.*

Olden, K., Director of NIEHS and NTP. 1992. Environmental health science research and human risk assessment. Remarks given at the Society of Risk Analysis 1992 Annual Meeting.

Science Advisory Board. 1990. *Reducing risk: Setting priorities and strategies for environmental protection,* September. SAB-EC-90-021.

Science Advisory Board. 1993. *An SAB report: Multi-media risk assessment for radon,* July. EPA-SAB-PAC-93-014.

Wilson, R., and E. Crouch. 1982. *Risk/benefit analysis*. Cambridge, Mass.: Ballinger Publishing, Co.

When and How Can You Specify a Probability Distribution When You Don't Know Much?

Yacov Y. Haimes,[1] Timothy Barry,[2] James H. Lambert[3]
Editors
Special Issue of *Risk Analysis: An International Journal,*
Pending Publication in 1994

Abstract

This paper presents an excerpt from the proceedings of the above-named workshop, and gives an overview of both the workshop and the proceedings. Five papers with diverse and complementary themes were reviewed and revised for publication.

Foreword

On January 25, 1993, nine members of the EPA/UVA Workshop's Steering Committee met in Washington, D.C., to formulate the agenda for the workshop and to identify topical areas for white papers. During the meeting some twenty-four basic questions were generated for consideration during the workshop. We subsequently grouped these questions into five white-paper topics:

- Dealing with quality assurance, skepticism, peer review, and closeness to reality
- Tails and extreme events
- Uncertainty versus variability

[1]Director, Center for Risk Management of Engineering Systems, University of Virginia, Charlottesville, VA 22903
[2]Chief, Science Policy Integration Branch, U.S. Environmental Protection Agency, 401 M Street, SW, Washington, DC 20460
[3]Ph.D. Candidate, Center for Risk Management of Engineering Systems, University of Virginia, Charlottesville, VA 22903

- What are the bases for our choices and when should we use probabilities?
- Uncertainty analysis for decision making

Some of these questions belong to more than one group, and we decided that some overlapping was advisable. On the basis of the recommendation made by the Steering Committee and prospective contributors, five white papers were commissioned. The contributors were asked to address the following topics:

A. Relevance to Workshop: Relate explicitly the themes of your white paper to the selection of probability distributions for Monte Carlo analysis in EPA problems.

B. Proposed Guidelines: Generate "straw man" guidelines for the selection of probability distributions that will be useful across a broad scope of regulatory problems. The guidelines resulting from the five white papers will be combined into a model that will lead to the expected results of the workshop.

C. Opposing View: Articulate and discuss the significance to the workshop of points of view that oppose or contradict those adopted in your paper.

D. State of the Art: Relate the state of the art of uncertainty analysis to your discussion.

The five commissioned papers and a number of others that came unsolicited from the participants were distributed in advance of the workshop to focus and stimulate the discussions.

Five papers with diverse and complementary themes were reviewed and revised for publication in this special issue. Hoffman and Hammonds, in "Propagation of Uncertainty in Risk Assessments: The Need to Distinguish Between Uncertainty Due to Lack of Knowledge and Uncertainty Due to Variability," suggest a pragmatic and rational direction to assessing variabilities along with lack of knowledge: one begins by deciding the endpoint of the assessment, then proceeds with an appropriate two-dimensional strategy. In a theoretical style, albeit just as grounded in the risk assessment experiences of EPA and others, Hattis and Burmaster, in "Assessment of Variability and Uncertainty Distributions for Practical Risk Analysis," show that variabilities and knowledge uncertainties, while both amenable to probabilistic descriptions, require different approaches to the assessment of distributions. Lambert, Matalas, Ling, Haimes, and Li, in "Selection of Probability Distributions in Characterizing Risk of Extreme Events," demonstrate that considering the tails of a variable's distribution and assessing the risk of extreme events present unique challenges and opportunities. Rowe, in "Understanding Uncertainty," gives a challenging framework for characterizing sources of uncertainty in measurements. Lastly, Finkel, in "Stepping Out of Your Own Shadow: A Didactic Example of How Uncertainty Can Inform and Improve Decision Making," gives a persuasive, elegant tutorial example in part to demonstrate that developing better tools are only an intermediate end for risk assessors--rather, the crucial issues are the quality of our societal judgment and its impact.

Background

Much of the environmental legislation administered by the federal government, while not specifically calling for assessments of risk, requires the agencies to set targets, or goals, that require *risk-like* analyses. In addition to legislative requirements, federal agencies are subject to a number of executive orders (e.g., Executive Order 12291) and federal policies that require them to conduct regulatory analyses. These analyses generally involve examinations of health and environmental effects, economic impacts, energy impacts, technical feasibility, barriers to implementation, and alternatives to direct regulation.

In the early years, the focus of regulatory analysis was on economic impacts and on fairly gross measures of environmental improvements, such as the number of tons of pollutant removed from the environment. Over the last several years, however, the use of quantitative risk assessment within the federal government and state governments has increased significantly. Risk assessment has become a powerful tool for supporting and shaping regulatory decisions, often equal to economic analysis in its usefulness. At its best, risk assessment adds a useful dimension by providing the decision maker with estimates of the public health and environmental benefits or effectiveness associated with regulatory alternatives. In concept, risk assessment provides common measures of what a regulation is expected to achieve, thus providing a level playing field to compare options critically and to ensure relative quality and consistency within and across programs.

New Risk Guidance at EPA

EPA has initiated a concerted effort to hold its regulatory analysis and research to the highest standards possible and practical. As part of this effort, EPA, on February 26, 1992, issued a policy titled *Guidance on Risk Characterization for Risk Managers and Risk Assessors,* along with technical guidance document titled *Guidance for Risk Assessment.* This policy is, in part, a response to a study conducted by the agency's senior risk managers. The study found that to improve risk assessments, agency staff should (1) present a full and complete picture of the environmental risks analyzed, (2) strive for greater consistency and comparability across risk assessments and across agency programs, and (3) recognize the critical role played by professional judgment in performing and interpreting risk assessments.

In August 1990, the Office of Management and Budget publicly criticized the way in which federal agencies, most notably the EPA, have implemented risk assessment and made a number of specific recommendations for improvement:

1. A renewed effort must be made to separate science from policy.

2. Decision makers should be provided with expected, unbiased risks, estimates of uncertainty, and outer ranges of potential risk.

3. Conservative risk management decisions and applications of margins of safety should only be made explicitly in the final regulatory decision.

4. Studies showing positive relationships should not outweigh those showing no relationship.

5. Regulatory decisions that increase net risk by eliminating one risk while increasing a more important risk elsewhere must be carefully avoided.

Problem Statement: The Analytic Challenge

To present a full and complete picture of risk, including a statement of confidence, the uncertainties within the risk assessments must be explicitly addressed. Sometimes, but rarely, environmental risk analysts have the luxury of extensive data to describe the quantitative relationships and their uncertainties. More often, however, data are very limited. How quantitative should the analysis be, both in estimating risks and representing their uncertainties, when data are limited? If a quantitative approach is selected, under which circumstances are Monte Carlo or other analysis methods appropriate? How should probability distribution functions be selected in these circumstances?

Workshop's Objectives

The objectives of this workshop were twofold:

1. To assess the state of the art in selecting input distributions with emphasis on their applications to environmental risks.

2. To establish theoretically sound and defensible foundations upon which to generate, in the future, a set of guidelines for EPA concerning the selection of probability distributions.

This second objective can be restated in terms of the following question:

When and how can you generate, specify, and use more defensible probability distributions?

Although this question can be addressed generically, special emphasis was placed on its relevance to the characterization, assessment, and quantification of environmental risk. Risk is defined as a measure of the probability and severity of adverse effects. Furthermore, evaluating the central tendency alone would not suffice. The workshop also addressed extreme events through upper bounds and/or other mechanisms, such as the conditional expectation of risk.

Organization and Format

The proceedings are edited notes generated by the workshop's participants through the use of nominal group technique and idea generation. The participants were divided into three groups; each group addressed a *Foundations* theme during two sessions and a *Synthesis* theme during another two sessions. With the conclusion of the three *Foundations* groups, the participants were re-organized into three new *Synthesis* groups. These six themes are correspondingly organized in these proceedings into two parts. Each group was assigned a facilitator and a rapporteur, who are listed here with the six groups. A listing of the participants in each group appears elsewhere in these proceedings.

Part 1: Foundations

A. Value of Uncertainty Analysis to Decision Making
Elisabeth Pate-Cornell (Facilitator), Nicholas Matalas (Rapporteur)

B. Uncertainty and Variability in Assuring the Quality of Uncertainty Analysis
William Rowe (Facilitator), Ronald Whitfield (Rapporteur)

C. Principles of Selecting Probability Distributions for Uncertainty Analysis
Owen Hoffman (Facilitator), Steve Bartell (Rapporteur)

Part 2. Synthesis

D. TOP-DOWN VIEW: Monte Carlo Analysis for the Needs of Decision Making
Adam Finkel (Facilitator), Dale Hattis (Rapporteur)

E. BOTTOM-UP VIEW: Recommended Tools for Selecting Distributions and Monte Carlo Analysis
Alison Taylor (Facilitator), David Burmaster (Rapporteur)

F. FORWARD VIEW: Plan of Action to Improve Monte Carlo Analysis
Richard Schwing (Facilitator), Stan Kaplan (Rapporteur)

These proceedings constitute a grassroots chronicle of the workshop's agenda from expert theoreticians, practitioners, academics, and other professionals with interest in the subject of uncertainties, variabilities, and risks. The text represents not only a technical document, but also a record of the organizational culture of the state of risk analysis in environmental regulatory decision making. Indeed, these proceedings reflect the integration and synthesis of a logical exploration of concepts and ideas of concern to the workshop's participants.

We agree with Stan Kaplan that in the edited notes there is a lack of consensus on the meaning of the term *Monte Carlo analysis*. In fact, the term is used in these proceedings, as it is elsewhere, with two very different meanings:

- Usage 1. *Monte Carlo* is a specific computational technique for doing calculations involving probability curves. As such, it is only one computational technique among many; it has it pros and cons, its areas of superiority and inferiority. However, it is just a mechanical technique for calculation and has no deeper philosophical substance or significance.
- Usage 2. In this usage, *Monte Carlo* is being used as a stand-in or synonym for "probabilistic analysis" itself. In this case, it has all the

very deep and profound philosophical and practical decision making, public communication, and trust-telling significance that probabilistic analysis has.

We encourage the readers of these proceedings to be alert for these two usages of *Monte Carlo analysis.*

Far from being of a single mind on the issues of uncertainty and variability in Monte Carlo analyses, the various subgroup sessions were as often contentious and questioning as they were conclusive, and it is appropriate that the edited notes reflect the spirit of these debates. The results from the sessions summarized in these proceedings of the EPA/UVA Worshop do not represent the opinions of the facilitators, rapporteurs, nor the proceedings editors. Indeed, there are statements in the edited notes with which individual facilitators, rapporteurs, and editors flatly disagree. The invaluable services of the facilitators and rapporteurs in managing the conference sessions and subsequently in guiding the editing process are much appreciated and applauded.

Workshop's Tasks

1. Review the various existing approaches to selecting distributions and assess the state of the art with emphasis on the applications to environmental risk assessment. Special issues to be discussed include the problems of poor and scant data; prediction of rare events; Bayesian priors and distribution selection; and empirical distributions, entropy, truncation, rejection of extreme variates, correlated inputs, efficient sampling designs.

2. Establish foundations upon which to generate a set of EPA guidelines concerning the selection of probability distributions. Special issues to be discussed include the criteria used to establish the guidelines; general rules of thumb, do's and don'ts; and a prioritized research agenda.

Methods for Risk and Reliability Analysis

Larry W. Mays[1]

Abstract

Recently, there has been considerable emphasis on the state of decay in the nation's infrastructure because of its importance to society's needs and industrial growth. Water distribution systems are one of the many kinds of infrastructure systems amenable to higher levels of serviceability. Conventional design methods often fail to recognize, analyze, and account for, systematically, the effects of the various uncertainties that prevail in the design and operation of water distribution systems. Because of the existence of design and operation uncertainties, water distribution systems have an associated risk or probability of failure and an associated reliability or probability of not failing. The definition of failure can take on several meanings for a water distribution system, ranging from not meeting required pressure heads to the mechanical failure of a pipe or component. For water distribution systems, there are no universally accepted definitions for risk or reliability. The purpose of this paper is to review methods for risk and reliability evaluation of water distribution systems.

Introduction

The urban water distribution system comprises three major components: pumping stations, distribution storage, and distribution piping. These components may be further divided into subcomponents that can in turn be divided into sub-subcomponents as shown in figure 1. For example, the pumping station component consists of structural, electrical, piping, and pumping unit subcomponents. The pumping unit can be further divided into sub-subcomponents: pump, driver, controls, power transmission, and piping and valves. The exact definition of

---

[1]Chair and Professor, Department of Civil Engineering, Arizona State University, Tempe, AZ 85287-5306

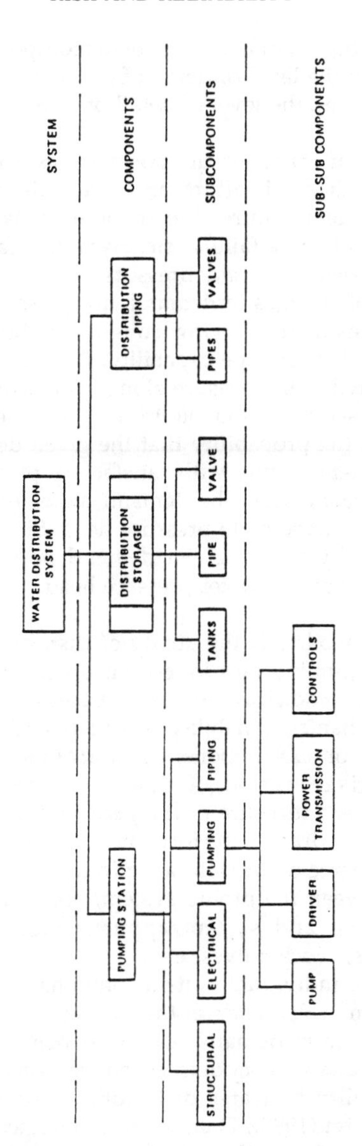

Figure 1. Hierarchical Relationship of System, Components, Subcomponents, and Sub-subcomponents for a Water Distribution System

components, subcomponents, and sub-subcomponents is somewhat fluid and depends on the level of detail of the required analysis and, to a somewhat greater extent, the level of detail of available data.

The reliability of water distribution systems is concerned with two types of failure: mechanical failure and hydraulic failure. Mechanical failure considers system failure due to pipe breakage, pump failure, power outages, control valve failure, etc. Hydraulic failure considers system failure due to demands and pressure heads being exceeded that could be the result of changes in demand and pressure head, inadequate pipe sizes, old pipes with varying roughness, insufficient pumping capacity, and insufficient storage capability. Because either the mechanical measure or the hydraulic measure alone is inadequate to measure the system reliability, it seems reasonable to unify this definition by specifying the reliability as the probability that the given demand nodes in the system receive sufficient supply with satisfactory pressure head. In other words, the failure occurs when the demand nodes receive either insufficient flowrate and/or inadequate pressure head. Similarly, a nodal reliability is the probability that a given demand node receives sufficient water flowrate with adequate water pressure head.

Mechanical reliability is the ability of distribution system components to provide continuing and long-term operation without the need for frequent repairs, modifications, or replacement of components or subcomponents. Mechanical reliability is usually defined as the probability that a component or subcomponent performs its mission within specified limits for a given period of time in a specified environment. Hydraulic reliability is a measure of the performance of the water distribution system. The hydraulic performance of the distribution system depends to a great degree on the following factors:

- interaction between the piping system, distribution storage, distribution pumping, and system appurtenances, such as pressure reducing valves, check valves, etc.
- reliability of the individual system components
- spatial variation of demands in the system
- temporal variation in demands on the system

Network reliability analysis models based on considering mechanical failure in the reliability of water distribution networks include Mays (1989), Hobbs and Beim (1988); Duan (1988); Quimpo and Shamsi (1987); Mays and Cullinane (1986); Wagner, Shamir, and Marks (1986, 1988a, 1988b); Duan and Mays (1987); and Tung (1985). The reliability of water systems due to the hydraulic failure resulting from mechanical failure was considered by Su et al. (1987) and Cullinane, Lansey, and Mays (1992) in an optimization model. None of these previous works actually quantifies a system reliability.

This paper focuses on new methodologies for water distribution system reliability; reliability analysis of pumping systems; and reliability-based optimization models for water distribution systems. The objective of this paper is to present the theoretical background for new methodologies. The publication by Mays (1989) also presents these methodologies along with others, including detailed example calculations and applications. The reader is referred to the following publications for a complete understanding of these methodologies: Bao and Mays (1990); Coals and Goulter (1985); Fujiwara and Tung (1991); Goulter and Bouchart (1987); Goulter and Coals (1986); Hobbs, Beim, and Gleit (1987); Kettler and Goulter (1985); Lansey and Mays (1987, 1989); Lansey et al. (1989); Mays and Cullinane (1986); Mays, Duan, and Su (1986); Mays (1989); Mays et al. (1989); Ormsbee and Kessler (1990); Quimpo and Shamsi (1987); Su et al. (1987); Tung (1985); Tung et al. (1987); and Woodburn, Lansey, and Mays (1987); Goulter and Kazemi (1988, 1989); Yen and Tung (1993); and Goulter, Davidson, and Jacobs (1993).

Methods for Reliability Analysis of Water Distribution System Components

Reliability concepts

The analysis of reliability and availability requires an understanding of some basic terms, which are defined in this section. The concepts represented by these terms will be used in later sections to quantify reliability and availability. The common thread in the analysis of reliability and availability is the selection of an appropriate failure density function. Failure density functions are used to model a variety of reliability-associated events, including time to failure and time to repair.

The reliability R(t) of a component is defined as the probability that the component experiences no failures during the time interval (o,t) from time zero to time t, given that it is new or repaired at time zero. In other words, the reliability is the probability that the time to failure T exceeds t, or

$$R(t) = \int_{t}^{\infty} f(t)\,dt \tag{1}$$

where f(t) is the probability density function of the time to failure. Values for R(t) range between 0 and 1. The probability density function f(t) may be developed from equipment failure data, using various statistical methods. In many cases, a simple exponential distribution is found appropriate. The unreliability F(t) of a component is defined as the probability that the component will fail by time t. Unreliability can be defined mathematically as

$$F(t) = \int_0^t f(t)\,dt = 1 - R(t) \tag{2}$$

The failure rate m(t) is the probability that a component experiences a failure per unit of time t given that the component was operating at time zero and has survived to time t. Note that the failure rate m(t) is a conditional probability. The relationship of m(t) to f(t) and F(t) is given as

$$m(t) = \frac{f(t)}{R(t)} \tag{3}$$

Sometimes the failure rate is called the hazard function. The quantity m(t)dt is the probability that a component fails during time (t, t + dt). Values for m(t)dt range from 0 to 1. Given the failure rate, the failure density function and the component reliability can be obtained as equations (4) and (5), respectively (Kapur and Lamberson 1977).

$$f(t) = m(t)\exp\left[-\int_0^t m(h)\,dh\right] \tag{4}$$

$$R(t) = \exp\left[-\int_0^t m(h)\,dh\right] \tag{5}$$

Time-to-failure analysis

Because the time to failure of a component is not certain, it is always desirable to have some idea of the expected life of the component under investigation. Furthermore, for a repairable component, the time required to repair the failed component might also be uncertain. This section briefly describes and defines some of the useful terminology in the field of reliability theory that is relevant in the reliability assessment of water distribution systems.

The mean time to failure (MTTF) is the expected value of the time to failure, stated mathematically as

$$MTTF = \int_{0}^{\infty} t\, f(t)\, dt \tag{6}$$

which is expressed in hours.

Similar to the failure density function, the repair density function, g(t), describes the random characteristics of the time required to repair a failed component when failure occurs at time zero. The probability of repair, G(t), is the probability that the component repair is completed before time t, given that the component failed at time zero. Note that the repair process starts with a failure at time zero and ends at the completion of the repair at time t.

Similar to the failure rate, the repair rate r(t) is the probability that the component is repaired per unit time t given that the component failed at time zero and is still not repaired at time t. The quantity r(t)dt is the probability that a component is repaired during time (t, t + dt) given that the components failure occurred at time t. The relation between repair rate, repair density, and repair probability function is

$$r(t) = \frac{g(t)}{G(t)} \tag{7}$$

Given a repair rate function r(t), the repair density function and the repair probability are, respectively,

$$g(t) = r(t) \exp\left[-\int_{0}^{t} r(h)\, dh \right] \tag{8}$$

$$G(t) = 1 - \exp\left[-\int_0^t r(h)dh\right] \tag{9}$$

The mean time to repair (MTTR) is the expected value of the time to repair a failed component. The MTTR is defined mathematically as

$$\text{MTTR} = \int_0^\infty t\,g(t)dt \tag{10}$$

where g(t) is the probability density function for the repair time.

The mean time between failures (MTBF) is the expected value of the time between two consecutive failures. For a repairable component, the MTBF is defined mathematically as

$$\text{MTBF} = \text{MTTF} + \text{MTTR} \tag{11}$$

The mean time between repairs (MTBR) is the expected value of the time between two consecutive repairs and equals the MTBF.

Availability and unavailability concepts

The reliability of a component is a measure of the probability that the component would be continuously functional without interruption through the entire period (o,t). This measure is appropriate if a component is nonrepairable and has to be discarded when the component fails. However, many of the components in a water distribution system are generally repairable and can be put back in service again. In that situation, a measure that has a broader meaning than that of the reliability is needed.

The availability A(t) of a component is the probability that the component is in operating condition at time t, given that the component was as good as new at time zero. The reliability generally differs from the availability because reliability requires the continuation of the operational state over the whole interval (0,t). Subcomponents contribute to the availability A(t) but not to the reliability R(t) if the subcomponent that failed before time t is repaired and is then operational at time t. As a result, the availability A(t) is always larger than or equal to the reliability

R(t), i.e., A(t) ≥ R(t). For a nonrepairable component, it is operational at time t, if and only if, it has been operational to time t, i.e., A(t) = R(t). As shown in figure 2, the availability of a nonrepairable component decreases to zero as t becomes larger, whereas the availability of a repairable component converges to a nonzero positive number.

The unavailability U(t) at time t is the probability that a component is in the failed state at time t, given that it started in the operational state at time zero. In general, the U(t) is less than or equal to the unreliability F(t); for nonrepairable components they are equal. A component is either in the operational state or in the failed state at time t, therefore,

$$A(t) + U(t) = 1 \tag{12}$$

Using exponential failure and repair density functions, the resulting failure rate μ and repair rate η, according to the definitions given previously, are constants equal to their respective parameters. For a constant failure rate and a constant repair rate, the analysis of the whole process can be simplified to analytical solutions. Henley and Kumamoto (1981) use LaPlace transforms to derive the unavailability as

$$U(t) = \frac{\mu}{\mu + \eta}\left[1 - e^{-(\mu+\eta)t}\right] \tag{13}$$

and the availability

$$A(t) = 1 - U(t) = \frac{\eta}{\mu + \eta} + \frac{\mu}{\mu + \eta}\; e^{-(\mu+\eta)t} \tag{14}$$

The steady state or stationary unavailability U(•) and the stationary availability A(•) for i approaches ∞ are, respectively,

$$U(\infty) = \frac{\mu}{\mu + \eta} = \frac{MTTR}{MTTF + MTTR} \tag{15}$$

and

$$A(\infty) = \frac{\eta}{\mu + \eta} = \frac{MTTF}{MTTF + MTTR} \tag{16}$$

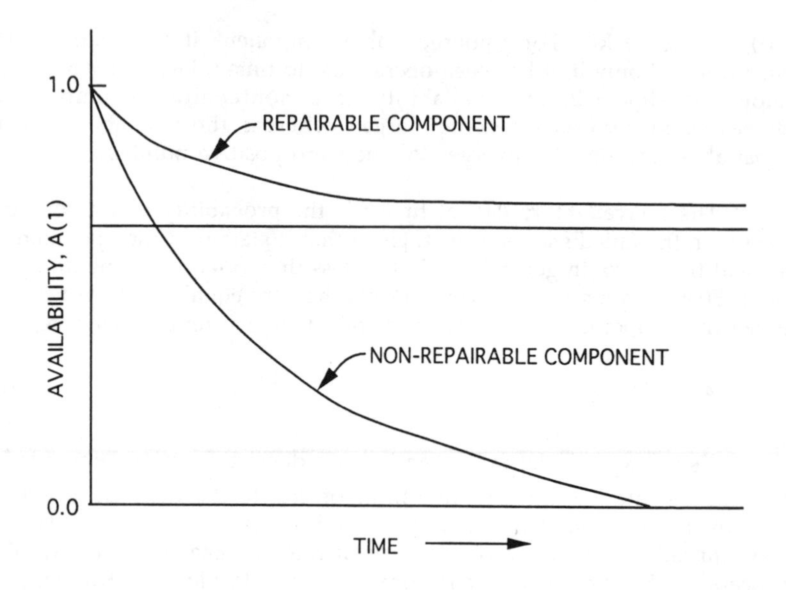

Figure 2. Availability for Repairable and Nonrepairable Components

As time gets larger, the steady state (or stationary) unavailability and availability for the pump can be calculated. The following relation is also true.

$$\frac{U(t)}{U(\infty)} = 1 - e^{-(\mu+\eta)t} \tag{17}$$

Model for Water Distribution System Reliability

The reliability of a water distribution system can be defined as the probability that the system will provide demanded flowrate at a required pressure head. Due to the random nature of pipes roughness, water demands, and required pressure heads, the estimation of water distribution system reliability is subject to uncertainty. Bao and Mays (1990) presented a methodology by which to estimate the nodal and system reliabilities of a distribution system accounting for such uncertainty using Monte Carlo simulation.

The hydraulic uncertainty is considered by treating the demand, pressure head, and pipe roughness as random variables. Assuming that the randomness of water demand (Q_d) and the pipe roughness coefficient (C) follows a probability distribution, a random number generator is used to generate the values of Q_d for each node and C for each pipe. For each set of values of Q_d and C generated, a hydraulic network simulator is used to compute the pressure heads at the demand nodes, provided that the demands are satisfied. The required pressure head (H_d) at given nodes can be treated as constant with both lower and upper bounds or as a random variable. The corresponding nodal and system hydraulic reliabilities are then computed.

The framework for the methodology is based on a Monte Carlo simulation consisting of three major components: random number generation, hydraulic simulation, and computation of reliability. The random number generator is the core of the methodology and is used to generate values of the random variables of demand (Q_d), pressure head (H_d), and the Hazen-Williams coefficient for pipe roughness (C). For each set of values of Q_d, H_d, and/or C generated, the University of Kentucky hydraulic simulation model (KYPIPE) is used to determine pressure heads for the nodes throughout the water distribution system. After a certain number of iterations, the nodal or system reliability is computed.

The nodal reliability (R_n) is the probability that a given node receives sufficient flowrate at the required pressure head. So theoretically the nodal reliability is a joint probability of water flowrate and pressure head being satisfied at the given nodes. However, it is difficult mathematically to derive and compute this joint probability. For instance, the flowrate and pressure head at a node are not independent. The approach used herein is to compute the conditional probability in terms of pressure head provided that the water demand has been satisfied or vice versa. This approach assumes that the water demand is satisfied (Q_s = Q_d). The nodal reliability can be defined as the probability that the supplied pressure head (H_s) at the given node is greater than or equal to the minimum required pressure head (H_d^l),

$$R_n = P(H_s > H_d^l \mid Q_s = Q_d) = \int_0^{\infty} f_s(H_s) \left[\int_0^{H_s} f_{dl}(H_d^l)\, dH_d^l \right] dH_s \quad (18)$$

Alternatively, both lower and upper bounds of required pressure head, H_d^l and H_d^u can be considered. In this case the nodal reliability is the probability that the supplied pressure head (H_s) at a given node is greater than or equal to the minimum required pressure head H_d^l and less than or equal to the maximum required pressure head (H_d^u),

$$R_n = P(H_d^u \geq H_s > H_d^l \mid Q_s = Q_d)$$

$$= \int_0^{\infty} f_s(H_s) \left[\left(\int_0^{H_s} f_{dl}(H_d^l)\, dH_d^l \right) \left(1 - \int_0^{H_s} f_{du}(H_d^u)\, dH_d^u \right) \right] dH_s \quad (19)$$

where $f_s(\ )$ represents the probability density functions of supply, and $f_{dl}(\ )$, $f_{du}(\ )$ represent the minimum and maximum requirements of pressure head at a given node, respectively. If the required pressure head (H_d) is considered as a constant with lower bound (H_d^l), the nodal reliability is given, respectively, below

$$R_n = P(H_s > H_d^l) = \int_{H_d^l}^{\infty} f_s(H_s)\, dH_s \quad (20)$$

or with lower and upper bounds, (H_d^l) and (H_d^u), the nodal reliability is

$$R_n = P(H_d^u \geq H_s \geq H_d^l) = \int_{H_d^l}^{H_d^u} f_s(H_s)\, dH_s \quad (21)$$

Although the nodal reliabilities depict a fairly complete reliability measure of the water distribution system, it is also convenient to use a single index such as "system reliability" to represent the composite effect of the nodal reliabilities. Such an index is difficult to define because of the dependence of the computed nodal reliabilities. Three heuristic definitions of the system reliability are considered (Cullinane 1989).

The system reliability (R_{sm}) could be defined as the minimum nodal reliability in the system,

$$R_{sm} = \min \left\{ R_{ni} \right\}, \; i = 1,2,\ldots,I \tag{22}$$

where R_{ni} is the nodal reliability at node i, and I is the number of demand nodes of interest. Another system reliability could be the arithmetic mean (R_{sa}), the mean of all nodal reliabilities,

$$R_{sa} = \frac{\sum_{i=1}^{I} R_{ni}}{I} \tag{23}$$

A third approach could be to define the system reliability as a weighted average (R_{sw}), which is a weighted mean of all nodal reliabilities weighted by the water supply at the node,

$$R_{sw} = \frac{\sum_{i=1}^{I} R_{ni} \bar{Q}_{si}}{\sum_{i=1}^{I} \bar{Q}_{si}} \tag{24}$$

where $\bar{Q}_{si}$ is the mean value of water supply at node i.

Algorithm

The procedure to evaluate the nodal and system reliability of a water distribution for hydraulic failure is illustrated in the flowchart in figure 3. The algorithm includes the following basic steps:

1. Assign distributions to Q_d, H_d, and/or C.

2. Generate Q_d, H_d, and/or C using Monte Carlo simulation.

3. Compute H_s at all nodes using the hydraulic simulator KYPIPE, assuming Q_d is satisfied ($Q_s = Q_d$).

4. Compute nodal and system reliabilities.

Reliability Analysis

A new methodology for the reliability analysis of pumping stations for water supply systems was developed by Duan and Mays (1990) that considers both mechanical failure and hydraulic failure. The reliability methodology models the available capacity of a pumping station as a continuous-time Markov process, using bivariate analysis and conditional probability approaches in a frequency and duration analysis framework. A supply model, a demand model, and a margin model are developed that are used to compute the expected duration of a failure, expected unserved demand of a failure, expected number of failures in the period of study, expected total duration of failures in the period of study, and expected total unserved demand in the period of study.

The frequency and duration analysis (FD), referred to herein as the FD approach, allows derivation of various reliability indices and is well suited for analyzing the reliability performance of a pumping system. Not only the failure probability, but also the failure frequency and the cycle time between failures can be analyzed by this approach. Hobbs (1985) was one of the first to recognize the importance and applicability of frequency and duration analysis to water supply systems. Some of his work includes the development of a methodology to compute the expected unserved demand and the inclusion of storage in the FD computations.

Duan and Mays (1990) presented a new methodology to analyze the reliability of pumping systems using a modified FD analysis to make the reliability analysis more realistic and complete. Both the mechanical failure and hydraulic failure of pumping systems are analyzed in computing the reliability parameters for the methodology. Hydraulic failure in this context refers to not meeting required demands and required pressure heads. The methodology has been programmed into a computer code called RAPS (Reliability Analysis of Pumping Systems). RAPS is used to determine the following eight reliability parameters for pumping systems: (1) failure probability; (2) failure frequency; (3) cycle time between failures; (4) expected duration of a failure; (5) expected unserved demand of a failure; (6) expected number of failures in the period of study; (7) expected total duration of failures in the period of study; and (8) expected total unserved demand in the period of study. RAPS has been tested on example problems ranging from two pumps and five demand states to ten pumps and twenty-five demand states.

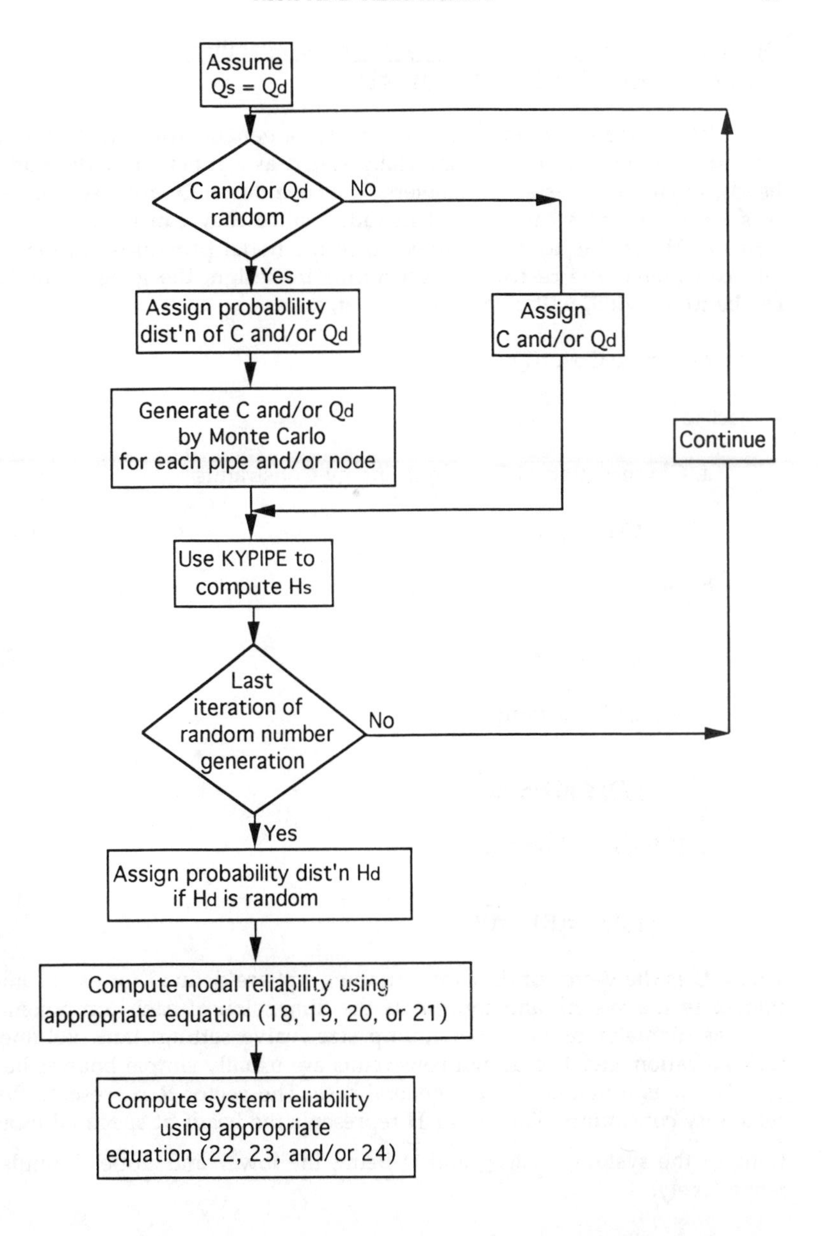

Figure 3. Flowchart of Algorithm to Evaluate System Reliability

Optimization Model for Reliability-Based (Availability) Design of Water Distribution Networks

The overall optimization problem for a general water distribution network design can be mathematically stated as a function of the nodal heads, **H**, and the design parameters. The pipe flows, **Q**, are a second set of system variables but are not included since they can be written in terms of **H** via the flow equations. Since the nodal pressures are generally considered the restricting constraints in design, the general model can be formulated, with respect to this set, as

$$\text{Minimize Cost } f(\mathbf{D},\mathbf{H}) \tag{25}$$

subject to

a. Conservation of Flow and Energy Constraints

$$\mathbf{G}(\mathbf{H},\mathbf{D}) = 0 \tag{26}$$

b. Head Bounds

$$\underline{\mathbf{H}} \leq \mathbf{H} \leq \overline{\mathbf{H}} \tag{27}$$

c. Design Constraints

$$\underline{\mathrm{j}(\mathbf{D})} \leq \mathrm{j}(\mathbf{D}) \leq \overline{\mathrm{j}(\mathbf{D})} \tag{28}$$

d. Reliability Constraints

$$\underline{\mathbf{r}(\mathbf{R})} \leq \mathbf{r}(\mathbf{R}) \leq \overline{\mathbf{r}(\mathbf{R})} \tag{29}$$

where **D** is the vector of decision variables that are defined for each component in the system and represents the dimension of each component, such as diameter of the pipes, pump size, valve setting, tank volume, tank elevation, etc. The design constraints are usually simple bounds but are shown as functions for the general case. The vector **R** represents the reliability constraints. The vector **H** represents the heads at specified locations in the system, with $\underline{\mathbf{H}}$ and $\overline{\mathbf{H}}$ being the lower and upper bounds, respectively.

The solution approach employs a technique whereby the problem is reduced to a form that is more manageable by large-scale nonlinear programming (NLP) codes. The technique reduces the problem by writing some variables called "state" variables that are dependent in terms of other "control" (independent) variables using equality constraints. This step results in a smaller, reduced problem with a new objective and a smaller set of constraints, many of which are simple bounds, that can now be efficiently solved by existing NLP codes. In this problem, the pressure heads, **H**, will be defined as the state or basic variables and written with respect to the design parameters, **D**, the control or nonbasic variables. This variables reduction technique has been successfully applied to problems in econometric control, oil and gas reservoir management, groundwater management, and large water distribution systems without a reliability constraint (Lansey 1987). A water distribution simulation model can be used to solve the network equations for the nodal heads given a set of design parameters. A computer program, WSAVOPT (Water Supply AVailability OPTimization), has been developed (Cullinane 1989; Cullinane, Lansey, and Mays 1992).

References

Bao, Y., and L. W. Mays. 1990. Model for water distribution system reliability. *Journal of Hydraulic Engineering,* ASCE 116 (September).

Coals, A., and I. C. Goulter. 1985. Approaches to the consideration of reliability in water distribution networks. In *Proceedings of the 1985 International Symposium on Urban Hydrology, Hydraulic Infrastructures and Water Quality Control.* The University of Kentucky, Lexington, Kentucky, July 23–25.

Cullinane, M. J., Jr. 1989. Methodologies for the evaluation of water distribution system reliability/availability. Ph.D. dissertation, Department of Civil Engineering, University of Texas, Austin, Texas.

Cullinane, M. J., Jr.; K. E. Lansey; and L. W. Mays. 1992. Optimization-availability based design of water distribution networks. *Journal of Hydraulic Engineering,* ASCE 118 (March).

Duan, N., and L. W. Mays. 1987. Reliability analysis of pumping stations and storage facilities. In *Proceedings of the American Society of Civil Engineers National Conference on Hydraulic Engineering,* edited by R. M. Ragan. New York: American Society of Civil Engineers.

Duan, N., and L. W. Mays. 1990. Reliability analysis of pumping systems. *Journal of Hydraulic Engineering,* ASCE 116 (February).

Duan, N. 1988. Optimal reliability-based design and analysis of pumping systems for water distribution systems. Ph.D. dissertation, Department of Civil Engineering, The University of Texas, Austin, Texas.

Fujiwara, O., and H. D. Tung. 1991. Reliability improvement for water distribution networks through increasing pipe size. *Water Resources Research* 27 (July): 1395–402.

Goulter, I., and F. Bouchart. 1987. Joint consideration of pipe breakage and pipe flow probabilities. In *Proceedings of the American Society of Civil Engineers National Conference on Hydraulic Engineering,* edited by R. M. Ragan. New York: American Society of Civil Engineers.

Goulter, I., and A. Coals. 1986. Quantitative approaches to reliability assessment in pipe networks. *Journal of Transportation Engineering,* ASCE 112 (March): 287–301.

Goulter, I., J. Davidson, and D. Jacobs. 1993. Predicting water main breakage rates. *Journal of Water Resources Planning and Management,* ASCE 119 (July/August).

Goulter, I., and A. Kazemi. 1988. Spatial and temporal groupings of water main pipe breakage in Winnipeg. *Can. Journal of Civil Engineering* 15 (1): 91–7.

Goulter, I., and A. Kazemi. 1989. Analysis of water distribution pipe failure types in Winnipeg, Canada. *Journal of Transportation Engineering,* ASCE 115 (2): 95–111.

Henley, E. J., and H. Kumamoto. 1981. *Reliability engineering and risk assessment*. Englewood, N.J.: Prentice Hall.

Hobbs, B. 1985. Reliability analysis of water system capacity. In *Proceedings of the American Society of Civil Engineers Specialty Conference, Hydraulics and Hydrology in the Small Computer Age,* edited by W. Waldrop. New York: ASCE.

Hobbs, B. F., and G. K. Beim. 1988. Analytical simulation of water system capacity reliability 1. Modified frequency-duration analysis. *Water Resources Research* 24 (9): 1431–44.

Hobbs, B., G. K. Beim, and A. Gleit. 1987. Reliability analysis of power and water supply systems. In *Strategic Planning in Energy and Natural Resources,* edited by B. Lev, et al. Amsterdam: North-Holland.

Kapur, K. C., and L. R. Lamberson. 1977. *Reliability in engineering designs*. New York: Wiley.

Kettler, A. J., and I. C. Goulter. 1985. An analysis of pipe breakage in urban water distribution networks. *Can. J. Civil Eng.* 12: 286–93.

Lansey, K. E. 1987. Optimal design of large-scale water distribution systems under multiple loading conditions. Ph.D. dissertation, The University of Texas, Austin, Texas.

Lansey, K. E., and L. W. Mays. 1987. Optimal design of large scale water distribution systems. In *Proceedings of the American Society of Civil Engineers National Conference on Hydraulic Engineering,* edited by R. M. Ragan. New York: American Society of Civil Engineers.

Lansey, K. E., and L. W. Mays. 1989. Optimization model for water distribution system design. *Journal of the Hydraulics Division,* ASCE 115 (October).

Lansey, K. E., N. Duan, L. W. Mays, and Y.-K. Tung. 1989. Water distribution system design under uncertainties. *Journal of Water Resources Planning and Management* ASCE 115 (September).

Mays, L. W., and M. J. Cullinane, Jr. 1986. A review and evaluation of reliability concepts for design of water distribution systems. Miscellaneous paper EL-86-1, U.S. Army Corps of Engineers, Environmental Laboratory, Waterways Experiment Station, Vicksburg, Mississippi, January.

Mays, L. W., N. Duan, and Y. C. Su. 1986. Modeling reliability in water distribution network design. In *Proceedings,* Water Forum 1986: World Water Issues in Evolution, edited by M. Karamouz, et al. New York: American Society of Civil Engineers.

Mays, L. W., ed. 1989. *Reliability analysis of water distribution systems.* Compiled by the Task Committee on Reliability Analysis of Water Distribution Systems. New York: American Society of Civil Engineers.

Mays, L. W., et al. 1989. Methodologies for the assessment of aging water distribution systems. Report No. CRWR 227, Center for Research in Water Resources, The University of Texas at Austin, July.

Ormsbee, L., and A. Kessler. 1990. Optimal upgrading of hydraulic-network reliability. *Journal of Water Resources Planning and Management,* ASCE 116 (November/December): 784-802.

Quimpo, R. G., and U. M. Shamsi. 1987. Network analysis for water supply reliability determination. In *Proceedings of the American Society of Civil Engineers National Conference on Hydraulic Engineering,* edited by R. M. Ragan. New York: American Society of Civil Engineers.

Su, Y. C., L. W. Mays, N. Duan, and K. E. Lansey. 1987. Reliability-based optimization model for water distribution systems. *Journal of Hydraulic Engineering,* ASCE 114 (12): 1539–56.

Tung, Y.-K. 1985. Evaluation of water distribution network reliability. In *Proceedings* of the ASCE Hydraulics Division Specialty Conference, Orlando, Florida.

Tung, Y.-K., K. E. Lansey, N. Duan, and L. W. Mays. 1987. Water distribution system design by chance-constrained model. In *Proceedings* of the 1987 National Conference on Hydraulic Engineering.

Wagner, J. M., U. Shamir, and D. H. Marks. 1986. Reliability of water distribution systems. Report No. 312, Ralph M. Parsons Laboratory, Massachusetts Institute of Technology, Cambridge, Mass.

Wagner, J. M., U. Shamir, and D. H. Marks. 1988a. Water distribution reliability: Analytical methods. *Journal of Water Resources Planning and Management,* ASCE 114 (3): 253–75.

Wagner, J. M., U. Shamir, and D. H. Marks. 1988b. Water distribution reliability: Simulation methods. *Journal of Water Resources Planning and Management,* ASCE 114 (3): 276–94.

Woodburn, J., K. E. Lansey, and L. W. Mays. 1987. Model for the optimal rehabilitation replacement of water distribution system components. In *Proceedings of the American Society of Civil Engineers National Conference on Hydraulic Engineering,* edited by R. M. Ragan. New York: American Society of Civil Engineers.

Yen, B. C., and Y.-K. Tung, eds. 1993. *Reliability and uncertainty analysis in hydraulic design.* Compiled by ASCE Subcommittee on Uncertainty and Reliability Analysis in Design of Hydraulic Structures of the Technical Committee on Probabilistic Approaches to Hydraulics of the Hydraulics Division of ASCE.

Reliability Analysis of Water Distribution Systems

Rafael G. Quimpo[1]

Abstract

The aging and deterioration of our water supply infrastructure have resulted in performance deficiencies. Limited budgets require that resources for maintenance and rehabilitation be allocated in the most efficient manner. If resources are insufficient to correct all the deficiencies, restoration maintenance must be prioritized. Prioritization using reliability measures has been suggested as a management tool. This paper examines the developments in this field, explains some of the analytical approaches, discusses the conceptual and computational difficulties encountered, and reviews some ongoing research to resolve some of the problems.

Introduction

Our research at the University of Pittsburgh has focused on developing a strategy for maintenance and rehabilitation of the water supply infrastructure. We have narrowed our interest to the reliability assessment of water supply. Although this aspect represents just one facet of the infrastructure problem (Walski and Pellicia 1982; Li and Haimes 1992), the cost of repairs, public inconvenience, business disruptions, and lost economic opportunities due to pipe breaks justify the commitment of scarce resources to improve decision making in this sector.

The water supply infrastructure consists of

1. *the water source*—This may be a river or a groundwater source. Considerations of risk and reliability regarding the

---

[1]Professor of Civil Engineering, Department of Civil and Environmental Engineering, University of Pittsburgh, Pittsburgh, PA 15261

water source may include aspects of hydrology and, hence, the probabilistic analysis of geophysical phenomena. It may also lead to risk con-siderations regarding the quality (potability) of the water source.

2. *the water treatment plant*—In most instances, raw water either from the river or the ground must be purified to be potable. Risk and reliability issues may relate to the analysis of the physical and chemical treatment processes and/or probabilities pertaining to the integrity of the process train.

3. *the distribution system*—This is the physical configuration of pipes, pumps, valves, tanks, reservoirs, and control structures that enable the distribution of the treated water to the end users.

Our reliability analysis will be limited to the distribution system. We will further assume that the reliabilities of the source of water and the treatment process are acceptable. This may not be the case, but we will leave these aspects to other investigators better equipped to deal with the associated problems. With this qualification, we proceed to analyze the reliability issues that arise.

There are several definitions of reliability. Our definition is based on the expectations of an individual water user who resides in a typical town or city. This individual's residence is supplied with water by a municipal authority or a private water supplier. Water comes from a source or sources of treated water and reaches the user's location through different possible paths. These paths depend on the configuration of the pipes and appurtenant structures that form the water distribution system. To this user, the reliability of water supply is equal to the probability that water will be available at sufficient hydraulic pressure to satisfy his or her requirements. This probability (reliability) would vary for different users at different locations in the system.

We therefore choose to characterize the reliability of the system not by a single numerical value but by a spatial distribution of point reliabilities, i.e., a "reliability surface." Such a surface is illustrated in figure 1. This approach accommodates the spatial variability as well as the disparate requirements of individual users. Knowledge of this surface can be an important tool for a decision maker. The reliability surface will give him or her a basis for allocating resources for repair and maintenance as well as capital improvements. Visual examination of this

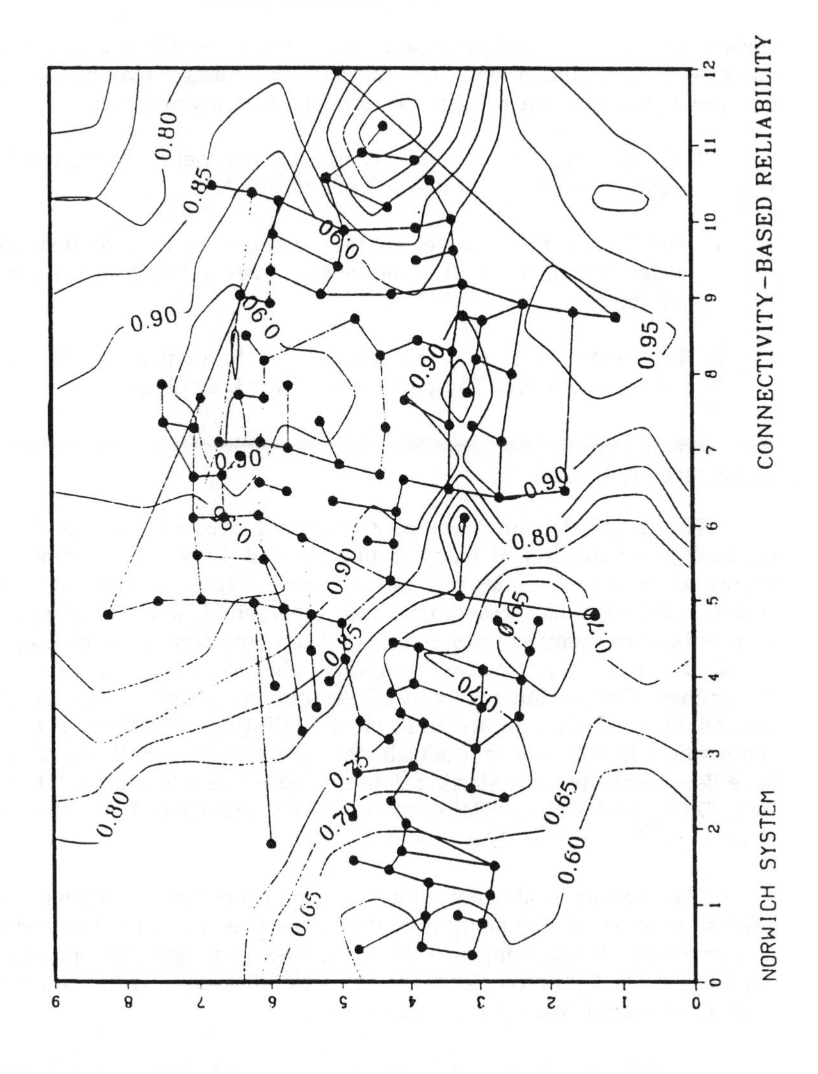

Figure 1. Reliability Surface for a Water Distribution System

surface will quickly identify critical areas of low reliability and, hence, areas needing rehabilitation. Thus, the reliability surface may be used to develop a prioritization strategy for infrastructure maintenance.

The calculation of the point reliabilities that define the reliability surface requires knowledge of two sets of probabilities:

1. The failure probabilities of the components that connect the water source to the demand point. These relate to component reliability.

2. The probabilities that the source and the demand points will remain connected. These relate to network topology.

Component Reliability

The physical connection between the source of water and the demand point consists of many components. A typical path from the source to the user includes pipes of different diameters and materials, control devices, tanks, reservoirs, and subassemblies of components, such as booster pump stations and control devices. Each of these components has failure (and, hence, reliability) characteristics that must be determined. Component reliability determination is complicated by the unavailability of data. Information on the failure probabilities of these components is not easy to attain. In the last decade, significant efforts have been expended in collecting data to obtain this information (O'Day 1982; O'Day and Staheli 1983; Andreou, Marks, and Clark 1987; Wagner, Shamir, and Marks 1988; Mays 1989).

The problem is also not limited to data collection. Even after data have been collected, the analysis of data itself presents some interpretation problems. Work along this line continues and significant progress has been made (Andreou, Marks, and Clark 1987; Goulter and Kazemi 1989; Al-Humoud, Wu, and Quimpo 1990).

If T denotes the time to failure of a system component and its distribution function is $F(t)$, Cox and Oakes (1984) define the survival function $S(t)$ as

$$S(t) = 1 - F(t) = P[T > t] \qquad (1)$$

This is related to the hazard function, $h(t)$ (Benjamin and Cornell 1970) by

$$h(t) = \frac{f(t)}{1 - F(t)} \,. \tag{2}$$

Integration of $h(t)$ yields

$$\int h(t)dt = \int \frac{f(u)du}{1 - F(u)} = -\ln[1 - F(t)] = -\ln S(t) \,, \tag{3}$$

so that

$$S(t) = e^{-\int h(u)du} \,. \tag{4}$$

Thus, if the hazard function is known, its survival function and, hence, its failure probability, may be determined. This may be done through the Proportional Hazards Model, which takes the form

$$h(t;z) = h_0(t)e^{\beta z} \,, \tag{5}$$

where z is a column vector of explanatory variables, β is a row vector of parameters, and $h_0(t)$ is an unspecified baseline hazard function.

This model was recently suggested by Andreou, Marks, and Clark (1987) for water distribution system components and examined by Al-Humoud, Wu, and Quimpo (1990). The model estimates the probability of breaking for each pipe as a function of a baseline hazard function that depends on time, and the multiplicative effects of several explanatory variables. A desirable feature of this model is its ability to separate the relationship between break failure and age of pipes from the other explanatory variables. The model is also able to use information from censored data. No prior distribution of the hazard function is assumed and, hence, it is nonparametric.

Since it is not central to the subject of the present paper, we will assume that the difficult task of component reliability analysis has been carried out and that the component reliabilities have been determined. Our work has now been narrowed to the problem of evaluating the probability that the source is connected to the user.

Network Reliability

In general, water distribution systems are redundantly connected. There are usually several paths from the source to the user. This redundance exists by design, so that if one path is disconnected, alternative paths are available to maintain the connection. This redundancy is, in fact, the essence of a "reliable" design. On the other hand, this redundancy is also the scourge of reliability analysts.

The tools and techniques developed in network theory can help one explore the connectivity between the source and the demand. The water distribution system may be thought of as a network of nodes interconnected by arcs, each of which has a probability of failure. If, for example, three arcs (pipes) in series connect the source to the demand point, the probability that the two will remain connected is simply the product of the reliabilities (complements of the failure probabilities) of the three arcs in series. Parallel (redundant) connections may also be evaluated easily for small networks.

For medium-sized networks, several techniques have been used to analyze the connectivity problem. Techniques such as the method of path sets, method of cut sets, and fault-tree analysis have been tried successfully (Tung 1985; Goodrich, Cullinane, and Goulter 1989). With block reduction techniques, wherein series-parallel blocks of the network are replaced by arcs of equivalent reliability, these methods have permitted small and medium-sized water distribution systems to be analyzed.

Consider the path-set method. A path set between nodes u and v, denoted by P_{uv}, is a set of arcs $\{e_1, e_2, \dots\}$ that connects u to v. A path set is minimal if it contains no subset that is also a path set. Let $P_1, P_2, \dots, P_m$ be m minimal path sets between the given node-pair and let E_i denote the event: {all arcs in P_i function}. Then, the node-pair reliability R_{ur} is

$$R_{u,v} = Pr\left[\bigcup_{i=1}^{m} E_i\right] \tag{6}$$

$$R_{u,v} = \sum_{i=1}^{m} \Pr[E_i] - \sum_{i<j} \Pr[E_i \cap E_j] + \sum_{i<j<k} \Pr[E_i \cap E_j \cap E_k] - \dots + (-1)^{m-1} \Pr[E_i \cap \dots \cap E_m] \tag{7}$$

where $Pr[.]$ denotes the probability. Quimpo and Shamsi (1991) used equation (7) to determine the node-pair reliabilities of all the demand points and to develop the reliability surface.

Although the method of path sets is suitable for small systems, the water distribution system of a medium-sized city has too many arcs to be easily analyzed by this or any of the above methods. One may reduce the number of arcs and, hence, lessen the computational burden by using a skeletonized system wherein only the major trunk lines are included in the network. This alleviates the problem somewhat, but often it will not suffice because interest may be on the smaller pipe sizes.

The connectivity problem for a general network has been analyzed and it has been shown that in general the problem is NP-hard, that is, the problem is intractable. It can be shown that, using any of the above methods, the computation time to analyze the reliability of a network of size n is of the order of 2^n. One will appreciate this exponential time requirement if we consider a value of $n = 60$. This yields $2^{60} = 1.153 \times 10^{18}$. If the reliability analysis of a system with $n = 60$ takes this many floating point operations (FLOPs) and each operation can be executed in one billionth of a second (the capability of a Cray computer), it will take 36.5 years to carry out the calculations! It is obvious that an algorithm with a shorter computation time is needed to solve the problem. Currently network analysts are trying to develop a polynomial-time solution to this exponential-time problem. Unless such a solution is found, the problem is considered NP-hard. Table 1 compares several polynomial and exponential time functions.

We have experimented with some reduction techniques in trying to develop a polynomial-time algorithm (which has fewer computation requirements than an exponential algorithm) to solve this problem. We have found that by using a combination of block reduction and factorization techniques (Satyanarayana and Wood 1985), we were able to analyze a 220-node network in slightly less than 4 hours using an IBM 486/33 machine. A 486/33 machine has a speed of 1.5M FLOPs per second. The approach appears promising and we are continuing to refine it.

Capacity-Weighted Reliability

However, this is just half of the problem. The above analysis pertains only to what is called a "connectivity-based reliability." This reliability measure calculates the probability that the water source is connected to the demand point. Although connectivity-based reliabilities often suffice in an initial examination, in water distribution systems the requirement is more stringent. Not only must the demand node be connected to the water source, but it is also required that a sufficient amount of water be delivered at the requisite hydraulic pressure.

Time Complexity Function	n			
	10	20	40	60
n	0.00001 second	0.00002 second	0.00004 second	0.00006 second
n^2	0.0001 second	0.0004 second	0.0016 second	0.0036 second
n^5	0.1 second	3.2 seconds	1.7 minutes	13 minutes
2^n	0.001 second	1.0 second	12.7 days	366 centuries
3^n	0.059 second	58 minutes	3855 centuries	1.3×10^{13} centuries

Table 1. Comparison of Several Polynomial and Exponential Time Complexity Functions (Source: Garey and Johnson 1979)

This compels us to revise our formulation to include the carrying capacity of the component pipes in the connectivity calculation. To do this, the hydraulics of the system must be incorporated into the analysis, which, up to this point, has been limited only to the topology of the network. If the capacity of each pipe is known, a reasonable approach is to weight the failure probability of a component pipe by the ratio of its carrying capacity to the required flow at the demand point. This capacity-weighted probability may then be used in the calculation of the point reliabilities described earlier. Strictly speaking, the point reliabilities calculated in this manner are no longer true probabilities. They may be considered as reliability indices.

The inclusion of the network hydraulics complicates the analysis because of how the capacity of a pipeline is defined. Besides its physical characteristics, such as diameter and friction factor, the discharge capacity of a pipe depends on the energy gradient. The energy gradient varies with flow conditions. Although software packages are available to calculate the discharge Q for pipes in a distribution system (Wood 1980), it is not practical to include detailed hydraulic analysis in the reliability assessment. Therefore, an assumption on maximum capacity of each component needs to be made. The flow in a pipe may be calculated by the Hazen-Williams equation

$$Q = 0.28\, C\, D^{2.63}\, S^{0.54}, \tag{8}$$

where Q is the discharge; C is a pipe roughness coefficient; D is the pipe diameter; and S is the hydraulic gradient. Wagner, Shamir, and Marks (1988) suggest to set the maximum flow in each pipe as that flow calculated by equation (8) if the pipe were installed at a hydraulic gradient of $S = 0.01$. Other alternatives may also be considered. Wu, Yoon, and Quimpo (1993) use this simplification as the basis for a capacity-weighted reliability analysis. In weighting the flow contributions, they also included partial fulfillment of the flow requirement in evaluating the reliability index.

The pipe capacities, calculated in this manner, may then be used as weights of the arc reliabilities in determining the connectivity between the source node and each demand point and the resulting "point reliability index" used to define the reliability surface. The remaining problem is to develop an algorithm that will calculate these reliabilities in polynomial, and preferably linear, time. The computational problem is still formidable. At the University of Pittsburgh, we are trying to use network reduction and factorization techniques to develop a polynomial-time algorithm.

Conclusions

If this attempt at developing a polynomial-time algorithm is successful, it will facilitate the reliability analysis of large water distribution systems and encourage water authorities to adopt the resulting decision-making tool for infrastructure assessment and rehabilitation.

References

Al-Humoud, J., S.-J. Wu, and R. G. Quimpo. 1990. Failure modeling of hydraulic systems. In *Hydraulic Engineering,* edited by H. Chang. New York: American Society of Civil Engineers.

Andreou, S. A., D. H. Marks, and R. M. Clark. 1987. A new methodology for modelling break failure patterns in deteriorating water distribution systems: Theory and applications. *Advances in Water Resources* 10 (March): 2–20.

Benjamin, J. R., and C. A. Cornell. 1970. *Probability, statistics and decision for civil engineers.* New York: McGraw-Hill.

Cox, D. R., and D. Oakes. 1984. *Analysis of survival data.* New York: Chapman and Hall.

Garey, M., and D. Johnson. 1979. Computers and intractability: A guide to the theory of NP-completeness. Bell Telephone Labs.

Goodrich, J., M. J. Cullinane, and I. Goulter. 1989. Water distribution system evaluation. In *Reliability analysis of water distribution systems,* edited by Larry W. Mays. New York: American Society of Civil Engineers.

Goulter, I. C., and A. Kazemi. 1989. Analysis of water distribution pipe failure in Winnipeg, Canada. *Journal of Transportation Engineering* ASCE 115: 2.

Li, D., and Y. Y. Haimes. 1992. Optimal maintenance-related decision making for deteriorating water distribution systems, 1 and 2. *Water Resources Research* 28: 1053–61 and 1063–70.

Mays, L. W., ed. 1989. *Reliability analysis of water distribution systems.* New York: American Society of Civil Engineers.

O'Day, K. 1982. Organizing and analyzing leak and break data for making main replacement decisions. *Journal of the American Waterworks Association* 74 (11): 588–94.

O'Day, D. K., and L. A. Staheli. 1983. *Information systems for facility maintenance decisions.* Rep. No. 2, Study for the Urban Infrastructure Network, The Urban Institute, October.

Quimpo, R. G., and U. M. Shamsi. 1991. Reliability-based distribution system maintenance. *Journal of Water Resources Planning and Management,* ASCE 117 (3): 321–39.

Satyanarayana, A., and R. K. Wood. 1985. A linear-time algorithm for computing K-terminal reliability in series-parallel networks. *SIAM Journal of Computing* 14 (November).

Tung, Y. K. 1985. Evaluation of water distribution network reliability. *Hydraulics and hydrology in the small computer age,* Vol. 1. New York: American Society of Civil Engineers.

Wagner, J. M., U. Shamir, and D. H. Marks. 1988. Water distribution reliability: Analytical methods. *Journal of Water Resources Planning and Management,* ASCE 114 (May).

Walski, T. M., and A. Pellicia. 1982. Economic analysis of water main breaks. *Journal of the American Waterworks Association* (March): 140–7.

Wood, D. J. 1980. *KYPIPES.* Engineering Continuing Education, University of Kentucky, Knoxville.

Wu, S.-J., Y.-H. Yoon, and R. G. Quimpo. 1993. Capacity-weighted water distribution system reliability. *Reliability Engineering & System Safety* 42 (1): 39–46.

Water Infrastructure Risk Ranking and Filtering Method

Tomas Dolezal,[1] Yacov Y. Haimes, and Duan Li[2]

Abstract

The availability of water distribution infrastructure is critical to the sustained support and growth of society. Water infrastructure consists of systems that carry water (pipe and irrigation systems) and systems that control natural water flow (dams and runoff systems). The smooth operation of water infrastructure has both social and economic ramifications. This paper extends and modifies a ranking methodology, the risk ranking and filtering (RRF) method, and applies it toward locating water distribution systems and subsystems with the greatest risks. The resulting short list of the water distribution (sub-)systems with the greatest risks is then used as an input to a budget-assignment problem with limited capital. The resulting methodological framework aids in effective allocation of scarce resources among various maintenance, rehabilitation, and expansion projects.

Introduction

The goals of the water infrastructure risk ranking and filtering (WIRRF) method developed in this paper are twofold: to introduce a risk ranking methodology for identifying water distribution (sub)systems with the greatest risks and to aid in effective allocation of scarce resources among various maintenance and rehabilitation projects.

The first phase of the WIRRF method is to aid decision makers in locating the riskiest pipe networks. Risk, a measure of the probability and

---

[1]Consultant, American Airlines Decision Technologies, P.O. Box 619616, M-D 4462, Dallas/Fort Worth Airport, TX 75261-9616

[2]Director and Associate Director, respectively, Center for Risk Management of Engineering Systems, University of Virginia, Charlottesville, VA 22903

severity of adverse effects (Lowrance 1976), is characterized for this paper by the following criteria: prior risk information, functionality, moderate-event risk, and extreme-event risk. These criteria in the WIRRF method define a risk hierarchy that encompasses the major risks associated with water pipe networks. These criteria are quantified for each pipe (sub) network. A filtering process is then used to locate those pipe networks with the greatest risks. Adopting an additive value function to incorporate multiple objectives, the collection of water pipe (sub)networks in the short list that resulted from the filtering process is ordered using the analytic hierarchy process (AHP) (Saaty 1980).

The WIRRF methodology is a derivative of the risk ranking and filtering (RRF) methodology (Haimes et al. 1993). The RRF is a decision tool initially developed to locate the top twenty components that contributed most to NASA's shuttle program risk. A methodological precedent for the WIRRF exists in the RIPS program designed for individual pipes (Muhlbauer 1990), where only individual pipes are ranked.

Building on the first phase of WIRRF, the second phase yields an ordered ranking for a set of maintenance, rehabilitation, and expansion projects. Each project is evaluated using benefit-cost analysis in a multi-objective framework. Benefits are defined in this paper as a decrease in the pipe (sub-)network's risk as measured in the first phase of WIRRF, when a project is enacted. The costs are defined as the dollar value required to finance each project. Resources are then allocated to the projects with the highest benefit-cost ratio, in descending order, until the funds are exhausted.

Background

The creation of the WIRRF methodology was motivated by three problem areas in the water infrastructure field: cost, complexity, and varied data formats. Each of the three areas has directly influenced the development of the WIRRF method.

The National Council on Public Works Improvement (NCPWI 1988) estimated that public works outlays range from $45 billion to upward of $70 billion annually. Further, the NCPWI noted that the percentage of gross national product (GNP) for public outlays was falling while the public works capital spending relative to private capital spending was also decreasing. Consequently, since there are fewer available funds, there is a need for a more efficient means of targeting their allocation.

The WIRRF methodology is applied to an area fraught with complexity, which makes it particularly difficult to make reliable decisions. The complexity is due to the type of information that is available and the constraints imposed on the decision process (O'Day et al. 1986). Appropriate information is often missing, not standardized, or incomplete. Information may be provided in several distinct formats: qualitative, quantitative, or mixed. The decision and risk analysis process is often driven by time constraints that inhibit a sound decision-making process.

The quality and quantity of data that characterize a water distribution system strongly influence the effectiveness and results of the decision process. The quality and quantity of available data can vary sharply from one water distribution system to another. In addition to this variability, there are also questions regarding the definition of terms (e.g., "leak" versus "break") and wide variety in the formats of the available stored data.

In general, because of more resources, larger water distribution systems have better databases that are well organized for easier access. Clark and Goodrich (1989), who studied water distribution systems, note that "it is a fact of life that in most utilities, distribution reliability is maintained by the knowledge of a few key personnel. Generally, these water treatment personnel use a good memory, repair records, a large wall map, and a hydraulic model of the larger transmission performance." Even when a sufficient database is available, there can still be many problems associated with it. The format of the data is often incompatible among different water distribution systems (O'Day et al. 1986).

The creation of and use of databases for water distribution systems are relatively new, and thus, there is much new work being done in this area (Walski et al. 1989). There are several current efforts motivated by the need to standardize and manage the database for the water distribution infrastructure (Van Blaricum and Hock 1991; Van Blaricum, Hock, and Kamur 1989; Person and O'Day 1986; Griggs 1988).

Phase 1 of the WIRRF: Risk Ranking and Filtering

The flow chart in figure 1 summarizes the first phase of the WIRRF methodology, which is divided into the following five major steps: (1) definition of the risk hierarchy, (2) quantification of the attributes and severities, (3) use of the telescoping filter, (4) use of the AHP in the prioritization phase, and (5) application of sensitivity analysis.

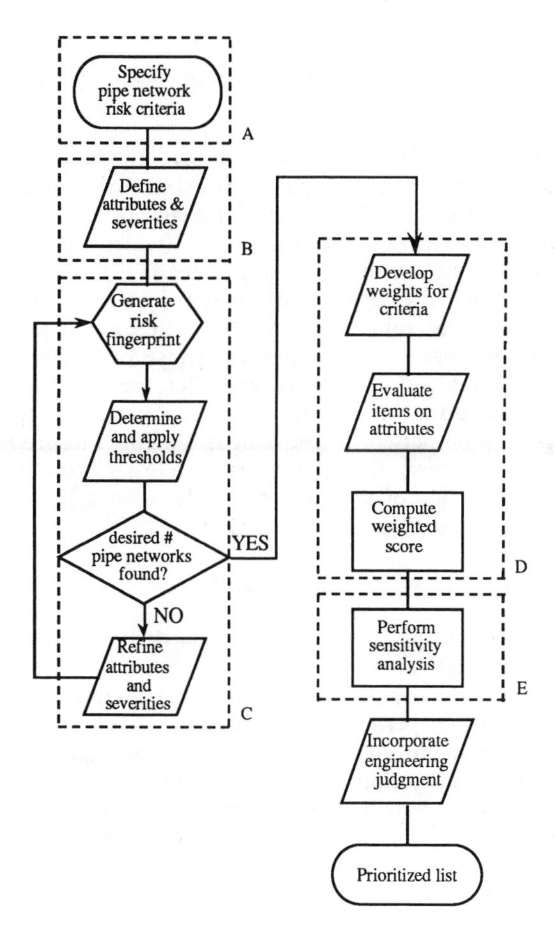

Figure 1. Five Steps in Phase 1 of WIRRF Methodology

The definition and specification of a risk hierarchy allow the objectives to be delineated explicitly. The four risk criteria of the pipe (sub-)network have been identified as (1) prior risk information, (2) functionality, (3) moderate-event risk, and (4) extreme-event risk. Prior risk information is the criterion that characterizes the historical aspect of the pipe networks. This allows the WIRRF methodology to take into account the previous operation of pipe networks. Functionality is the criterion that measures the pipe network's ability to operate in the presence of defects. This provides a measure of how robust the pipe (sub-)network is in remaining operational. Moderate-event risk is the criterion that characterizes the most probable failure events. Extreme-event risk goes further than moderate-event risk to provide the decision maker with information about low-probability/high-consequence events or worst-case scenarios.

The use of measurable attributes allows for the quantification of the pipe risk criteria. The assessment process for each risk criterion defines mappings from criteria scales (for example, the probability of a specified event) to discrete scales of severity (which range from 1 to 5, where 1 denotes very low and 5 denotes very high). A risk fingerprint, comprising a graphical interpretation of risk for all the risk criteria, is created for each pipe (sub-)network. Thresholds are then specified to reduce the original list of pipe networks. Those pipe networks that do not exceed the threshold are not retained for further evaluation. This step is repeated by adjusting the threshold level until a specified number or percentage of pipe networks is attained. When the desired number of pipe networks is reached, the AHP is used to order the remaining pipe networks. The final step of this process involves a sensitivity analysis of the criterion weights derived from the AHP.

Ranking hierarchy and criteria quantification

The first two steps in phase one of the WIRRF methodology are the definition of a risk hierarchy and the quantification of the risk criteria. The ranking hierarchy provides a framework within which to select the pipe networks that possess the greatest risk. The quantification step uses procedures to quantify the four risk criteria through the use of proxy attributes. The quantification of the pipe (sub-)network risk criteria is facilitated by the use of attributes that are representative and have magnitudes that are readily measurable. Figure 2 illustrates the ranking hierarchy employed within the WIRRF methodology.

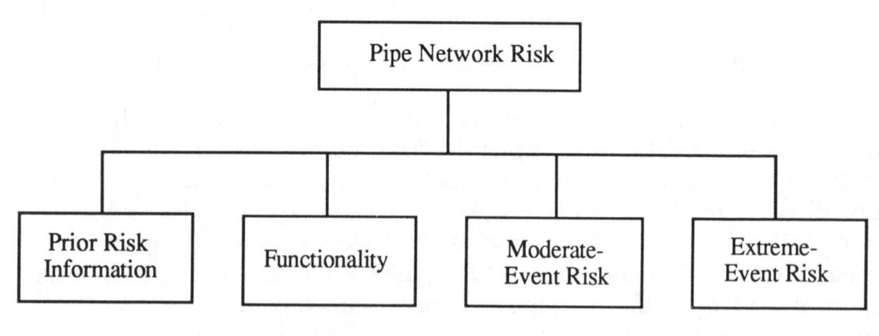

Figure 2. Ranking Hierarchy in WIRRF Methodology

The four risk criteria for the pipe (sub-)network of the WIRRF methodology are quantified by first associating a "proxy attribute" to each risk criterion. A proxy attribute, as defined by Keeney and Raiffa (1976), is a measurable quantity that represents the degree to which the criterion is satisfied (or unsatisfied). For a pipe network, for example, a proxy attribute may have units of probability, the number of miles of pipe in a water distribution (sub-)system, or the number of pipe failures. We note that no correspondence between a criterion and a proxy attribute is perfect. The goal is to find a satisfactory representation that is complete, measurable, and as representative as possible. A risk criterion is considered measurable if there exists a proxy attribute that represents the criterion; that is, the magnitude of the proxy attribute can be well defined and can be determined for all pipe networks.

For a risk criterion of the pipe (sub-)network, such as functionality, the possible magnitudes ascribed to the proxy attribute are part of a representative measurement scale. To compare risk severity over the four risk criteria, the measured attributes are transformed into a severity scale. The severity scale is defined as follows:

very low - 1
low - 2
medium - 3
high - 4
very high - 5

The decision maker is required to assess the ranges that correspond to each of the severity scores for the four risk criteria. This assessment needs to be performed only once, early and objectively in the decision process, and is facilitated by the use of a graph. The motivation and quantification behind the four risk criteria and their definitions are explained below.

(i) Prior risk information

The first of the four risk criteria is prior risk information. Prior risk information deals with attributes that can quantify factors and characterizes historical trends that affect the pipe networks. For example, the prior risk information criterion can be used to identify trends in failure rates and critical areas. It can also aid in locating areas within the pipe (sub-)network that require attention and resources, as well as highlighting areas of the pipe (sub-)network that have achieved better than average performance.

The value of the records and/or databases will influence the quality (and quantity) of information represented by the prior risk information criterion. Research has indicated that there is a high probability that historical information is often incomplete or simply absent (Van Blaricum and Hock 1991). In this case, the decision maker will be required to employ his or her engineering judgment.

Figure 3 illustrates the prior risk information hierarchy. The associated attributes used to quantify this risk criterion include, but are not restricted to, pipe type, pipe age, number of previous breaks, and associated trends. A combination of pipe type and pipe age can often provide a decision maker with a better understanding of failure data. The number of previous breaks has also been found to influence the failure rate (Andreou, Marks, and Clark 1987). From a historical perspective, it is also important to know who laid the pipe and how it was laid, since studies have indicated that when examined separately, the age and type of pipe laid is often a poor indicator of failure (O'Day et al. 1986).

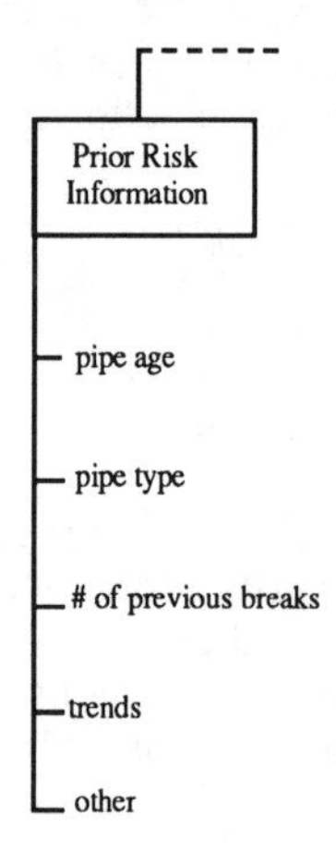

Figure 3. Prior Risk Information Hierarchy

As an illustrative example of converting proxy attributes to a severity scale, we examine two attributes for the prior risk information criterion: pipe age and pipe type. Both pipe age and pipe type are independent variables that contribute to the failure rate for a given pipe. This criterion juxtaposes the two attributes and is used to generate the severity score. This mapping is accomplished through the use of a matrix procedure developed by the U.S. Air Force and McDonnell Douglas Corp. (U.S. Air Force 1988). By using this matrix, the ages and types of pipes are combined so that the output is within the five-point severity scale. Figure 4 is an example of this type of matrix. When the number of proxy attributes is larger than two, the decision maker needs to create a higher dimensional matrix or create multiple two-dimensional matrices to determine the severity level.

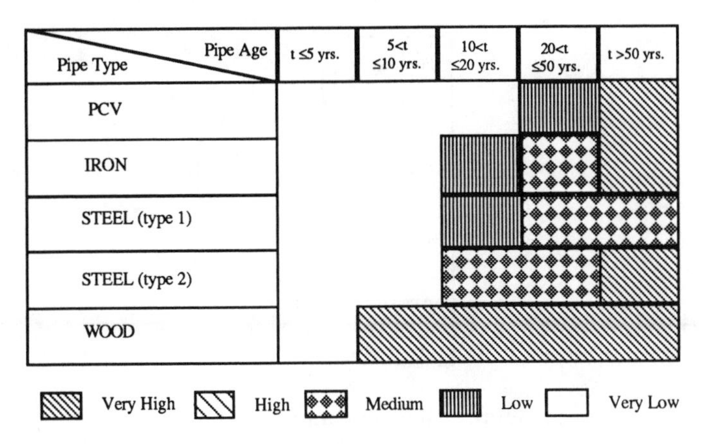

Figure 4. Sample of Prior Risk Information Matrix

(ii) Functionality

The term *functionality* (or alternatively, fault tolerance) is defined as the ability of a pipe (sub-)network to maintain an operational state in the presence of defects and anomalies. Functionality can be measured through various techniques of redundancy (Johnson 1989). The assumption made for this criterion is that a pipe (sub-)network that does not have a high degree of redundant pathways is considered more of a potential problem than a similar pipe (sub-)network with a higher degree of redundancy. Figure 5 illustrates the functionality risk criterion hierarchy.

The risk criterion of functionality is designed to take into account what will happen when an inevitable single failure occurs in a pipe network (sub-)system (Wagner, Shamir, and Marks 1988a, 1988b). Consequently, if the pipe network can continue to operate at an acceptable level with the occurrence of a failure, the risk of the pipe network will be considered less than a nonredundant network.

The attributes that measure functionality risk for pipe networks include, but are not restricted to, the level of redundancy, the number of sources/sinks, and the percentage that the pipe network is gravity-fed versus pump-fed. The level of redundancy can be measured in a number of ways, including the number of paths from a given set of sources to a given set of demand nodes and the number of incoming feeds to each demand node.

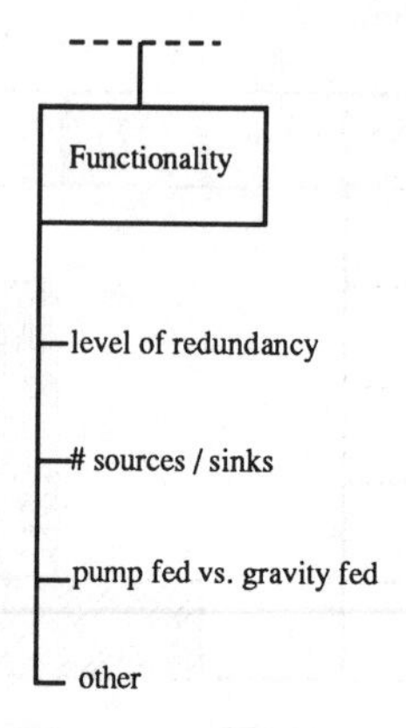

Figure 5. Functionality Hierarchy

(iii) Moderate-event risk

The criterion of moderate-event risk gives the decision maker a feel for the most probable loss that can be expected. This criterion provides information about the center portion of the loss distribution.

As seen in figure 6, moderate-event risk can be partitioned into two parts: the likelihood of the most probable failure events in a pipe

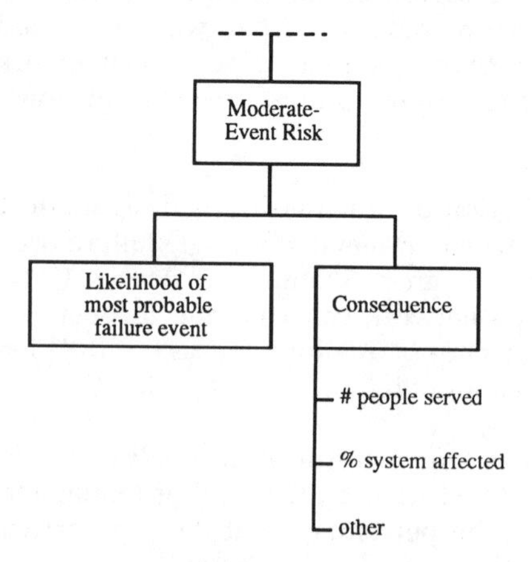

Figure 6. Moderate-Event Risk Hierarchy

(sub-)network and the most likely effect. The most likely effect describes the resultant consequence of a (sub-)network failure, such as the number of people not served, the percentage of the system affected, the economic loss from industries that rely on constant water supply, and the fire loss due to the loss of water pressure or the lack of adequate water quantity.

The likelihood of the most probable failure events can be approximated by the network's unreliability. The quantification of the moderate-event risk can then be obtained using a matrix design that combines the two attributes—the reliability and the most likely effect. Reliability can be quantified by probability ranges; the most likely effect can be quantified through the percentage loss of service for the pipe network and the number of people not served. This approach provides a ready transition to the five-point severity scale. Figure 7 illustrates an example matrix.

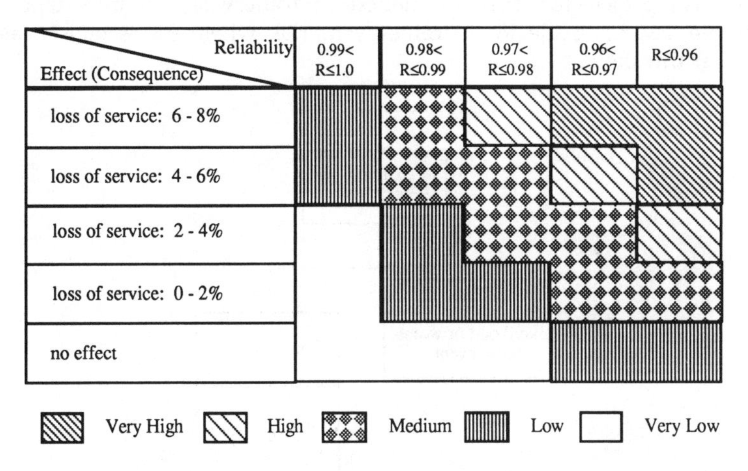

Figure 7. Sample of Moderate-Event Risk Matrix

(iv) Extreme-event risk

The concept of extreme events has received much attention in the past decade (Asbeck and Haimes 1984). Extreme events often have a low probability of occurrence, but a high consequence if they do occur. They thus have a large impact on public reaction and on the community's perception of risk. The extreme-event risk criterion combines the worst-case scenario with the likelihood of the worst case. The measure of extreme events provides the decision maker with more information, which gives the decision maker a more comprehensive view of the risks than is provided by a measure of the moderate-event risk alone.

Uncertainty is an inherent feature when assessing risk in deteriorating water distribution (sub-)systems. To a certain extent, the extreme-event risk criterion could be used to measure the uncertainty through capturing the characteristics of the variance.

As seen in figure 8, extreme-event risk is partitioned into two parts—the likelihood of worst-case event (of the pipe network) and the consequence of the worse-case scenario effect. The worst-case effect can be quantified by the number of people served and the percentage of the system affected. Similar to the procedure used in the moderate-event risk, the pair of the likelihood and the effects of worse-case scenarios can be mapped onto a five-level scale of extreme-event risk severity. Since extreme events are concerned only with low-probability/high-consequence events, there are only two effect categories: critical (when the percentage of customers who do not receive service during system failure exceeds a preselected level) and noncritical (otherwise). Figure 9 depicts a sample severity scale for extreme-event risk where 8% is chosen as a critical threshold.

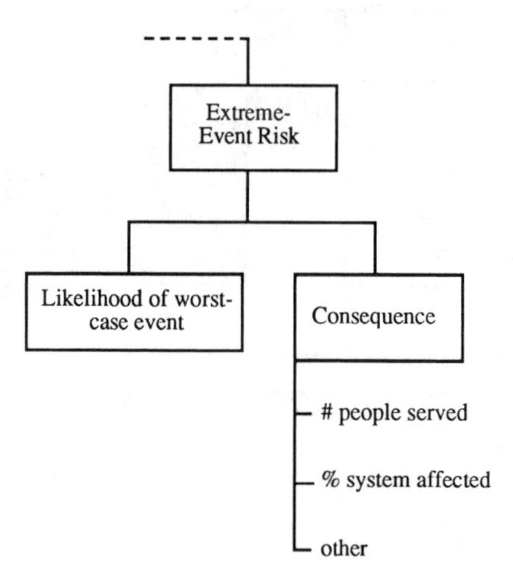

Figure 8. Extreme-Event Risk Hierarchy

The hierarchy of these four criteria constitutes the structure for the ranking process and provides a basis for all the steps in phase one of the WIRRF methodology. Although the above attributes may not provide exhaustive coverage of the universe of risk attributes, they represent a broad base from which to create new comparison methods for assigning severity scores. The four risk criteria were designed to be as independent as practically possible. Note, however, that several sources of information may be used as inputs to more than one risk attribute. The information, however, is processed and extracted in a different fashion in accordance to the content of the specific risk criterion.

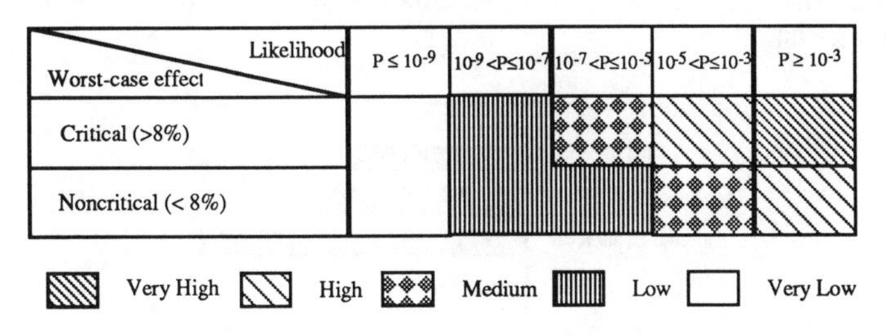

Figure 9. Sample of Extreme-Event Risk Matrix

Telescoping filter phase

The telescoping filter phase of the WIRRF methodology is an iterative procedure that pares down the number of pipe networks under consideration. A threshold acts as a filter that is defined by the decision-maker. The threshold allows only the riskiest pipe (sub-)networks (or, alternatively, the pipe (sub-)networks that require the most attention and resources) to be retained.

The telescoping filter phase of the WIRRF methodology back-loads the expenditures on data collection and analysis. The filtering process does not require exhaustive data collection and analysis. The filter allows the decision maker to remove quickly those pipe networks that are evidently not top candidates through rough scanning with minimum data collection and analysis. The speed and effort required in the "first cut" of the decision analysis is often an important consideration. First, the availability of data (i.e., the absence of quantified data) is often an issue, and second, the task of data collection is time and resource consuming.

A distinguishing feature of the WIRRF methodology is the use of the risk fingerprint of severity scales, an intuitive and graphical tool used to represent the pipe network risk. When the severity scales are defined for each of the four risk criteria, two scores can be generated for each attribute: (1) a percentile score (compared to the pipe network that has the highest score in a given specific risk criterion) and (2) an absolute severity score (defined on the 1 to 5 scale). Risk fingerprints are represented on a bar graph where the criteria are plotted along the vertical axis and either the percentile or the severity score is plotted on the horizontal axis. Figure 10 is a graphical representation of a percentile risk fingerprint that represents the contribution of a pipe network's risk relative to the other pipe networks. Figure 11 is a graphical representation of the severity risk fingerprint that represents the absolute contribution to pipe network risk.

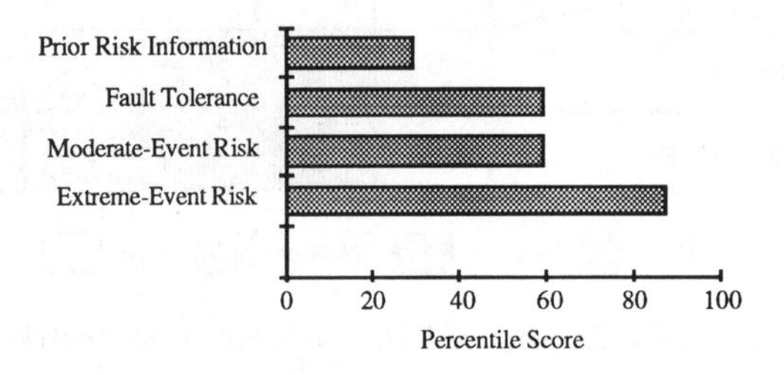

Figure 10. Example of a Percentile Risk Fingerprint

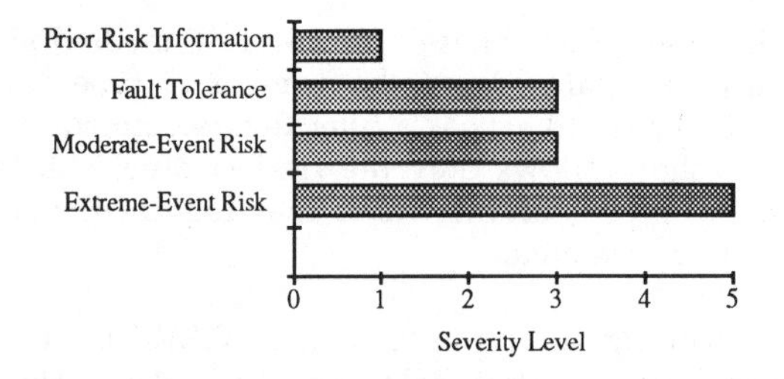

Figure 11. Example of a Severity Risk Fingerprint

The next step in the filtering process is the creation of a threshold. A threshold is a decision rule on either of the two types of fingerprints (percentile or severity). Those pipe networks with a risk criterion greater than the threshold will not pass through the filter. This process is applied repeatedly—in telescoping fashion (the decision rule or threshold typically becomes more exclusive at a later iteration)—to narrow the list of prospective (sub-)networks. In each successive iteration, either the threshold rule is made more restrictive or more information is gathered. Consequently, the measurement becomes more refined throughout the process.

Finally, it is important to note the versatility of the telescoping filter phase. A subset of the four risk criteria could be used in the early stages of the telescoping filter, since the level of available information and restrictions on further data collection may dictate the options available to the decision maker.

Prioritization phase

Assessing the risk of failures in a pipe network requires the judicious use of engineering judgment. The WIRRF methodology combines engineering judgment with a graphic risk fingerprint to put risk information into representative formats. Once the initial list of pipe (sub) networks is pared down, using data quantification and the telescoping filter, the decision maker is left with an unordered short list of pipe networks that possess the highest risk.

The AHP (or some other ranking methodology) can be used to establish a final ranking for the networks in the short list that resulted from the filtering process. The AHP currently used in the WIRRF methodology generates weights for each of the four risk criteria: prior risk information, functionality, moderate-event risk, and extreme-event risk. The weights are derived through a pairwise comparison among the four criteria. The goal of this pairwise comparison is to weight the relative importance of each criterion against the other. The severity scores of networks in the short list are then multiplied by the derived weights and a weighted sum is used to arrive at a final ranking. The impact of the selected weighting coefficients on the outcome of the ranking order can be evaluated through a sensitivity analysis.

Phase 2 of WIRRF: Budget Allocation

The WIRRF uses the concept of risk in phase 1 to rank the pipe (sub-)networks in decreasing order of risk. The second phase in the WIRRF methodology is to rank maintenance, rehabilitation, and expansion projects that are needed for the pipe (sub-)networks of the highest risks on the short list. Consequently, the goal in the second phase of WIRRF is to assign resources effectively to a number of competing projects by incorporating budgetary considerations. This goal is achieved through benefit-cost analysis in a multiobjective framework.

If the allocated budget is greater than or equal to the combined cost of the proposed projects, then the allocation solution is trivial; all projects are funded. However, if the available budget is constrained, that is, the available budget is less than the total required for all the proposed projects, then the funds need to be allocated as efficiently as possible. Therefore, the problem is how to assign limited resources while minimizing four risk criteria (prior risk information, functionality, moderate-event risk, and extreme-event risk).

Benefit-cost analysis is used in the second phase of WIRRF to identify the most efficient projects. The cost of a given improvement project is obtained from the contractor's or firm's bids to execute the improvement. Denote the cost of a proposed project by $C. The benefits of a given improvement project are defined as the decrease in the total risk

$$\Delta R = w_P(R_P - R_P') + w_F(R_F - R_F') + w_M(R_M - R_M') + w_E(R_E - R_E') ,$$

where (R_P, R_F, R_M, R_E) and (R_P', R_F', R_M', R_E') are risk measures of prior risk information, functionality, moderate-event risk, and extreme-event risk without and with the proposed project, respectively. Consequently, the benefit-cost ratio (B/C) for the proposed improvement project is

$$B/C = \frac{\Delta R}{\$C} .$$

The benefit-cost ratio is calculated for all proposed projects and projects are then chosen in decreasing order of benefit-cost ratio until the budget is exhausted.

Since the severity in each of the four risk criteria is assessed by a five-level scale, it is possible that in certain situations the impact of a project is not large enough to differentiate the risk measure with a project from the risk measure without a project, i.e., (R_P', R_F', R_M', R_E') remains the same as (R_P, R_F, R_M, R_E). The proposed benefit-cost analysis methodology does not recognize the outcome of these projects with small risk-reduction impacts. In most situations, however, this phenomenon does not constitute a limitation for the proposed methodology, since the goal of the budget allocation phase is to identify those projects with the highest benefit-cost ratios.

Epilogue

Societies are critically dependent on the physical infrastructure that provides shelter, water, waste disposal, and transportation. However, the investment in and the maintenance of the infrastructure is a constant and expensive process that is often neglected in favor of the more attractive and often short-term political goals (Griggs 1988).

This paper provides a framework for a methodology that can potentially quickly locate and rank the riskiest water distribution (sub) systems and the most cost-effective projects. The WIRRF methodology provides the decision maker with a flexible approach that structures the decision-making process. The WIRRF methodology makes use of measurement theory, and clearly delineates the various objectives. The risk hierarchy allows the objectives and goals in risk assessment to be stated clearly. The incorporation of fault tolerance (functionality) and extreme events in the risk hierarchy provides the decision maker with a more complete picture of the involved risk. The graphical risk fingerprints clearly represent the contribution of each criterion to pipe network risk. The second phase of WIRRF extends the benefit-cost analysis, an economic tool that is used extensively in industry and for public policy decisions, into a multiobjective framework. It provides the decision maker with a budgeting methodology that places particular emphasis on minimizing the risk while allocating scarce resources in an efficient manner.

Acknowledgments

This paper was supported, in part, by the National Science Foundation under grant BCS-8912630, titled "Integrating the Statistics of Extremes with Conditional Expectation." We gratefully acknowledge the comments and suggestions of B. F. Hobbs and R. G. Quimpo.

References

Andreou, S. A., D. H. Marks, and R. M. Clark. 1987. A new methodology for modelling failure patterns in deteriorating water distribution systems: Theory. *Adv. Water Resour.* 10: 2–10.

Asbeck, E. L., and Y. Y. Haimes. 1984. The partitioned multiobjective risk method (PMRM). *Large Scale Systems* 6, (1): 13–38.

Clark, R. M., and J. A. Goodrich. 1989. Developing a data base on infrastructure needs. *Journal AWWA* (July): 81.

Griggs, N. S. 1988. *Infrastructure engineering and management.* New York: Wiley and Son, Inc.

Haimes, Y. Y., J. H. Lambert, V. Tulsiani, D. Li, J. Pet-Edwards, D. Tynes, and R. Weinstock. 1993. Risk ranking and filtering method. Submitted for publication in *Risk Analysis.*

Johnson, B. W. 1989. *Design and analysis of fault tolerant digital systems.* Reading, Mass.: Addison Wesley.

Keeney, R. L., and H. Raiffa. 1976. *Decisions with multiple objectives.* New York: John Wiley and Sons.

Lowrance, W. W. 1976. *Of acceptable risk: Science and the determination of safety.* Los Altos, Calif.: W. Kaufmann.

Muhlbauer, W. K. 1990. RIPS, a pipeline evaluation system. Paper presented at the Pipeline Engineering Symposium of the Petroleum Division, ASME, New Orleans, January 14–18.

NCPWI, National Council on Public Works Improvement. 1988. Final report by J. M. Giglio, chairman, February.

O'Day, D. K., R. Weiss, S. Chiavari, and D. Blair. 1986. *Water main evaluation for rehabilitation/replacement.* AWWA and USEPA, April 30.

Person, R. A., and D. K. O'Day. 1986. A computerized infrastructure management program. *Public Works* (February): 65–9.

Saaty, T. L. 1980. *Analytic hierarchy process.* New York: McGraw-Hill.

U.S. Air Force. 1988. Software risk abatement. In Software risk management, a tutorial edited by B. Boehm. Washington, D.C.: IEEE Computer Science Press.

Van Blaricum, V. L., and V. F. Hock. 1991. A computerized maintenance management system for water distribution networks. U.S. Army Corps of Engineers Research Lab, *Corrosion91*, paper 515.

Van Blaricum, V. L., V. F. Hock, and A. Kamur. 1989. Analysis of aged water distribution systems. U.S. Army Corps of Engineers Research Lab, *Corrosion89*, paper 592.

Wagner, J. M., U. Shamir, and D. H. Marks. 1988a. Water distribution reliability: Analytical methods. *Journal of Water Resources Planning and Management* 114 (May).

Wagner, J. M., U. Shamir, and D. H. Marks. 1988b. Water distribution reliability: Simulation methods. *Journal of Water Resources Planning and Management* 114 (May).

Walski, T. M., R. Wade, J. W. Sjostrom, and D. Schlessinger. 1989. Conducting a pipe break analysis for a large city. U.S. Army Corps of Engineers, Waterways Exp. Station.

Reliability-Based Assessment of Corps Structures

Mary Ann Leggett[1]

Abstract

Due to the limited amount of funding available for maintenance, rehabilitation, and replacement of aging Corps structures, the director of civil works, Headquarters, U.S. Army Corps of Engineers (HQUSACE) requested a regional assessment of future modernization needs in 1989. In 1990, the Inland Waterways User's Group began formulating this information into an economic traffic-based national evaluation model. As work on this model progressed, the user's group realized that a structure's condition should also be considered. A multidisciplinary team was assembled by the structures branch of the engineering division at HQUSACE to determine the feasibility of modeling a structure's condition. This team established the basis of a methodology for a reliability-based engineering assessment model for Corps structures. Output from the formulated procedure is now required as input into a risk-based economic model that is used to assess Corps structures under consideration for major rehabilitation.

Introduction

Many Corps structures have exceeded their design life and/or capacity, and major maintenance work is necessary to keep them operational. As structures age, their maintenance costs increase. Locks 1 and 2 on the Kentucky River are over 150 years old, while Point Marion Lock and Dam opened in 1993. Although the median age of Corps structures is approximately 40 years, more than 40% of the structures exceed their 50-year design life. Performance below satisfactory levels puts further strains on overloaded systems. Structures within each particular water-

[1]Computer Scientist, Information Technology Laboratory, U.S. Army Engineer, Waterways Experiment Station, Vicksburg, MS 39180

way were generally built at the same time, thus compounding the problem for some Army engineer divisions. Also, the tows currently in use are bigger than the chamber space within most of the older locks.

Major capital requirements will be needed for rehabilitation and efficiency improvement on Corps structures. Presently, revenue for major projects is provided by allocation in the federal budget and, for projects on the inland navigation system, by usage of the fuel taxes placed in the Inland Waterways Trust Fund. With the current emphasis on decreasing the federal budget, available funding will not be sufficient to meet all the system's requirements.

Solution Objective

A long-term investment strategy is needed for identifying and prioritizing critical needs within the Corps of Engineers. When considering investment decisions for these structures, three main areas need to be addressed:

- Current operation and maintenance work
- Planning for future rehabilitation work
- Design for future improvements

The initial purpose of this study was to establish a methodology for evaluating an engineering system's performance or the likelihood (probability) of a problem occurring. Then, the results from this engineering, condition-based evaluation can be used in a risk-based benefit-cost model to determine the optimum course of action for the decision problem under the restriction of limited resources. Additionally, the engineering, condition-based evaluation would provide consistency in comparing the relative conditions of different components of a structure, the relative performances of alternative rehabilitation designs, and the overall conditions of different structures.

Reliability Method

Reliability is defined as the mathematical probability that a system will operate as required. Methods of reliability have been utilized to determine the lifetimes of systems and to relate these lifetimes to factors that influence performance. Civil engineers have begun to use reliability analysis to make a reasonable quantification of the expected condition of a system, such as in the recently developed Load and Resistance Factor Design method for steel structures. In probabilistic analysis, the parameters used in calculations are treated as random variables, where they are represented by probability distributions instead of explicit values. Results from this probabilistic analysis may be expressed in the form of a reliability index.

Traditionally, a factor of safety (FS) is used to measure the safety of structures and their components. For any performance mode, a capacity (C) and a demand (D) function may be expressed as the factor of safety or safety ratio (SR),

$$FS = SR = C/D \; . \tag{1}$$

The capacity function could be a material's strength or ultimate stress, while applied load or applied stress is represented by the demand function. A structure will exhibit unsatisfactory performance when the demand placed on it exceeds the structure's capacity. When capacity equals demand, the limit state for SR is attained, which can be stated as

$$C/D = 1 \; . \tag{2}$$

In traditional deterministic analysis, the components are designed such that the ratio in equation (2) exceeds unity by some acceptable minimum value, which depends on the problem and the performance mode being investigated.

In probabilistic analysis, uncertainty in capacity and demand can be expressed as a probability distribution for each variable, as shown in figure 1, with associated limit states. Probability distributions are constructed by allowing one or more of the independent variables from a deterministic analysis to be random variables and by performing the calculations using the random variables rather than a single value. Then the probability of unsatisfactory performance, $P(u)$, can be expressed as

$$P(u) = P(C < D) = P(C/D<1) \; , \tag{3}$$

and the reliability, R, as

$$R = 1 - P(u) \; . \tag{4}$$

Generally, the distributions of capacity and demand are unknown, since the distributions of the input parameters on which they are based are unknown. But if the mean, μ, and standard deviation, σ, of the input parameters can be determined, then the means and standard deviations of capacity and demand may be calculated. Then, a reliability index, β, may be used to express reliability as a function of the means and standard deviations of capacity and demand. This reliability index may be defined in terms of $E[\ln C/D]$ or $\mu_{\ln C/D}$, and $\sigma_{\ln C/D}$, which are probabilistic moments of the lognormally distributed function (see figure 2), and measure how much the expected average value of the safety ratio exceeds the limit state.

Theoretically, moments can be obtained by integrating the performance function with the distribution range of the random variables. In practice, the integration is usually approximated using simulation methods, Taylor series approximations, or point-estimate methods. Probability distributions of the random variables are replaced by means and standard deviations of the random variables in these methods. Approximate integrations by the point-estimate methods or the Taylor series approach are achieved by performing repetitive, deterministic analyses using the various possible value combinations of the random variables.

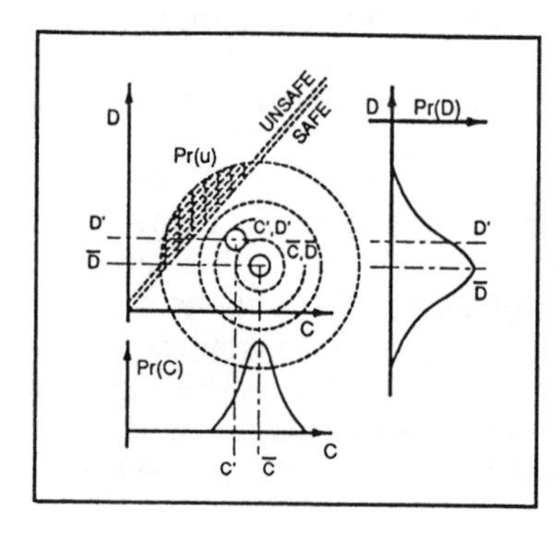

Figure 1. Joint Probability Distribution of C and D

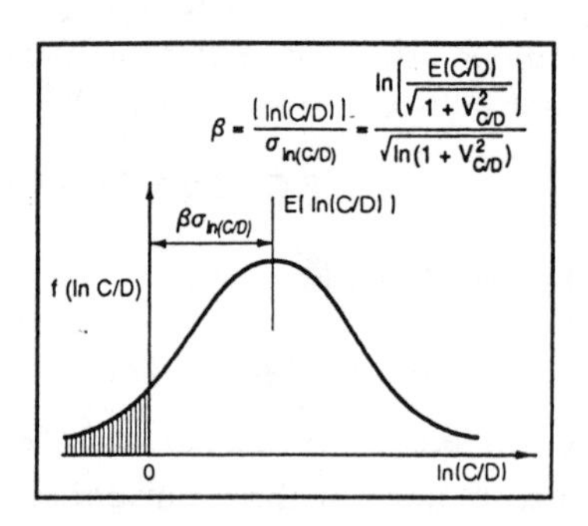

Figure 2. Transformed Lognormal Distribution of C/D

Miter Gate Example

This example illustrates the application of reliability assessment procedures to vertically framed miter gates. Emsworth Lock and Dam is a 110-foot lock located on the Ohio River. Analysis of the vertical beam, a main structural member, of the original upper gate and the replacement gate installed in 1979 is presented. This example demonstrates the method of reliability determination and its associated output, which is used as input into the economic risk-based model and is not intended as an assessment of the gates. A complete report on the reliability assessment of these gates and the upper gate at Monongahela River Lock and Dam 2 is available (Mlakar and Stough 1991). Because the old Emsworth gate and its replacement are basically the same, this example demonstrates a comparison between the assessment of a gate affected by corrosion and fatigue and a gate in an as-constructed condition. Miter gates are subjected to hydrostatic water loads that represent normal conditions, and extreme maintenance conditions, where the gates are used to retain upper pools during lock dewatering. Other loads include resistance to moving the gates through the water and surge loadings. For a single time period, the loading conditions on the gate can be represented by the event tree shown in figure 3.

Vertical beams, analyzed for bending and shear, are assumed to be supported simply at the top and bottom girders and carry water loads that act on the skin plate. Figure 4 shows a cross section of the vertical beam. The dimensions of the gates and the component parts of the vertical beam used in the deterministic analyses of stress are presented in table 1, with the statistical parameters for structural member thickness, variable loads, and limit-state steel stresses. Inspection records and observations were used to obtain the thickness of the vertical beam in the old gate and its standard deviation. The limit-state steel stresses are based on the yield strengths of the steels, which are expected means based on statistical averages. The old gate at Emsworth was fabricated from American Society for Testing Materials (ASTM) A-7 steel with a yield strength of 33,000 psi, and to account for the effects of fatigue due to 500,000 lockages, a 44% reduction in the limit-state yield strength was used. ASTM A-36 steel was used to fabricate the replacement gate. The pool levels are assumed to have fixed values.

Computer spreadsheets (tables 2 and 3) have been used to compute reliability indices using the appropriate loads, statistical parameters, and limit-state stresses. The reliability index for shear in the vertical beam is 2.7.

$$\beta = \frac{\mu_{\ln C/D}}{\sigma_{\ln C/D}} = \frac{0.49596}{0.18231} = 2.7\,, \tag{5}$$

where the logarithm of the mean safety ratio is

$$\begin{aligned} \mu_{\ln C/D} &= \ln \mu_{C/D} - \frac{\sigma^2_{\ln C/D}}{2} \\ &= \ln 1.66959 - \frac{(0.18231)^2}{2} = 0.49596\,, \end{aligned} \tag{6}$$

and the logarithm of the safety ratio's standard deviation is

$$\begin{aligned} \sigma_{\ln C/D} &= \sqrt{\ln\left[1 + (\sigma_{C/D} / \mu_{C/D})^2\right]} \\ &= \sqrt{\ln\left[1 + (0.30693 / 1.66959)^2\right]} = 0.18231\,. \end{aligned} \tag{7}$$

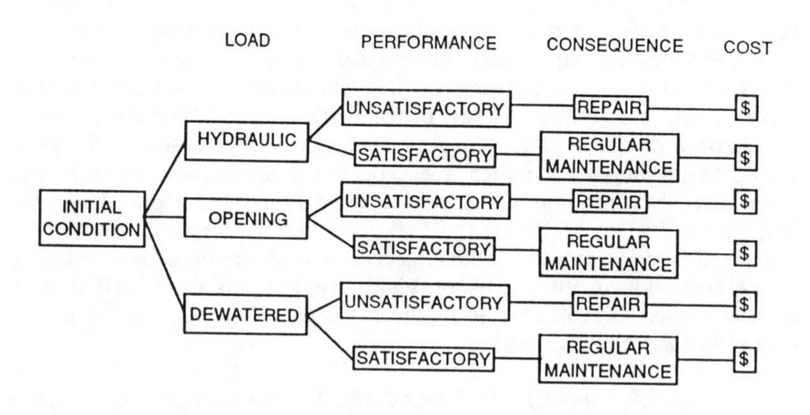

Figure 3. Single Time Event Tree for Emsworth Miter Gates

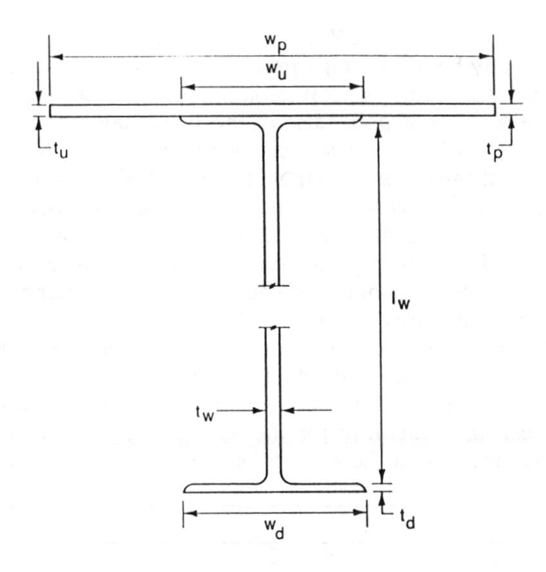

Figure 4. Vertical Beam Cross Section

Standard deviation, $\sigma_{C/D}$, of the safety ratio's (SR) probability distribution is derived from the standard deviations of the random variables. These derivatives $(SR\text{-}SD)$ are approximated by a finite difference expression in the Taylor series method.

	Parameter	Original gate		Replacement gate	
	Constant Values				
H	Height of gate	319.0 in		319.0 in	
v_s	Vertical member spacing	58.0 in		58.0 in	
l_w	Web length	24.0 in		30.0 in	
γ	Water density	0.0362 lb/in^3		0.0362 lb/in^3	
E_{ch}	Chamber elevation	689.67 ft msl		689.67 ft msl	
E_u	Upper pool elevation	710 ft msl		710 ft msl	
E_l	Lower pool elevation	692.67 ft msl		692.67 ft msl	
h_u	Upper pool height	243.96 ft msl		243.96 ft msl	
h_l	Lower pool height	36.04 ft msl		36.04 ft msl	
	Random Variables				
		Mean	Std. Dev.	Mean	Std. Dev.
t_w	Web thickness	0.25875 in	0.0275 in	0.522 in	0.0261 in
F_v	Allowable shear stress	12163.2 psi	1824.47 psi	23694.5 psi	3354.17 psi

Table 1. Values for the Emsworth Miter Gate

A reliability index of 2.7 for shear in the old gate's vertical beam indicates a slightly below average probability of satisfactory performance even under normal loadings. Corrosion, fatigue, and an added load that results from a seven-foot pool raise in the 1930s have resulted in a level of performance at which outages for repair may occur. It is not known whether this pool raise was anticipated in the original design. Before the gate was replaced, four vertical beams developed fractures in the downstream flanges under normal loads. Similarly, a reliability index of 13.3 can be calculated for the replacement gate under the same conditions. This reflects a high probability of satisfactory performance under reasonably anticipated impact loadings. To add torsional stiffness to the replacement gate, the vertical beams were sized to place the skin plate as far upstream as buoyancy would allow. A high reliability index may therefore be attributed to low stresses in the vertical beams. The difference between the reliability for the new gate (13.3) and the old gate (2.7) is a good indicator of the amount of deterioration in the old gate.

Constants	H	l_w	γ	v_s	E_{ch}	E_u	E_l
	319	24	0.0362	58	689.67	710	692.67

RV	t_w	F_v		*SR*	ln(*SR*)	h_u	h_l
Mean	0.25875	12163.2		1.66959	0.49596	243.96	36.04
Std Dev	0.0275	1824.47		0.30693	0.18231		
SR-SD	0.17744	0.25044					
			Beta	2.7			
	0.28625	12163.2		1.84703		243.96	36.04
	0.23125	12163.2		1.49214		243.96	36.04
	0.25875	13987.6		1.92003		243.96	36.04
	0.25875	10338.7		1.41915		243.96	36.04

Table 2. Old Emsworth Vertical Beam Hydraulic—Shear

Conclusion

A reliability-based engineering analysis is preferred over factors of safety as a procedure for assessing the performance level of Corps structures. Factors of safety relate to the reserve capacity between expected loads and structural failure, and do not directly relate to serviceability or performance. In traditional structural analysis, design values are conservatively chosen based on experience and judgment, whereas probabilistic analysis yields reliability indices from the average value and variability of the input variables.

The major emphasis of this study was to provide a method for assessing the condition of Corps structures by means of a reliability-based model that is mathematically sound. This work has established the basis for general criteria development (USACE 1992, 1993) and several supporting models to assess aging structures. Methods developed here have been incorporated into an economic risk-based model that was used to support funding for major rehabilitation at several Corps structures.

Constants	H	l_w	γ	v_s	E_{ch}	E_u	E_l
	319	30	0.0362	58	689.67	710	692.67

RV	t_w	F_v		SR	$\ln(SR)$	h_u	h_l
Mean	0.522	23694.5		8.20182	2.09201	243.96	36.04
Std Dev	0.0261	3554.17		1.29682	0.15714		
SR-SD	0.41009	1.23027					
			Beta	13.3			
	0.5481	23694.5		8.61191		243.96	36.04
	0.4959	23694.5		7.79173		243.96	36.04
	0.522	27248.6		9.43209		243.96	36.04
	0.522	20140.3		6.97155		243.96	36.04

Table 3. New Emsworth Vertical Beam Hydraulic--Shear

Acknowledgments

The author acknowledges the Operations and Engineering Divisions, HQUSACE, for sponsoring this study. The USACE Institute for Water Resources developed the economic risk-based model with which these reliability values are used.

References

Mlakar, P., and B. Stough. 1991. Condition analysis and evaluation of navigation structures: Miter gates. JAYCOR, Vicksburg, Miss.

U.S. Army Corps of Engineers. 1992. Reliability assessment of navigation structures. *Engineer Technical Letter* 1110-2-532, Washington, D.C.

U.S. Army Corps of Engineers. 1993. Guidance for majoı rehabilitation projects for fiscal year 1996. Memorandum CECW-OM-O, dated 17 August. Washington, D.C.

A Theoretical Framework for Risk Assessment

George Apostolakis[1]

Abstract

This paper puts forth a theoretical framework for the development of any risk model. The focus is on the distinction between model and parameter uncertainty.

The Model of the World

The "model of the world" is the mathematical model that is constructed for the physical situation of interest, such as a physical phenomenon's occurrence and impact on a system. The "world" is defined as "the object about which the person is concerned" (Savage 1972). Occasionally, we will refer to the model of the world as simply the model or the mathematical model. Constructing and solving such models is, of course, what most physical scientists and engineers do.

The model of the world may or may not contain uncertain quantities. A simple example of a model without uncertainties is the Darcy equation for ground water flow in saturated media,

$$q(x/k,M) = -k\frac{\partial q}{\partial x}, \qquad (1)$$

where $q(x/k, M)$ is the specific discharge in the x direction, h is the hydraulic head, and k is the hydraulic conductivity. The specific discharge is a function of x and is conditional on the assumption that the

[1]Professor of Engineering and Applied Science, Mechanical, Aerospace & Nuclear Engineering Dept., 38-137 Engineering IV, University of California, Los Angeles, CA 90024-1597

numerical value of the hydraulic conductivity is known and on the assumption that the set of hypotheses, M, that has led to this equation is reasonable for the situation at hand.

Many important phenomena cannot be modeled by deterministic expressions, such as equation (1). For example, the occurrence of earthquakes exhibits variability that we cannot eliminate; it is impossible for us to predict when the next earthquake will occur. We then construct models of the world that include this uncertainty. A simple example that will help us make the discussion concrete is the Poisson model for the probability of occurrence of r events in a period of time, t, that is,

$$h(r/\lambda, t, M) = \frac{e^{-\lambda t}(\lambda t)^{I}}{I!}. \tag{2}$$

The argument of this expression shows explicitly that this probability is conditional on our knowing the numerical values of two parameters, namely, the occurrence rate, λ, and the period, t, as well as on our accepting the assumption that M = {the times between successive events are independent}. As with all probabilities, it is understood that this probability is also conditional on the totality of our knowledge and experience.

The uncertainty in the model of the world is sometimes referred to as "randomness," or "stochastic uncertainty." Other analysts prefer to refer to the probabilities in the model of the world, such as that of equation (2), as "frequencies" (Kaplan and Garrick 1981), reserving the word "probability" for a different kind of uncertainty, which will be discussed later. Stochastic models of the world have also been called *aleatory* models. Webster's defines *aleatory* as "of or depending on chance, luck, or contingency." We will also use this terminology because of its simplicity and because it is not used in other contexts, to avoid confusion.

To generalize the preceding example, let us assume that we are interested in the probability of an event, A, and that this probability can be calculated under a number of mutually exclusive and exhaustive sets of hypotheses, M_i, $i = 1, \ldots, n$. Each M_i has its own set of parameters that are represented by the vector ϕ_i. Then, the required set of probabilities will be given by expressions of the form $h_i(A/\phi_i M_i)$, which form a set of n mutually exclusive and exhaustive aleatory models.

Similarly, when the model of the world contains no uncertainties, we write $q_i(x/\phi_i M_i)$, where we show explicitly that the calculated (possibly by a computer code) quantity q_i is a function of the independent variable vector x and is conditional on the parameters ϕ_i of the ith set of hypotheses M_i.

The Epistemic Model

As stated in the preceding section, each model of the world is conditional on the validity of its assumptions, M_i, and on the numerical values of its parameters, ϕ_i. Since there may be uncertainty associated with these conditions, we introduce the *epistemic* probability density function (pdf) $\pi(\phi_i, M_i)$, which represents our knowledge regarding the numerical values of the parameters and the validity of the model assumptions. Webster's defines epistemic as "of or having to do with knowledge." In risk assessment, this pdf is also referred to as the "state-of-knowledge" pdf. Since the set of models is usually discrete, we can also write $\pi_i(\phi_i/M_i)p(M_i)$, where $p(M_i)$, $i = 1, \dots, n$ is the probability mass function (pmf) of the set of models M_i, and $\pi_i(\phi_i/M_i)$ is the conditional pdf of the parameter vector ϕ_i of the model M_i.

The probability $p(M_i)$ needs more discussion. It is interpreted as the probability that the model M_i is "valid" or "acceptable." Of course, for this interpretation to be meaningful, we must define acceptability. Theoretically, we could identify conditions that, if met, would allow us to declare that a model is valid. These conditions would vary depending on the objectives of the model. For example, in many parts of a probabilistic risk assessment (PRA), the practitioners know that the relevant uncertainties are large and they act accordingly without ever really defining what large means. Thus, a model that may predict a quantity within a factor of two or three is sometimes considered acceptable (in severe nuclear reactor accident analyses, for instance), while such accuracy would be completely unacceptable in other areas (e.g., in reactor physics calculations).

A simple and, perhaps, familiar example demonstrating the importance of the objectives of a model involves the "goodness-of-fit" problem in statistics (Benjamin and Cornell 1970; Winkler and Hays 1975). Given a set of observations, we wish to evaluate the validity of the hypothesis that these observations have arisen from, say, the normal distribution. Our decision whether to accept this hypothesis may be based on the so-called Pearson χ^2 test, which is quantitative and relies on estimating the χ^2 statistic from the observations and, then, comparing it to fractiles of the χ^2 distribution given in standard tables. Alternatively, if great accuracy is not our aim, we may choose simply to plot the observations on normal probability paper and decide to accept the hypothesis of normality, if the data points plot approximately on a straight line.

The preceding example shows that the context within which the decision regarding the validity of a model is made is important. It does not, however, tell us anything about the probability that the model is valid, which is what we have introduced in our formalism. This is because these goodness-of-fit tests are based on relative-frequency-based statistics, where one accepts a hypothesis (when a decision rule, such as the Pearson test, leads to that conclusion) and then acts as if that hypothesis were true. In our formalism, there is a probability that is associated with the validity of each model and this is a fundamental difference. This leads us to say a few words about the philosophical point of view that we have adopted.

Probability is interpreted as a measure of degree of belief (De Finetti 1974). We accept that it is meaningful to say that one event is more, equally, or less likely than another event. Probability is simply a numerical expression of this likelihood. When we say that event A has probability 0.6, we mean that A is more likely than all events whose probability is less than 0.6 and it is less likely than all events whose probability is greater than 0.6. This primitive notion of likelihood, along with several other axioms, allows the development of a rigorous mathematical theory of subjective probability (De Groot 1988) that encompasses the standard mathematical theory based on the Kolmogorov axioms.

This interpretation of probability is to be contrasted with the widely accepted interpretation of probability being the limit of relative frequencies. The issue of which interpretation to accept has been debated in the literature and is still unsettled, although, in risk assessments, there has not been a single study that has been based on relative frequencies. The practical reason is that the subjective interpretation naturally assigns probability distributions to the parameters of models (the vector ϕ of our earlier discussion). The large uncertainties typically encountered in probabilistic risk assessments make such distributions an indispensable part of the analysis. Although some authors have proposed to treat probabilities in the aleatory model as frequencies (Kaplan and Garrick 1981; Parry 1988) and the probabilities in the epistemic model as subjective, this distinction, from a conceptual point of view, is unnecessary and may lead to theoretical problems (Apostolakis 1990). Finally, it is worth noting that Paté-Cornell and Fischbeck (1992) argue that people may have aversion toward epistemic uncertainty and that this attitude should be included in the decisionmaking process. (Formal decision theory does not allow the attitude toward probabilities to be part of the analysis.)

Returning to the probability of accepting a model, we rewrite it as $p(M_i/O)$ to make it explicit that it depends on the objectives of the analysis (as stated earlier, this, and all probabilities, are also conditional on the totality of our knowledge and experience). These objectives can be stated in rigorous terms, in which case it is clear when we have met them, or, as is often the case, they remain unquantified ("everyone knows about them"). This situation reflects the ambiguity that is present in the assessment.

We note that the probability distributions that reflect epistemic uncertainty are the ones that are updated as empirical evidence is gathered. This is done using Bayes's theorem, as shown in numerous references, e.g., Lee (1989), Benjamin and Cornell (1970), Winkler and Hays (1975), and Apostolakis (1990) (the latter also addresses the updating of the epistemic probabilities for models). The use of this theorem guarantees that probabilities are updated coherently (De Finetti 1974). It can be shown that, when the empirical evidence is very strong, the epistemic distributions become delta functions about the exact numerical values of the parameters and the model that emerges is the only valid one. At this point, no epistemic uncertainty exists and the only uncertainty in the problem is the aleatory uncertainty in the model of the world. The latter can never be removed. For example, when all epistemic uncertainty is eliminated, we will know that the Poisson model of equation (2) is the one that applies to a particular phenomenon and we will know the numerical value of the rate λ. However, we will still be unable to predict how many events will occur in a period $(0, t)$ and we will only be able to calculate the aleatory probabilities, as equation (2) dictates.

The Case of One Model of the World

In the preceding formulation, we have considered explicitly a number of models M_i, $i = 1, \ldots, n$. This is the general case, where the uncertainty about which model (hypotheses) is applicable to a physical situation is reflected in $p(M_i/O)$. There are important situations, however, where only one model of the world is available. Even though this model, which may be deterministic or aleatory, may represent the best available technology, we still know that there are significant model uncertainties associated with this model. In other words, this model alone does not meet our explicit or implicit objectives.

To include this situation in our formalism, we introduce a parameter, E, into the model of the world, which may be multiplicative or additive. For example, for an aleatory model we may write $h(A/\phi, M)E$; in this case, there is only one set of hypotheses, M, and the parameter E is multiplicative.

We now consider two cases. In the first case, E is treated as a deterministic parameter of the model of the world. However, its numerical value may be uncertain; therefore, in the epistemic model we represent this uncertainty by a pdf $g(E/O)$. In the second case, E itself exhibits aleatory variability, in which case we introduce its aleatory pdf $f(E/\theta)$. The parameter vector θ of this pdf has its own epistemic pdf, $w(\theta/O)$.

This formulation has been employed in fire risk assessment. The time-to-damage $T_{D,DRM}$ of an object in a compartment where there is a fire is calculated by a deterministic computer code. (The subscript DRM stands for "deterministic reference model.") The actual damage time is considered to be given by the product $T_{D,DRM}E$, where the factor E represents a correction due to the fact that the deterministic reference model is inadequate. In earlier applications, this factor was treated as deterministic with an epistemic pdf reflecting the analysts' uncertainty regarding the amount of systematic over- or underestimation of the damage time by the computer code (Siu and Apostolakis 1981). In later studies, it was recognized that E itself may be random and an aleatory distribution was introduced (Siu and Apostolakis 1985). This distribution was the lognormal distribution with parameters μ and σ. The epistemic model in this case involves a pdf over the vector (μ,σ).

References

Apostolakis, G. 1990. The concept of probability in safety assessments of technological systems. *Science* 250: 1359–64.

Benjamin, J. R., and C. A. Cornell. 1970. *Probability, statistics, and decision for civil engineers*. New York: McGraw-Hill.

De Finetti, B. 1974. *Theory of probability*, vols. 1 and 2. New York: Wiley.

De Groot, M. H. 1988. Modern aspects of probability and utility. In *Accelerated life testing and expert opinions in reliability*, edited by C. A. Clarotti and D. V. Lindley. Amsterdam: North-Holland.

Kaplan, S., and B. J. Garrick. 1981. On the quantitative definition of risk. *Risk Analysis* 1: 11–27.

Lee, P. M. 1989. *Bayesian statistics.* New York: Oxford University Press.

Lindley, D. V. 1985. *Making decisions.* 2d ed. London: Wiley.

Parry, G. W. 1988. On the meaning of probability in probabilistic safety assessment. *Reliability Engineering and System Safety* 23: 309–14.

Paté-Cornell, M. E., and P. S. Fischbeck. 1992. Aversion to epistemic uncertainties in rational decision making: Effects on engineering risk management. In *Risk-based decision making in water resources V,* edited by Y. Y. Haimes, D. A. Moser, and E. Z. Stakhiv. New York: American Society of Civil Engineers.

Savage, L. J. 1972. *The foundations of statistics.* New York: Dover Publications.

Siu, N., and G. Apostolakis. 1981. Probabilistic models of cable tray fires. *Reliability Engineering* 3: 213–27.

Siu, N., and G. Apostolakis. 1985. On the quantification of modeling uncertainties. In *Proceedings of the 8th International Conference on Structural Mechanics in Reactor Technology SMiRT.* Amsterdam: North-Holland.

Webster's New World Dictionary. New York: Simon & Schuster.

Winkler, R. L., and W. L. Hays. 1975. *Statistics.* New York: Holt, Reinhart and Winston.

Risk Analysis: Wet Weather Flows in S.E. Michigan

Jonathan W. Bulkley, F. ASCE[1]

Abstract

Institutional aspects of risk analysis are examined in two separate activities in the state of Michigan. The state of Michigan's relative risk project, patterned after the U.S. EPA's risk reduction report, provided an institutional mechanism to rank environmental issues facing the state of Michigan in terms of relative risk in 1991–92. This project identified twenty-four important environmental issues facing the state. These were categorized into four primary groups ranging from highest priority to medium priority. Seventeen of the identified issues are directly related to water resources. Combined sewer overflows are identified as the primary focus remaining for point-source discharges. The second institutional aspect of risk analysis considered in this paper is the development of a demonstration program for combined sewer overflow in the Rouge River Basin, which will help establish the trade-off between increased cost and risk reduction from such overflows.

Introduction

This paper examines the institutional aspects of risk analysis through two separate activities in the state of Michigan. The first activity is the effort undertaken by the state of Michigan to consider environmental problems facing the state and to assess and rank these problems by the relative risk associated with the several environmental problems. The second activity is to provide a brief report on the National Wet Weather Demonstration Project that is currently under way in the Rouge River Basin in southeast Michigan. The emphasis on the second

[1]Professor of Natural Resources and of Civil and Environmental Engineering; Director of the National Pollution Prevention Center for Higher Education, University of Michigan, 2506 B Dana, Ann Arbor, MI 48109-1115

activity is not on the full range of activities under way for the demonstration project. Rather, the paper focuses on the concept of risk-based decision making between the regulators and the regulated communities and how the work currently under way will provide valuable insights not only for the state of Michigan but for other communities facing difficult choices for control of combined sewer overflows (CSOs).

Part I: Michigan's Environment and Relative Risk

In September 1991, one year after the completion and distribution of the U.S. EPA Science Advisory Board's risk reduction report (U.S. EPA 1990), the state of Michigan undertook a relative risk project to rank environmental issues facing the state. This project and four similar projects in the states of Vermont, Washington, Louisiana, and Colorado were funded by the U.S. EPA; they were derived from the major effort of the U.S. EPA Science Advisory Board previously cited. The state of Michigan was fortunate that Professor William Cooper of Michigan State University had been a leader of the EPA Science Advisory Board's effort, and he provided leadership for the state's relative risk assessment project.

Within the state, the purpose of the relative risk project included the concept of using the information to balance the goals of environmental protection with available funds from both public and private sources with the identified environmental needs. By ranking the environmental needs in terms of relative risk, the decision makers would have guidance on the most important environmental problems facing the state. These rankings then could be matched against available funds in order to assure that the most pressing needs would be addressed according to priority rankings. It should be noted that the issue of population growth was not directly addressed. Population growth includes important aspects that will affect the severity of the environmental concern. The project did adopt the joint declaration of the National Academy of Sciences and the Royal Academy, "If current predictions of population growth prove accurate and patterns of human activity on the planet remain unchanged, science and technology may not be able to prevent either irreversible degradation of the environment or continual poverty for much of the world" (Michigan Department of Natural Resources 1992).

It should be noted that the Michigan Relative Risk Project considered that all issues ranked are important. Furthermore, the project considered residual risks, i.e., risks as they exist with current controls in place including regulations, programs, and in-place infrastructure. Three committees were established to undertake the identification of key environmental issues and then to perform the relative risk ranking of the issues. These committees were a Citizens Committee, an Agency Committee, and a Scientists Committee.

These committees functioned from September 1991 until July 1992. Initially, all three committees met together in a two-day workshop. Subsequently, the committees met separately until it was time to combine the rankings of the three groups. The overall process included the following:

- identify issues
- go to the public (public hearings) and refine the issues
- prepare white papers on each issue
- rank the issues into one of four categories; combine the rankings

Relative environmental risk: State of Michigan

The final effort of the three committees culminated in a joint meeting that produced final relative risk rankings of the twenty-four identified environmental issues facing the state of Michigan. The final rankings produced the following four categories into which all of the identified environmental issues were placed. The four categories included High-High (the category of highest relative environmental risk), High, Medium-High, and Medium (the category of lowest relative environmental risk). The final rankings are shown below (* denotes environmental issues that are closely related to water resource issues). Issues are listed alphabetically within each category.

HIGH-HIGH:
* Absence of land use planning that considers resources and the integrity of ecosystems
* Degradation of urban environments
* Energy production and consumption: practices and consequences
* Global climate change
  Lack of environmental awareness
  Stratospheric ozone depletion

HIGH:
* Alteration of surface water and ground water hydrology, including the Great Lakes
* Atmospheric transport and deposition of air toxics
* Biodiversity and habitat modification
  Indoor air pollutants

* Nonpoint-source discharges to surface water and ground water, including the Great Lakes
* Trace metals in the ecosystem

MEDIUM-HIGH:
* Contaminated sites
* Contaminate surface water segments
* Generation and disposal of hazardous waste
* Generation and disposal of high-level radioactive waste
  Generation and disposal of low-level radioactive waste
* Generation and disposal of municipal and industrial solid waste
  Photochemical smog
* Point-source discharges to surface water and ground water, including the Great Lakes. (Primary focus now: CSOs and aging infrastructure)

MEDIUM:
* Accidental releases and responses
* Acid deposition
  Criteria and related air pollutants
  Electromagnetic field effects

Observations

Of the twenty-four ranked environmental issues, seventeen (or more than 70%) are directly related to water resources. The single most important issue as ranked initially by the three separate committees and that later was ranked most important by the three committees meeting together is the issue, "Absence of land-use planning that considers resources and integrity of ecosystems."

In terms of undertaking these relative risk assessments, the three committees worked with the following common definitions:

1. Risk: Any involuntary exposure to harmful substances or conditions outside of the work place.

2. Residual risk: Risk remaining after current regulatory control and other control measures are considered.

Accordingly, the relative risk ranking performed in the state of Michigan was based on residual risk. The Michigan Department of Natural Resources is considering the results of this relative ranking process.

The process of establishing environmental priorities in the state of Michigan is, as in most states, a complicated activity. The state legislature acts to set policy through funding decisions and through the passage of state laws to constrain the limits of environmental abuse. Beyond the state legislature, environmental policy is implemented through a number of specialized agencies, including the Michigan Department of Natural Resources, the state Health Department, the state Department of Agriculture, and the National Resources Commission. Recently, a special commission has been reviewing legislative acts and statutes for the purpose of consolidating the legislation into a more uniform environmental code.

It should be noted that the discharges from CSOs were ranked medium-high by the Michigan relative risk project. The committees concluded that the discharges from CSOs constituted the primary residual risk associated with point-source discharges for the state of Michigan. This finding reflects the presence of a number of major cities and certain service areas in the state that have combined sewer collection systems. Wet weather events including both rainfall and snowmelt can produce major flow events that result in the discharge to surface waters of a mixture of storm water and sanitary sewage at designated overflow locations. This wet weather problem associated with CSOs has been identified by the U.S. EPA as the water quality need area requiring the greatest combination of financial resources for the country as a whole over the next twenty years. Consider the information in table 1 from the 1992 U.S. EPA Needs Survey (Bureau of National Affairs 1993).

It should be noted that personnel at the municipal level in this country believe that the $41.2 billion for CSO is an *underestimate* of what is actually needed. From a national perspective, CSO is the single most demanding water quality problem when viewed in terms of financial resources required. The other reality associated with all of these financial needs is the fact that the era for federal grants has ended. Federal funds have been utilized to capitalize the state revolving funds (SRFs); these are *loan* funds and not *grant* funds. Accordingly, the local units of government are responsible for paying for the construction as well as the operation and maintenance of all of these facilities. Thus, it is imperative that the funds utilized to correct CSO problems be utilized in the most cost-effective and efficient manner possible. The use of risk reduction provides a viable means to identify such efficient and effective solutions.

$ Amount (Billions)	Activity	% of Total
41.2	Combined Sewer Overflows	30.05
31.3	Secondary Treatment	22.83
17.9	New Collection Systems	13.06
15.5	Advanced Waste Water Treatment Plant Facilities	11.31
14.7	New Interceptor Sewers	10.72
8.8	Nonpoint-Source Agricultural and Silviculture Needs	6.42
3.6	Replacement/Rehab–Sewers	2.63
2.8	Correction of Inflows	2.04
1.2	Groundwater, Estuaries, and Wetlands	.88
.1	Storm Water Controls	.07
TOTAL $137.1 Billion		100%

Table 1. U.S. Water Quality Need Areas

Part II: Wet Weather Flows in S.E. Michigan Institutional Aspects: Risk Analysis

Background

The three-county area of southeast Michigan includes the city of Detroit plus more than seventy-eight additional communities outside the city. All these units of government are served by a very extensive wastewater collection system and the single wastewater treatment plant operated by the Detroit Water and Sewage Department. This treatment plant treats the sewage from a population of nearly three million people and over ten thousand industries in the greater Detroit metropolitan area. The entire city of Detroit is served by combined sewers. Certain areas in both Wayne County and Oakland County outside of the city are served by combined sewers. Other communities in Wayne and Oakland counties are served by separated sanitary and storm sewer systems.

In 1985, CSOs were identified as a major cause of water quality degradation in the Rouge River watershed of southeast Michigan. The Rouge River drains portions of Oakland County and Wayne County as well as a small portion of Washtenaw County. Its drainage is generally the western portion of the city of Detroit and the northwest and western suburban areas. Overall, the Rouge River watershed includes 438 square miles in the three counties and includes all or part of forty-eight separate municipalities in its watershed. The land use in the Rouge River watershed ranges from heavy industry in the southeast portion of the basin to undeveloped or agricultural land in the north and west. More than 50% of the watershed is urbanized. About 25% of the Rouge River Basin is served by combined sewer systems. It is estimated that the CSOs in the Rouge River Basin annually discharge six billion gallons of storm water, including two billion gallons of sanitary sewage to the Rouge River. These discharges take place at 168 discharge locations in the basin.

In 1987, the International Joint Commission (IJC) for the Boundary Waters United States–Canada identified forty-two areas of concern in the Great Lakes. These areas of concern are locations where toxics or other pollutants contribute to poor water quality. The Rouge River is one of these designated areas of concern. There are fourteen areas of concern identified by the IJC in the state of Michigan. In response to the identification of CSOs being a major cause of water quality degradation in the Rouge River, and also in response to the designation by the IJC of the Rouge River's being an area of concern, a remedial action plan (RAP) was developed for the Rouge River by the regional planning agency with input and support from the Michigan Department of Natural Resources. The Rouge River RAP was completed in 1988; the RAP was subsequently adopted by the Water Resources Commission (WRC) of the state of Michigan in 1988. This Rouge River RAP called for the building of more than forty CSO basins in the watershed as well as the separation of sewers in certain municipalities within the basin.

On October 19, 1989, the WRC issued NPDES permits to nine entities within the Rouge River watershed including both Wayne County and the city of Detroit. These permits specified control of the CSOs in each of the designated areas. The permit for the city of Detroit also included the operation of the wastewater treatment plant. In the view of the WRC, these NPDES permits issued in October 1989 reflected the goals of the Rouge River RAP—specifically the elimination of raw sewage discharges to the Rouge River and the protection of the public health by 2005.

However, the permittees did not agree. They felt that the NPDES permits went far beyond the RAP developed for the Rouge River Basin. In particular, the city of Detroit, Wayne County, and Oakland County requested that the federal district court take jurisdiction over the Detroit permit and the Rouge River CSO permits. The federal court determined it had pendent jurisdiction over the "time and manner" needed to meet permit standards, but it did not take jurisdiction over the entire permit. The court appointed a monitor to attempt to negotiate a settlement between the parties. The Michigan Department of Natural Resources never agreed to the jurisdiction of the federal court in this matter, but the department did agree to work with the court and the monitor to settle the permittees' contested cases.

With regard to the CSO issues in the contested NPDES permits, the permittees were very concerned with the specified requirement that the CSO basins be sized to capture and provide at least a thirty-minute detention, the CSO overflow resulting from a once-in-ten-year storm with a storm duration of one hour. In southeast Michigan, a storm of this magnitude would deliver about two inches of rainfall in one hour. The permits also required a three-phase approach to correcting the CSO problem in the Rouge River watershed. Phase I is to operate, repair, and maintain the existing facilities to minimize the discharge of raw sewage. Phase II requires that CSOs be controlled to eliminate the discharge of raw sewage and protect the public health by 2005. Phase III requires that additional controls be applied to comply with water quality standards at times of discharge.

In accordance with the direction from the federal court, the monitor initiated the negotiations to resolve the CSO conflicts between the regulatory agency and the permittees. This conflict-resolution process lasted from August 1990 to June 1991. The resulting settlement reflects institutional aspects of risk analysis applied in the context of control of CSO.

Insights

It needs to be recognized that a certain level of CSO control already exists in southeast Michigan. For example, there are nine CSO facilities currently operating to control CSO/excess inflow in southeast Michigan. The volumes of these facilities range from 4.5 MG to 94.7 MG. The average control level for the existing facilities is 4.52 MG/1000 acres of CSO area. An example (Hinshon 1990) has been developed that illustrates the level of control that would be provided if the permittees implemented the CSO permits as issued. Consider the following:

1. Capture and treat ten-year, one-hour storm for 30 minutes.

$Q = CIA$ $\quad$ C = Coefficient of runoff
I = Rainfall intensity (inches/hour)
A = Area (acres)

For an area 40% impervious, C = 0.34.

Q = CIA = .34 x 2"/hr x 1000 acres = 680 cfs

680 cfs x 30 min x 60 sec/min x 7.48 gal/cu ft =
9.16 MG/1000 acres

The size of the basin would be 9.16 MG for every 1000 acres of CSO area to be served.

2. The MDNR policy document for CSO indicated that a facility designed to retain for transport and treatment the flow from a one-year, one-hour storm will also provide roughly thirty minutes of treatment for flows generated by a ten-year, one-hour storm. As part of the CSO policy, the MDNR was also requiring the complete capture of the one-year, one-hour storm. In southeast Michigan, the one-year, one-hour storm has a rainfall intensity of 1"/hour. Accordingly, to capture and retain a one-year, one-hour storm, consider the following:

1"/12"/ft x 1000 acres x 43,560 sq ft/acre x 7.48 gal/cu ft =
27.2 MG/1000 acres

If runoff volume is 55%,

27.2 MG/1000 x .55 = 15.0 MG/1000 acres.

Clearly, these are very rough calculations; however, the results indicate that these proposed permits would require a much higher level of control than exists at present. Further, the assumption that if one sized a basin for the ten-year, one-hour storm with a thirty-minute detention, it would capture and retain the one-year, one-hour storm does not appear valid.

Another aspect that troubled the permittees was the significant divergence between what the NPDES permits specified for CSO control and what the Rouge River RAP had incorporated. For example, the RAP utilized a one-year, one-hour storm with a twenty-minute detention time to size the CSO basins. Accordingly, consider the following:

Q = CIA = .34 x 1"/hr x 1000 x 1000 acres = 340 cfs

340 cfs x 20 min x 60 sec/min x 7.48 gal/cu ft =
3.05 MG/1000 acres

It is clear that the NPDES permits as issued in October 1989 symbolize a major change for handling CSOs in the Rouge River Basin. The resolution of the conflict over the size of the CSO basins is ongoing. Institutional aspects of risk reduction have played and will continue to play an important role in these activities.

To conclude the insights portion of this presentation, note that the control of CSOs in the Rouge River Basin itself would set the pattern for correction of CSO throughout the greater Detroit service area. It is estimated that in the area served by the Detroit Water and Sewage Department, including a portion of the Rouge River watershed, the total area served by combined sewers is roughly 115,000 acres. From the previous calculations shown, it is clear that the ten-year, one-hour storm with a thirty-minute detention and the complete capture of the one-year, one-hour storm would require significant numbers of storage basins. Also, both of these requirements far exceed what the communities thought had been agreed to in the Rouge River RAP. This range of impact in terms of the number of 30 MG overflow basins is shown below.

1. Rouge River RAP

   3.05 MG/1000 acres x 115* = *350 MG,*
   or roughly 12 basins = 30 MG each

2. TenYear, One-Hour Storm = Thirty-Minute Detention

   9.16 MG/1000 acres x 115* = 1058 MG,
   or roughly 35 basins = 30 MG each

3. One-Year, One-Hour Storm = Complete Capture

   15.0 MG/1000 Acres x 115* = 1725 MG,
   or roughly 58 basins = 30 MG each

*It is estimated that the service area includes 115,000 acres of combined sewer area or $\frac{115{,}000 \text{ acres}}{1000 \text{ acres}} = 115.$

Conflict resolution

The immediate problem required exploration of whether or not the regulatory agency, Michigan Department of Natural Resources, and the communities, the city of Detroit and communities represented by

Oakland County and Wayne County, could agree on a mutually satisfactory program to enable the NPDES permits to be issued and actual implementation of the CSO program to begin. The negotiations extended from August 1990 to June 1991. While these negotiations under the monitor appointed by the federal court were under way, a separate and independent Contested Case Hearing on these NPDES permits was initiated through the MDNR. This Contested Case Hearing took place before a hearing officer and basically was a trial. All of the parties were represented by attorneys; expert witnesses were called to testify before the hearing officer. The Contested Case Hearing on the CSO issue was adjourned once the negotiation process came to closure.

In contrast, the monitor worked primarily with the technical representatives of the regulatory agency and the communities. Lawyers were present as needed. The negotiations among the parties resulted in a settlement agreement for the CSO issues of the NPDES permits that were approved by the parties on June 28, 1991. There are a number of important elements in this settlement agreement, including the following:

1. The CSO program in the Rouge River watershed is to be a *regional* program.

2. There will be a phased effort to control the CSO overflows.

3. There will be prompt construction of a series of CSO control facilities in the Rouge River Basin, including basins, tunnel storage, and sewer separation.

4. The CSO basins to be contracted will be of different sizes. The sizes include
   - one basin to provide the ten-year, one-hour storm with a thirty-minute detention
   - four basins to provide the one-year, one-hour storm with a thirty-minute detention
   - three basins to provide the one-year, one-hour storm with a twenty-minute detention
   - one basin will be as large as possible to provide as much treatment as possible for a one-year, one-hour storm (This site has very limited space to build a basin.)

5. This negotiated approach will help determine the optimum basin design needed to meet the Rouge River RAP and to meet water quality standards at the time of discharge.

6. The demonstration projects have been designated as Phase II A projects.

7. The facilities built after the demonstration projects have been completed and have undergone performance evaluation will be designated as Phase II B projects.

8. This cooperative effort should provide significant information and much better information than is presently available on the "correct" or "right" size of a retention basin in an urbanized community such as Detroit.

9. In lieu of two basins for one community, the community has elected to build a tunnel storage project.

The design and implementation of Phase II A is currently under way.

Concluding Observations

One important point to emerge is that the regulatory agency had very little actual information available regarding setting standards for sizes of retention basins. The negotiated settlement provided a viable mechanism whereby the demonstration project would be undertaken in a timely fashion. The actual operations of this set of demonstration projects would provide solid information of value not only to other communities in the Rouge River watershed but for other CSO areas in southeast Michigan. It is anticipated that these results will be valuable to other communities throughout the country. The basic question addressed throughout the negotiation process was one of risk reduction. No one could say how much better a large basin with longer detention time would be in terms of CSO control. As a consequence, the demonstration projects are being designed to develop the information needed to provide for the optimum-sized basin to control CSOs in urban settings. Accordingly, the trade-off between increased size (cost) and associated risk reduction from the overflow of combined sewage will be more clearly established. A final note: the Congress of the United States has taken note of the unique situation existing in the Rouge River watershed and a National Wet Weather Demonstration Project has been established that will extend these demonstration projects for CSO by including storm water discharges, contaminated sediments, and the full range of issues contributing to adverse impacts on surface water quality as a consequence of wet weather events.

References

The Bureau of National Affairs. 1993. *Environmental Reporter* October: 997.

Michigan Department of Natural Resources. 1992. *Michigan's Environment and Relative Risk,* Michigan Relative Risk Analysis Project. July.

Hinshon, R. 1990. Working paper. Hinshon Environmental Consultants. October.

U.S. Environmental Protection Agency, Science Advisory Board. 1990. *Reducing risk: Setting priorities and strategies for environmental protection*. Washington D.C. September.

Poorer Is Riskier: Opportunity for Change

Richard C. Schwing[1]

Abstract

With regard to water resources, it is appropriate to build upon two interacting themes of the late Aaron Wildavsky (1988). The first of Wildavsky's themes is the idea of *resiliency*. The second theme, which interacts with resiliency, is *"richer is safer."*

Existing empirical data are provided to support some of Wildavsky's views. There is a universal tendency in the advanced nations of the world for three important statistical variables to increase fairly uniformly as time progresses. They are safety, as measured by life expectancy; quality of life, as measured by real GNP per person; and energy use per person.

Causation cannot be proven by these correlations. However, the data indicate that increased available energy leads to increased development, which in turn leads to increases in average life span. A final section provides several examples of microlevel and macrolevel mechanisms for income to cause an increase in longevity.

Richer Is Safer

Contrary to the thesis of this paper, one can imagine a number of ways in which wealth leads to poor health. Wealth that leads to a lack of exercise, overeating, job stress, long working hours, and substance abuse is detrimental. On the other hand, access to health care, nutrition, improved housing, healthier and safer working conditions, and a more

[1]Principal Research Engineer, Operating Sciences Department, Research & Development, General Motors Corporation, 30500 Mound Road, Warren, MI 48090-9055

secure and less stressful life experience contributes to better health. While the good and the bad are inextricably mixed, the data indicate that the benefits of wealth or income overwhelm the detriments.

As a result, we support Wildavsky's view that almost all reductions in the wealth of a citizenry can lead to measurable impacts on their well-being. Specifically, their health is degraded and their lives are shortened. Morbidity and mortality data from several sources demonstrate that the impacts are significant.

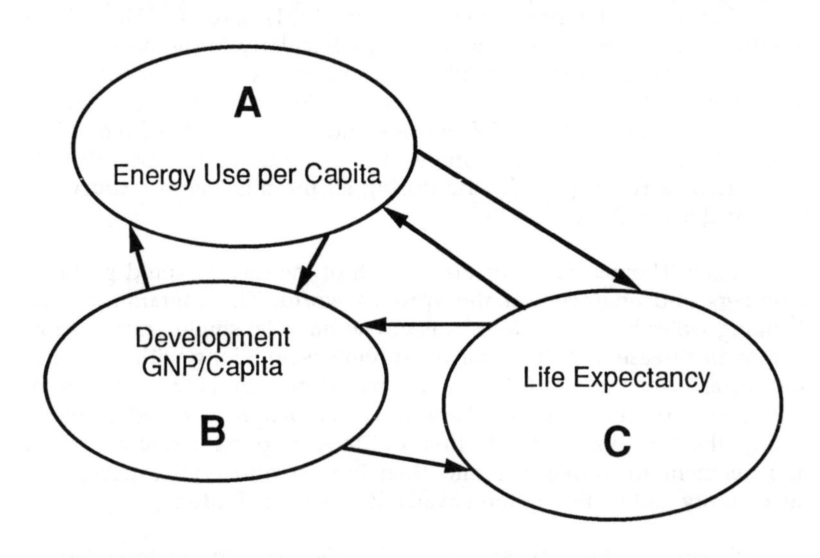

Figure 1. Relationship Between Longevity, Energy, and Development (Starr and Searle 1990)

Because the impacts are indirect, causality cannot be proven. However, evidence from several sources now demonstrates that the relationships are robust and consistent. Furthermore, regarding mortality rates or longevity, the evidence is manifold. The impacts are diagrammed in figure 1 to illustrate their interactive nature. The following section includes evidence from several studies to illustrate the correlations.

Energy use versus development: Path A-B

In an elegant paper, Starr and Searle (1990) analyzed the relationship between development (standard of living as measured by GNP per capita) and either primary or electrical energy.

Starr and Searle provided three datasets demonstrating how tightly gross national product is coupled with primary energy or electrical energy. Correlations for the U.S. are presented in figures 2(a) and 2(b). Cross-section data from several nations are provided in figure 3. Because the correlations are so strong, one would expect development to be *directly* influenced by energy use.

Life expectancy versus development: Path B-C

The medical writer Lewis Thomas (1984) asserts, "There is no question that our health has improved spectacularly in the past century, but ... one thing is certain, it did not happen because of medicine, or medical science, or the presence of doctors. ... Medical care itself—the visits by doctors in the homes of the sick and the transport of the sick to hospitals—could have had no more than marginal effects on either the prevention or reversal of disease during all the nineteenth century and the first third of the twentieth."

Later Thomas surmises that "Much of the credit should go to the plumbers and engineers of the Western world. The contamination of drinking water by human feces was at one time the single greatest cause of human disease and death for us; it remains so, along with starvation and malaria, for the Third World. ... Long before plumbing, ... during the 17th and 18th centuries, we became richer people and were able to change the way we lived. The first and most important change was an improvement in agriculture and then human nutrition, especially in the quantity and quality of food available to young children."

Sagan and Afifi (1978) in analyzing the longevity of some 99% of the world's population also find medical technology far less important than economic growth. They conclude that "economic development has been shown to add approximately thirty years to life expectancy."

Life expectancy versus energy consumption and electricity consumption: Path A-C

Energy consumption, just shown to be a correlate of development, has been investigated by Nathwani, Siddall, and Lind (1992) as a predictor of life expectancy. These authors have compiled data from around

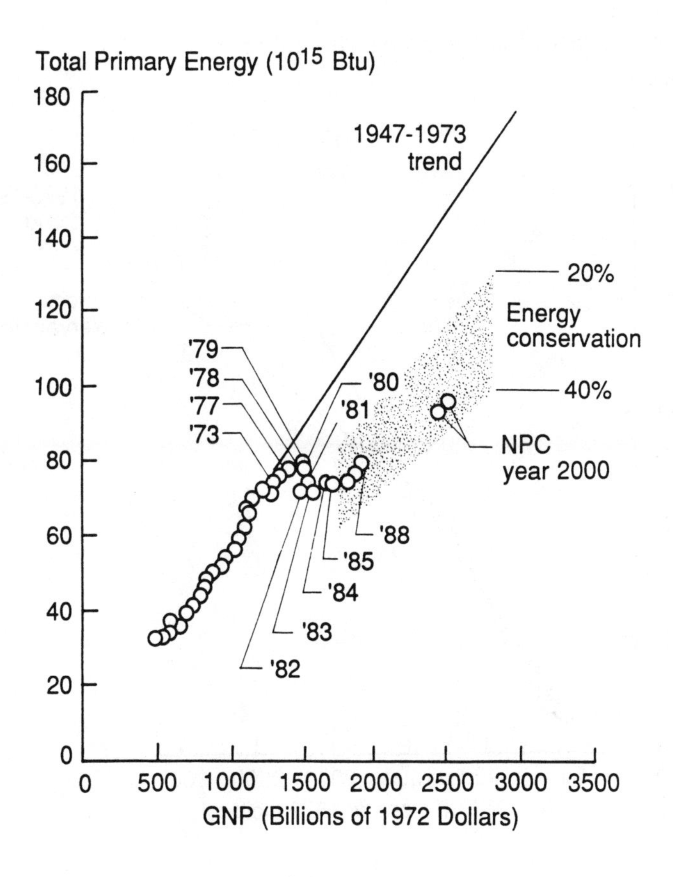

Figure 2(a). Correlation Between Primary Energy Use and Development (Starr and Searle 1990)

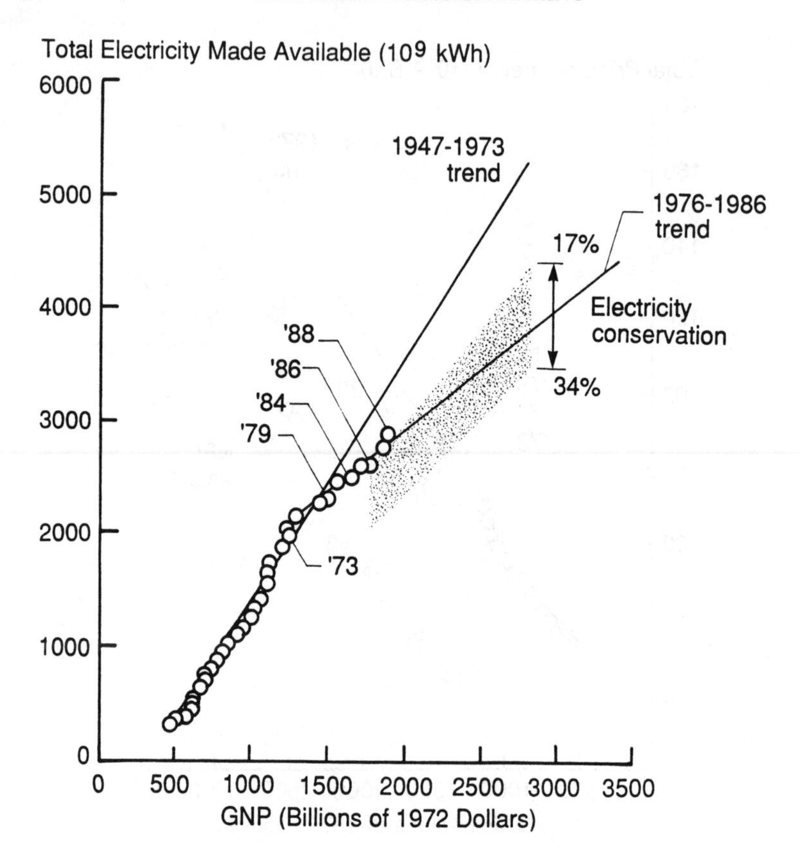

Figure 2(b). Correlation Between Electrical Energy Use and Development (Starr and Searle 1990)

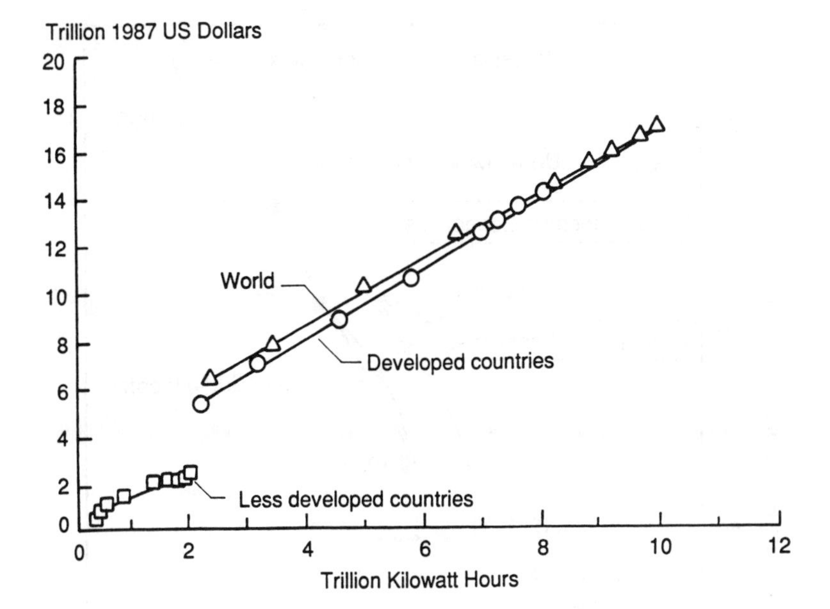

Figure 3. Correlation Between Development and Energy Use; Cross-Section Data for Developed and Less Developed Countries (Starr and Searle 1990)

the world and have derived robust correlations. Figure 4, among several in their study *Energy for 300 Years,* shows life expectancy versus energy consumption per capita and demonstrates the powerful impact of energy consumption on life expectancy.

Of course, the more efficiently energy can be utilized to accomplish its purpose, the more directly energy consumption will influence well-being or life expectancy. As discussed by Mills (1991), electricity is usually the most efficient method to deliver energy to its intended purpose.

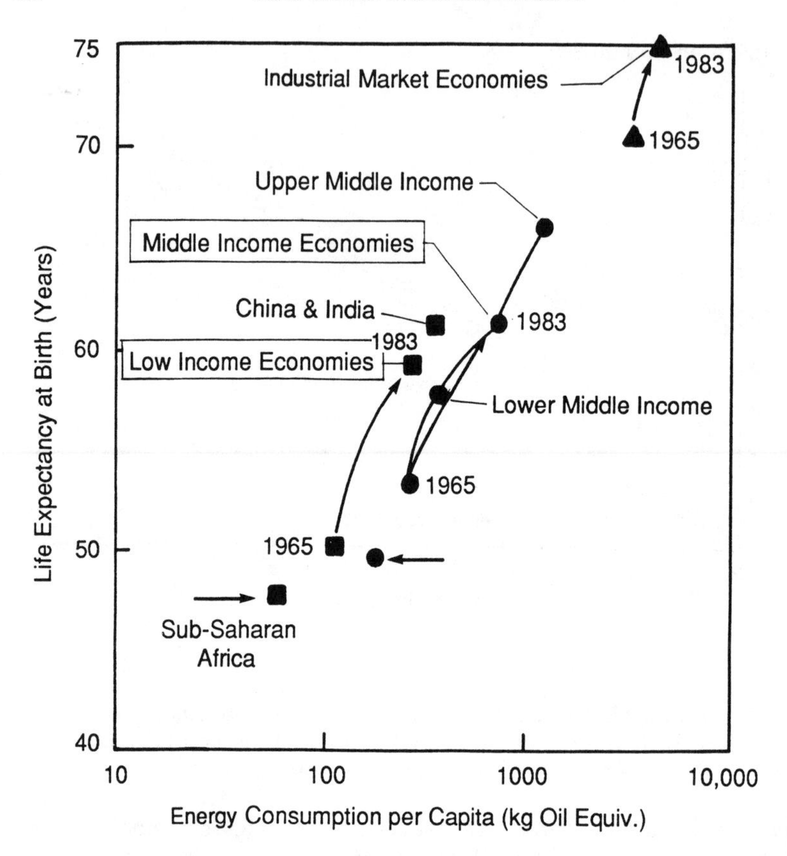

Figure 4. Correlation Between Life Expectancy and Energy Consumption for Different Economies (Nathwani, Siddall, and Lind 1992)

Among dozens of correlations contained in the study by Nathwani, Siddall, and Lind (1992), a correlation containing data for several well-developed economies is included here. Figure 5 shows the relationship between life expectancy and electricity consumption in Canada, the U.K., the U.S., and Japan. Again, a robust and positive impact on life span can be seen.

In summarizing the information that shows the positive effects of energy and/or electricity consumption on life expectancy, Nathwani, Siddall, and Lind (1992) argue that "energy *is* wealth, and it clearly plays an important part in the creation of more wealth. Wealth leads among other things to better nutrition, better housing, increased medical knowledge and technology, and the (ability of) people and institutions to apply them. Energy also contributes directly to health through cleanliness, warmth and light."

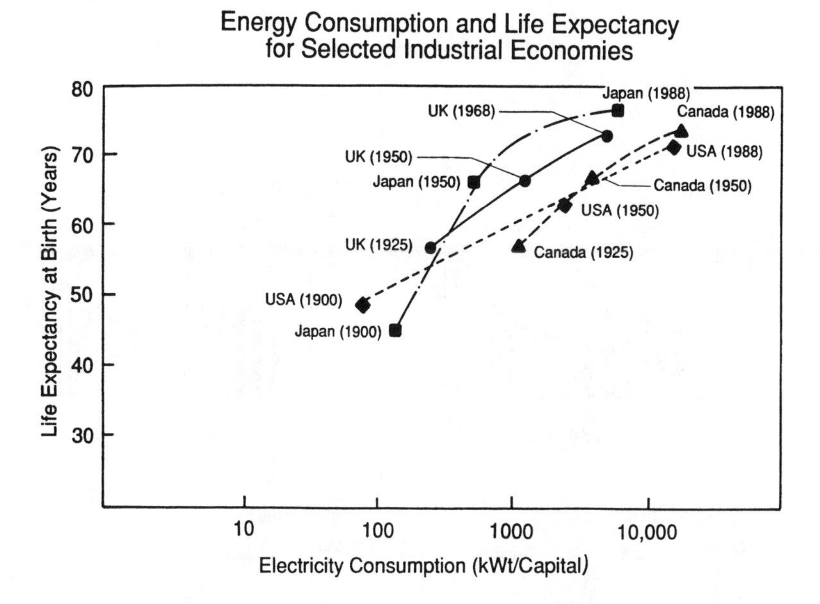

Figure 5. Correlation Between Life Expectancy and Electricity Consumption for Selected Industrial Economies (Nathwani, Siddall, and Lind 1992)

Life expectancy versus average income, social class, family income, and employment

In his book *Searching for Safety*, Wildavsky (1988) boldly states, "Increasing the income for countries or classes, as we shall see, increases their safety far more than all regulations to reduce risk."

Dozens of studies verify these general relationships, and in a survey article Antonovsky (1967) states, "Despite the multiplicity of methods in the thirty-odd studies cited, and despite variegated populations surveyed, the inescapable conclusion is that [economic] class influences one's chances of staying alive. Almost without exception, the evidence shows that classes differ on mortality rates."

More recent data from Britain, reported by Hart (1986), reveal the magnitudes by age, sex, and class (see figure 6). These data, shown in bar graphs for the four age groups, indicate the strongest gradients with class are imposed on youth.

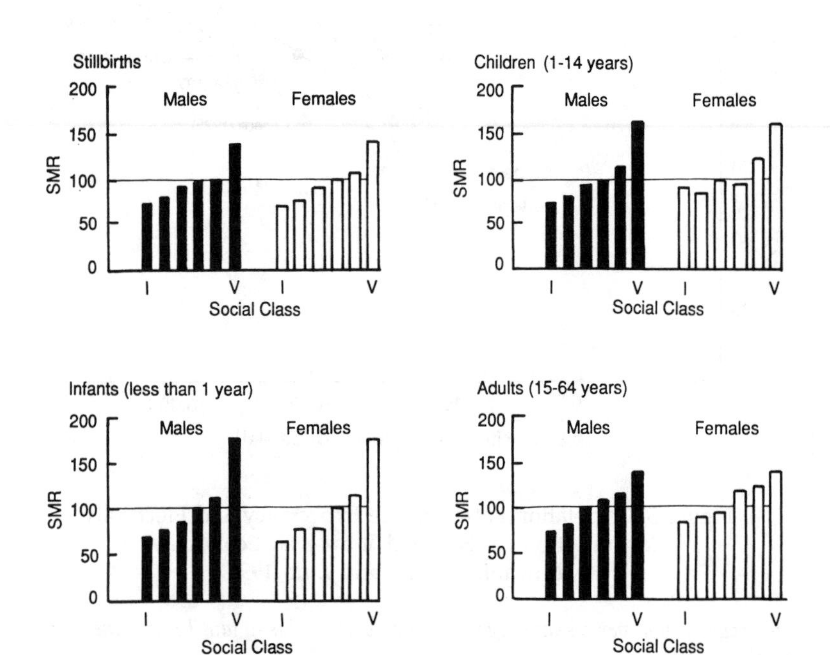

Figure 6. Standard Mortality Ratios for Males and Females versus Social Class (Hart 1986)

John Fox (1977) of the Office of Population Censuses and Surveys in Great Britain provides similar data for infant deaths, broken down by diseases and accidents (see figure 7). For infant deaths, steeper gradients are recorded for causes such as acute bronchitis, bronchopneumonia, and accidents than for conditions associated with pregnancy and birth or with congenital anomalies.

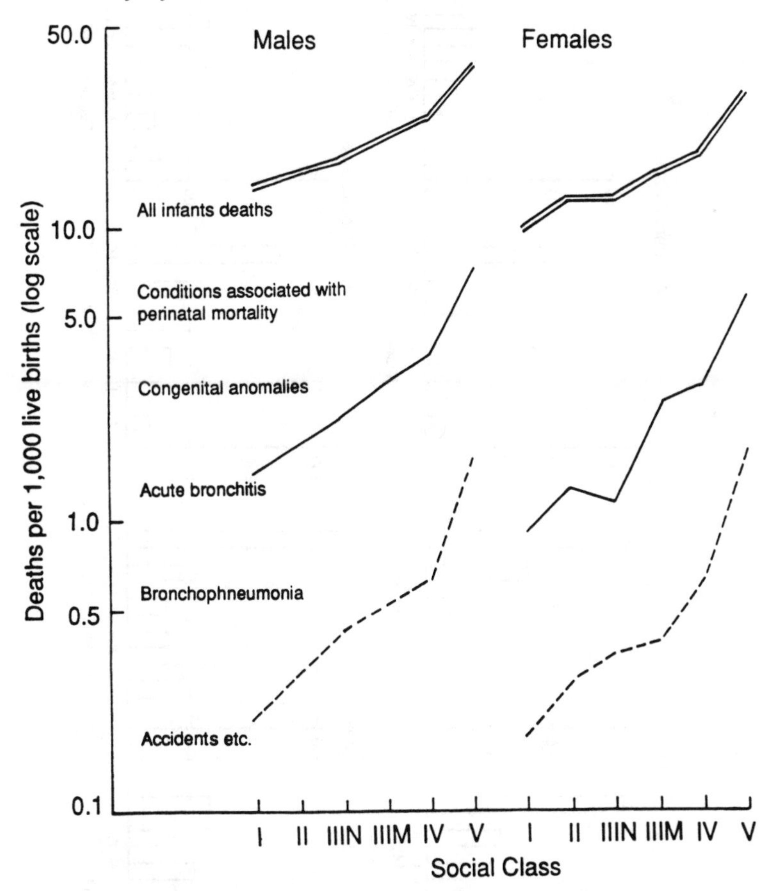

Figure 7. Infant Mortality versus Social Class (Fox 1977)

The most tragic impact of social class in Great Britain is shown in graphs for children under fourteen years of age (Office of Population Censuses and Surveys 1978). In figure 8, I have plotted these data for accidental falls, pedestrians, fires, and poisonings. In the category of accidental fires for males in 1970–72, the mortality for Class V is sixteen times greater than the mortality for Class I. A similar set of data, analyzed by Alison Macfarlane (1979), is provided for children aged one to fourteen in *home* accidents as shown in figure 9.

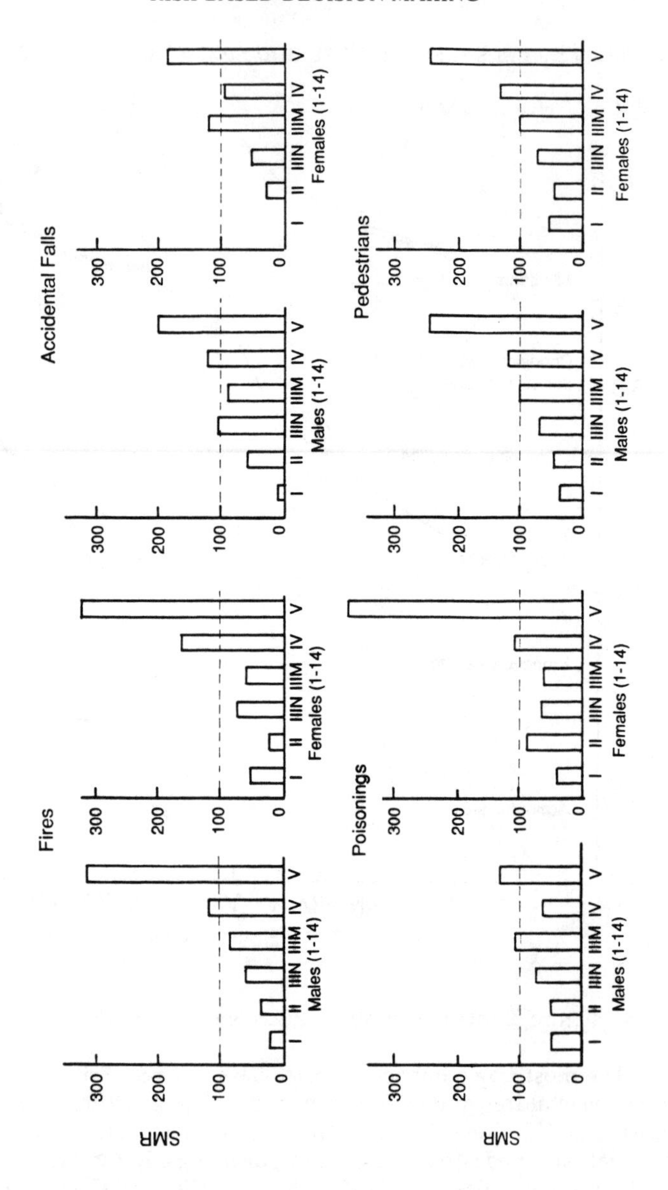

Figure 8. Accident-Specific Standard Mortality Ratios for Males and Females Aged 1–14 (Office of Population Censuses and Surveys 1978)

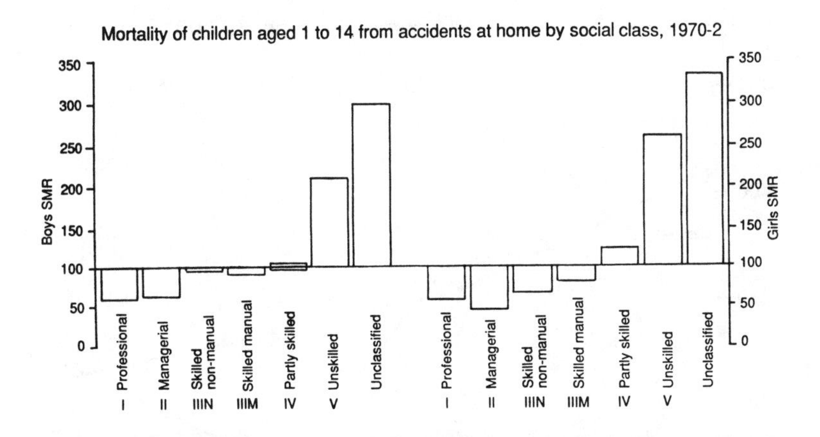

Figure 9. Home Accident Mortality Ratios for Children Aged 1–14 (Macfarlane 1979)

Because of taboos in labeling social class, similar data do not exist for the United States and Canada. However, one study by Claudia R. Baquet and colleagues at the Division of Cancer Prevention of the National Cancer Institute in Bethesda, Maryland, (1991) does indicate that socioeconomic status has a measurable impact on the incidence rates of certain cancers. The net effect of these impacts can be seen in figure 10.

Finally, as development is impeded to the extent that unemployment increases, changes in self-esteem, depression, compulsiveness, and anxiety take a toll on well-being. Linn, Sandifer, and Stein (1985) have shown that the unemployed visit their physicians five times as often and spend five times as much time in bed. Another database, analyzed by Tabor (1982) and Brenner (1973), indicates that a 1% increase in unemployment in the U.S. over a five-year period leads to as many as 36,900 deaths, including 19,000 excess heart attacks and 1,100 suicides. According to the calculations of Keeney and von Winterfeldt (1986), a 1% change in unemployment induces one excess death for every twenty-seven persons losing their jobs.

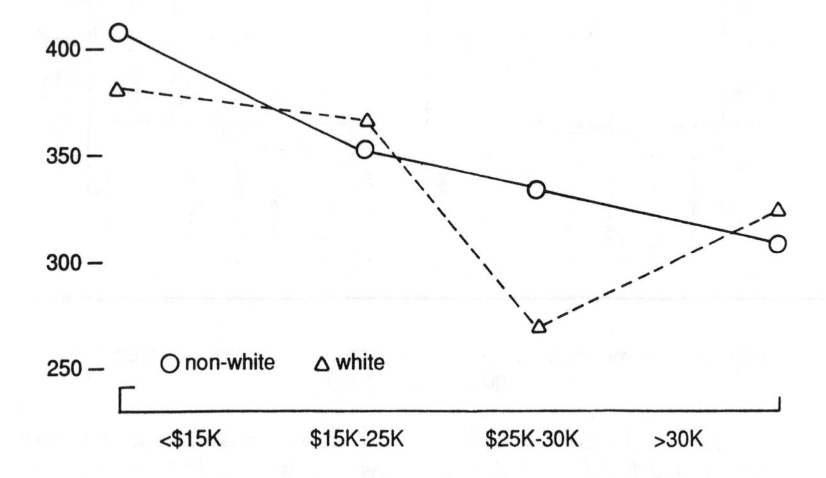

Figure 10. Cancer Incidence by Race and Income Class (per 100,000 population) (Baquet 1991)

Models of Causation

In this infinitely complicated world, it would be impossible to prove that wealth directly improves longevity. However, plausible mechanisms exist at the micro (personal) and macro (policy) level to link longevity to income.

Microlevel mechanisms

To improve the odds of extending my own life span, I could consider a number of expenditures, including living in a masonry home instead of frame; renting a "cherry-picker" instead of a ladder to trim my trees; equipping a number of my receptacles with ground-fault detectors; obtaining security lighting; linking my home with police and fire services via available electronic devices; purchasing a heavier vehicle with dual airbags and antiskid brakes; and purchasing more frequent and elaborate health tests to detect the emergence of a threatening disease. The list could go on and on, but a time constraint and a budget constraint have prevented or postponed these options for my family and me.

Clearly, using actuarial criteria, many of these options are good choices. However, my ultimate demise will be due to something I have neglected, I have inadequately protected against, or which cannot yet be prevented.

Needless to say, most of us already have a long list of options already exercised but not yet available to everyone. It is usually an income constraint that makes the difference. I find it both tragic and ironic that the "quality child care" lobby in the U.S. does not use the data in figure 9 to bolster its case.

Macrolevel mechanisms

Policy analysts in the Office of Management and Budget in the United States have examined fifty-three policies designed to save lives. The results of their study are contained in the 1992 U.S. Budget Document (U.S. Government 1991). I have published (Schwing 1991, 1992) the plot shown in figure 11, which depicts the cost of saving lives on the vertical axis and the number of lives saved on the horizontal axis.

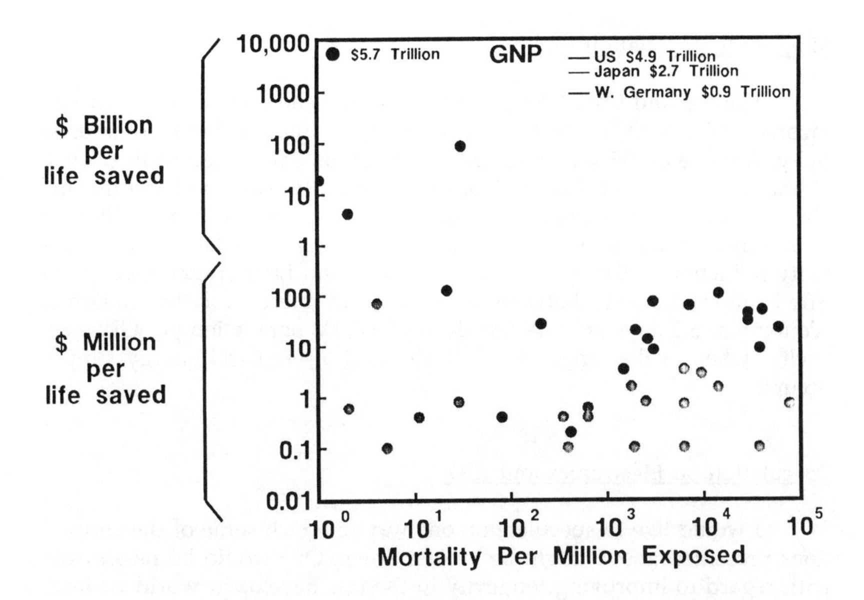

Figure 11. Risk Reduction Data for Costs of Health and Safety (Schwing 1991)

With regard to the "richer is safer" theme, there are three elements in figure 11 to emphasize. They are

- existing policies trade longevity for substantial costs
- policy efficiencies vary over a wide range
- by trading efficient policies for inefficient policies, we could extend the population's longevity even further

The possibility exists that some of the most efficient options for risk reductions exist at the individual micro level and that macro national policies can make us less safe unless they too are efficient. Surely some of the programs that are shown in figure 11 exceed the ability of most economies to abate risk. *As the portfolio of risk abatement options grows, it is paramount to know that the list we choose to implement is efficient in terms of the available resources.*

Ethicists are uncomfortable imposing an expenditure constraint on life-saving issues. However, when we realize that expenditures for the highest points in figure 11 match or rival the GNP of several nations, the reality gives us pause. Society must acknowledge that, even in life-and-death matters, resources are not infinite.

Magnitude of Relationship

Hadley and Osei (1982) have quantified the relationship between income and health in the United States by estimating the income elasticity. A value of .05 was obtained using census data at the county level. Graham, Chang, and Evans (1992) have recently estimated the income elasticity in a more complex model to be 0.034. For the combination of family income and unearned income, a 1% change translates into a mortality reduction of 0.05%. Nathwani, Siddall, and Lind (1992) have quantified the relationship between mortality and energy for the Canadian economy as 6.2 days per GW (.85 deaths/100,000 population per GW electricity) when in the range of a 20% shortfall of electrical energy supply occurs.

Speculation on Electronics and Risk

I would like to speculate on one way in which some of the correlations presented above might be broken down. One would be pessimistic with regard to improving longevity in the less developed world without a huge transfer of resources, given the strong correlation between longevity and development.

With regard to natural hazards, however, there might be reason for optimism. The cost of computing power is halved every eighteen months (see figure 12). Although my telephone bill does not show it, I believe a similar relationship exists in communicating data and text.

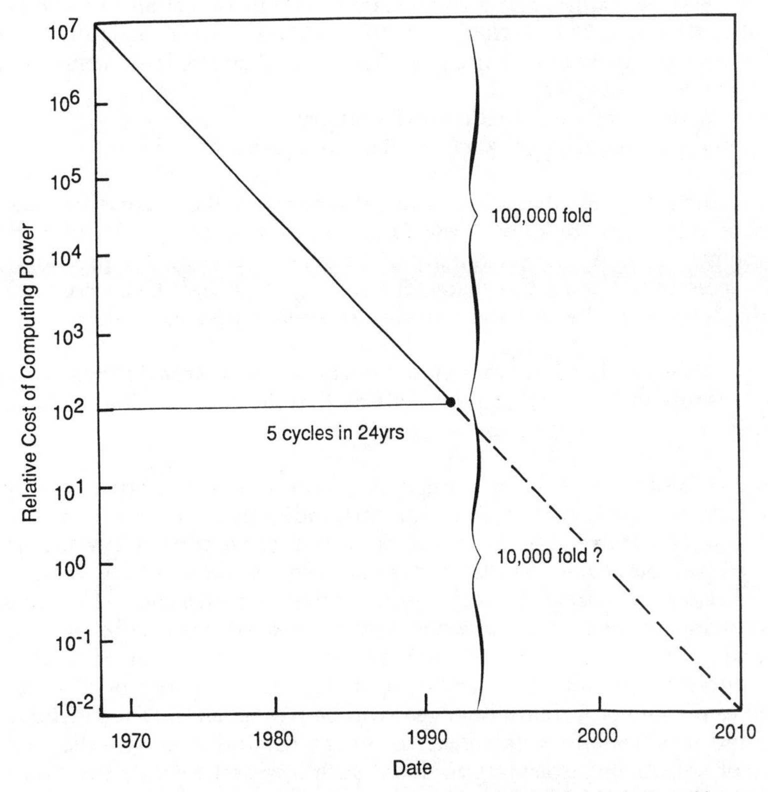

Figure 12. Relative Cost of Computing Power (Zachary and Yoder 1993)

As there are many aspects of hazard abatement that depend on communication and computer technologies, we might be optimistic about providing cheap means to reduce natural hazard risks even though society may not be able to afford to abate other significant categories. Once again, we must prioritize to obtain the maximum benefit from available resources, as we will not totally decouple computer or communication power from costs. These technologies may well move to the top of our priorities established by any cost-effectiveness metric.

Recapitulation

Stanley Kaplan and B. John Garrick (1981), in Volume 1, Number 1 of the journal *Risk Analysis: An International Journal,* articulate the fundamental tenets of risk analysis. Their "set of triplets idea" consists of

- What can happen?
- How likely is it that that will happen?
- If it does happen, what are the consequences?

Attempts to abate risk often flounder on the second of these tenets. It is crucial to know "how likely," or probable, any particular scenario might be. As the probability of an event decreases, the probability of successfully aiming the abatement strategy decreases. Conversely, the costs decrease as the abatement strategies become more focused.

Though there are no *high-consequence/low-probability events* represented in figure 11, it is quite likely that the costs per life saved for many scenarios will be extremely expensive.

It follows that focused preparations for low-probability events are likely to be costly, as there is a high probability they will never be used. As low-probability events proliferate, and prevention scenarios are developed, resources are drained away. We should carefully examine Wildavsky's contrast between *anticipation* and *resilience.* He states: "Knowing so little about whether a given risk will materialize (though we know that the chance is low), we are in great danger of wasting resources in the pursuit of wrong leads (the vast majority of all leads, due to the low probability involved, will be wrong leads). The inevitability and irreducibility of this uncertainty is what makes unwise the strategy of anticipatory spending to avoid such low-probability risks. Something better is needed."

To my knowledge, no one predicted the AIDS tragedy, however, any solution to this disastrous increase in human suffering will ultimately result from our previous and continued investments in the biological sciences.

With this as a model, I would advocate that planners invest in strategies for *resilience.* That is, strategies that allow us to deal with surprise. There is no room for *hubris.* We do not know what the future holds for us.

Besides investments in medical science, investments should be enhanced in at least three kinds of infrastructures synonymous with developed societies. These infrastructures allow us *to move, to communicate, and to learn. Mobility* provides one means for populations to avoid tragedy. As indicated by others in this conference, the ability to *communicate* enhances mitigation. Finally, investments in *education and the scientific enterprise* will provide the most resilient tools to prepare for future unknowns.

Ultimately, as stated in the introduction, the richer society is the safer society because it is more resilient.

References

Antonovsky, A. 1967. Social class, life expectancy and overall mortality. *Milbank Memorial Fund Quarterly* 48: 13.

Baquet, C. R., et al. 1991. Socioeconomic factors and cancer incidence among black and white. *Journal of the National Cancer Institute* 83 (8): 551–7.

Brenner, M. H. [see *Mental Illness and the Economy*. Cambridge, Mass.: Cambridge University Press.]

Fox, J. 1977. Occupational mortality 1970–72. *Population Trends. Great Britain*. 9: 8–15. Information taken from *The Registrar General's Decennial Supplement 1931 England and Wales*, Part IIA, Occupational Mortality. London: Government Statistical Service.

Graham, J. D., B. H. Chang, and J. S. Evans. 1992. Poorer is riskier. *Risk Analysis* 10 (3): 333–7.

Hadley, J., and A. Osei. 1982. Does income affect mortality? An analysis of the effects of different types of income on age/sex/race-specific mortality rates in the United States. *Medical Care* 20 (9): 901–14.

Hart, N. 1986. Inequalities in health: The individual versus the environment. Royal Statistical Society. Part 3: 228–46.

Kaplan, S., and B. J. Garrick. 1981. On the quantitative definition of risk. *Risk Analysis: An International Journal* 1 (1): 11–27.

Keeney, R., and D. von Winterfeldt. 1990. Why indirect health risks of regulations should be examined. *Interfaces* 16: 13–17. [See also R. L.

Keeney. 1990. Mortality risks induced by economic expenditures. *Risk Analysis* 10: 147–59.]

Lewis, H. W. 1992. *Technological risk.* New York: Norton.

Linn, M. W., R. Sandifer, and S. Stein. 1985. Effects of unemployment on mental and physical health. *American Journal of Public Health* 75 (5): 502–6.

Macfarlane, A. 1979. Child deaths from accidents 2: Place of accident. *Population Trends* (Spring): 10–15.

Mills, M. P. 1991. *Ecowatts: The clean switch.* Science Concepts Inc., (April).

Nathwani, J. S., E. Siddall, and N. C. Lind. 1992. *Energy for 300 years.* Institute for Risk Analysis, University of Waterloo. [See also N. C. Lind, J. S. Nathwani, and E. Siddall. 1991. *Managing risk in the public interest.* Institute for Risk Research, University of Waterloo.]

Office of Population Censuses and Surveys. 1978. Deaths of children (1–14) in England and Wales registered 1970–72. *Occupational Mortality 1970–72,* Series DS No. 1. London: HMSO.

Sagan, L. A., and A. A. Afifi. 1978. Health and economic development II: Longevity research memorandum 78-41. International Institute of Applied Systems Analysis 18 (August).

Schwing, R. C. 1991. Risky business: Improving decision making. *Regulating risk: The science and politics of risk.* National Safety Council, June.

Schwing, R. C. 1992. Conflicts in health and safety matters: Between a rock and a hard place. *Proceedings of the Fifth Engineering Foundation Conference 1992.* New York: Engineering Foundation.

Starr, C., and M. Searle. 1990. Global energy and electricity futures: Demand and supply alternatives. *Energy Systems and Policy* 14: 53–83.

Tabor, M. 1982. The stress of job loss. *Occupational Health and Safety* 51 (6): 20–6.

Thomas, L. 1984. Scientific frontiers and national frontiers. *Foreign Affairs* (Spring): 980–1.

U.S. Government. 1991. Reforming regulation and managing risk-reduction sensibly. In *Budget of the United States government for fiscal year 1992,* section IX.C. , part two: 367–76.

Wildavsky, A. 1988. *Searching for safety.* New Brunswick (U.S.A.) and London (U.K.): Transaction Publishers.

Zachary, G. P., and S. K. Yoder. 1993. Computer industry divides into camps of winners and losers. *Wall Street Journal* 74 (January 27): A1.

Measuring the Benefits of Flood Risk Reduction

Leonard Shabman[1]

Abstract

Property damages avoided, land price analysis, and contingent valuation were techniques used to estimate the economic benefits of flood risk reduction for residential land parcels in Roanoke, Virginia. Evidence of benefits also was available from voting behavior in a bond referendum. The benefit estimates from the different techniques diverged significantly, but in ways that can be explained by the limitations of the techniques. Benefit estimates should be treated as a first approximation of community and landowner interest in flood risk reduction, but final investment decisions will be justified in the political arena.

Introduction

Beginning with the Mississippi River Commission in the late 1800s, and continuing to the planning efforts of the Miami Conservancy District of Ohio in the 1920s, benefits from reduced flood risk were estimated as the expected difference in land prices with a flood control project versus land prices without one. These estimates were a starting point for negotiations between local flood control and levee districts and benefiting landowners on the tax payments that would be made to pay for project construction (Shabman 1989).

Benefit estimation practices deviated from the land price method in the 1930s, when the 1936 Flood Control Act shifted flood control project construction and financing responsibility to the federal government. The federal flood control project construction program was expected to

[1]Professor, Environmental and Resource Economics, Department of Agricultural and Applied Economics, Virginia Tech, Blacksburg, VA 24061-0401

provide relief from human suffering and to advance national economic prosperity. Such broad social objectives were not adequately reflected in private land market transactions, and new approaches to benefit estimation were sought (Maass 1970; White 1936). Nonetheless, despite the broadly stated social goals of the act, benefits were computed as the present value of real property damages avoided (PDA). The property damages avoided technique relied on the hydrologic information that was routinely developed for project planning. Also, the PDA approach had a compelling investment logic: if current expenditures for a flood control project were less than the present value of avoided future property repair costs, then the project was justified.

However, the narrow perspective of the method was recognized in federal benefit estimation guidelines (Federal Inter-Agency River Basin Committee 1950; Water Resources Council 1969, 1973, 1983). Many of these guidelines, while not using the term directly, implied that landowners' and society's "willingness to pay" for flood hazard reduction was the standard by which benefits should be established. Determinants of willingness to pay would include not only property damages avoided, but also avoided individual and community disruption, avoided post-flood trauma (medical bills, worker productivity losses), and avoided preflood anxiety. Therefore, in the political arena where project investment priorities were set, the simple calculations of PDA benefits were part of a broader consideration of the "nonproperty" effects that would be mitigated by the project (Shabman 1973).

Economists recently have developed a survey-based benefit estimation technique that is capable of reflecting nonproperty concerns (Mitchell and Carson 1989). The contingent valuation method (CVM), which has been used to value environmental assets, could ask questions to determine individuals' willingness to pay for flood protection. However, the CVM method has not been used to estimate flood control benefits.

To date there has been no research that compares the results of these different techniques. An opportunity to do such a comparison did present itself in the city of Roanoke, Virginia. The city of Roanoke has experienced ten floods of varying magnitudes since the turn of the century, with major floods in 1972 and 1978. However, it was the record flood of 1985 that motivated city officials to press for a U.S. Army Corps of Engineers project of flood walls, bridge replacements, and river channel improvements to increase the capacity of the river to carry floodwater (*Roanoke Times & World News* 1989; U.S. Army Corps of Engineers 1984).

In a city of about 100,000 people, the project would reduce the probability of flooding for only 2,265 single- and multiple-family residential structures. Also, the project would provide flood protection to about $250 million invested in businesses and industry employing about 5,000 people. A special referendum was called April 11, 1989, asking voters to approve a $7.5 million bond issue to finance the city's cash share of the federal project's cost. The citywide utility tax would be raised from 10% to 12% to repay the bond, increasing the average household's utility bill by approximately two dollars per month. Because alternative flood reduction proposals were considered politically infeasible, rejection of the bond issue implied that the city would be without any flood protection measures for years to come. The bond issue passed with over 56% support (4,271 to 3,273), with 19.6% of the registered city voters casting ballots.

The techniques of property damages avoided, land price analysis, and contingent valuation were used to estimate the economic benefits of flood risk reduction for the same residential land parcels in Roanoke, Virginia. Evidence of the benefits the community attributed to the project also was available from voting behavior in the bond referendum. Individual study procedures and results have been reported elsewhere (Driscoll, Dietz, and Alwang 1994; Shabman and Thunberg 1991; Shabman and Stephenson 1992; U.S. Army Corps of Engineers 1984). For purposes of comparison, the CVM and PDA studies of benefit estimation are classed as hypothetical choice techniques because the benefits are estimated from speculations made by analysts (PDA) or a survey respondent (CVM). The studies also included an application of the land price technique and an interpretation of voting behavior in a bond referendum. For this report, techniques that interpret land market prices and voting behavior are termed revealed choice approaches because the benefit estimates are grounded in actual choice behavior. After briefly describing each of the four studies, the next two sections of the paper report on how the estimates compare and the possible reasons for differences. Implications for measurement of flood risk reduction benefits are also provided.

The Property-Damages-Avoided Estimates

The analyst using the PDA technique computes the repair costs to real property with and without a flood hazard reduction project for a given river flow event. The difference between repair costs is the estimated annual benefit of the project for that flow. The analyst then weights each flow by its likelihood of occurrence and sums over all flows to calculate the expected benefit of the project for any year. Extending the probability-weighted annual estimates over project life and discounting the estimates back to present value yields the expected present value of avoided damages to real property.

PDA will equal individual and social willingness to pay only if the following assumptions hold: 1) the property owner is willing and able to pay to restore the damaged property; 2) the property owner is only concerned with the real property effects of flooding; 3) the property owner uses the same technical information in the same way as the agency technical expert; 4) the income of property owners is not an influence on willingness to pay and the property owner has no access to disaster aid or flood insurance, that is, has no alternative to the project for reducing the consequences of flood loss; 5) the property owner expects the expected repair cost reductions to be capitalized into the resale value of the property; and 6) the property owners are the only members of the community who are willing to pay for the project.

PDA estimates were obtained from the U.S. Army Corps of Engineers (Wilmington District) for single-family homes in a moderate-income neighborhood in a residential area in Roanoke. The area had experienced serious flood damage in the 1985 flood, but all the properties had been repaired. The homes were typical of those in the city in age and construction. These estimates are reported later in the paper.

The Contingent-Value Estimates

The CVM technique relies on a sample survey to elicit statements of willingness to pay, termed a "bid," for a project. A statistical analysis of the sample data can expand the survey results to the whole population and can be used to explain the factors that were significant in the stated expressions of willingness to pay. Also, in developing the survey and in the statistical analysis, the several assumptions about the PDA estimates can be directly addressed. Consider each assumption: 1) the property owner's bid should reflect the likelihood that he or she will repair certain property damages; 2) the property owner's bid should reflect concerns about real property as well as nonproperty and community effects of flood risk; 3) the property owner can be given the same technical data as the agency expert, however, his or her bid may reflect a different interpretation of the data; 4) the property owner's bid can be statistically adjusted for the availability of flood insurance, disaster aid, and for the property owner's income; 5) the property owner's bid can be statistically adjusted for land price expectations; and 6) bids may be elicited from people who do not own flood-prone property but who may be willing to pay toward the project out of concern for the community.

The contingent value study for Roanoke was completed in the fall of 1989 (Shabman and Thunberg 1991). In-person interviews were conducted with seventy-four landowners (55% of the total population of landowners in the case study area) using a prepared CVM survey as the interview protocol. Because the project's effect was to reduce the probability of flooding, a special effort was made to ensure that the landowners understood the risk reduction they were being asked to pay for. First, the reduction in the probability of floodwater entering the landowner's first floor at least once in a period of ten years was described to the landowner as follows (the number in parentheses is an example of the calculation for one house):

> *Floods both larger and smaller than the 1985 flood can occur in the future. All these possible floods are considered in planning a project to protect all properties along the river. Therefore, the Corps has calculated the chance of floodwaters entering the first floor or basement of your residence both before the project is built and after it is built. If no flood control project is built there would be a (40)% chance that floodwaters would enter the first floor or basement of this residence at least one time in ten years. After the project is built the chance that floodwaters would enter the first floor or basement of your house will be reduced to (20)%.*

As each respondent was informed of these flood probabilities, a pair of pie charts illustrating the chance of floodwater entering their home with and without the project was displayed. Several follow-up questions were asked to ensure that the respondent understood the project's effect.

The interviewer also needed to elicit a statement of willingness to pay from each respondent. First the landowner was informed that the project cost would be shared between the city and the federal government, but that the actual financial arrangements had yet to be made. Respondents were then presented with the following scenario:

> *Suppose the cost of the project will be paid by property owners (commercial and residential) as a one time only city special assessment as soon as the project is built.*

Respondents were then presented with a card listing dollar amounts from $0 to $5,000 and asked to state how much they would be willing to pay for project construction. The respondent was then asked how much more than this initial bid he would be willing to pay. The initial and incremental bids were then summed and the respondent was asked to confirm whether the total bid was a maximum. The bids were inter preted as the present value of the benefits so they could be compared with the PDA (and land price comparison) estimates.

Over 50% of the respondents bid zero. All zero bidders were asked to state a reason for the bid. If they stated that they could not afford more or that the project was not worth anything to them, the bids were distinguished from "protest bids." Protest bidders were those who bid zero because they (for some reason) objected to being asked their willingness to pay. For example, protesters may have felt that they had already paid once for the flood problem through damages incurred in the 1985 flood, and so they found it unfair that they would be asked to pay again. Of the seventy-four completed interviews, twenty-two individuals made a protest bid. The "protest" bids were considered to be legitimate expressions of zero willingness to pay and were included in the analysis.

Several other variables measured in the survey were used to estimate a statistical equation for describing the CVM bids. Preflood anxiety and concerns about community disruption from flooding were elicited using Likert scaling procedures (Allee et al. 1980; Blocker and Rochford 1986) and the scale was included in the estimated equation. Real property concerns were represented in the equation by the structural value of the home and the expectations individuals stated that they had for an increase in land prices. Other factors expected to affect willingness to pay included ownership of flood insurance and household income. The statistical analysis found that the variables having the most influence on the size of the CVM bids, in order of importance, were concerns for community disruption, expectations of real property damage, expected increases in land prices, and reductions in preflood anxiety.[2]

The Land-Price Estimates

Land price analysis, technically termed hedonic price analysis, is used to separate the effects of flood risk statistically from all other factors that may influence the sale price of land (Donnelly 1989; Park and Miller 1982; Thompson and Stoevener 1983).[3] A land transfer price represents an agreement between willing buyers and willing sellers on the worth to

[2]A separate analysis that excluded protest bids did not alter these results.

[3]Willingness to pay cannot always be interpreted as the simple difference in land prices. This has been explained elsewhere and illustrated with the Roanoke data (Driscoll, Alwang, and Dietz in press). For purposes of this paper, the simple land price differences are reported and described.

them of a land parcel. Nonproperty concerns of traders may affect willingness to pay and to sell. Also affecting the terms of trade will be the traders' expected access to disaster aid and flood insurance and the traders' income levels. However, unlike both the PDA and CVM techniques, the information available to the land market traders may not be the same as that available to experts. If a flood risk is present and recognized by the traders, buyers' willingness to pay and land prices may be depressed. If a recognized flood risk is removed, sellers will seek a higher price, but willingness to pay may also rise.

A hedonic price equation was estimated for over three hundred land sales during the years 1959 to 1991 (Driscoll, Dietz, and Alwang 1994). The price equation was estimated for the same Roanoke study area where the PDA benefits were computed and where the CVM survey was administered. Land price differences were determined primarily by house and lot characteristics. The hedonic price model also included the location of the property in the flood plain, termed flood zone, to represent the expected frequency of flooding; and a variable that measured the severity of, and time since, the previous floods.

Land prices were depressed after the 1985 flood, suggesting that there would be benefits from a project to reduce flood risk. However, there were many floods of smaller magnitude before 1985 that had no statistically significant effect on land market prices. Therefore, if the hedonic price study had been done before 1985, the conclusion would have been that there would be no benefits of flood protection.

The Voter Behavior Results

The 1989 bond referendum election was held solely for giving the city the authority to issue bonds to pay the city's cash share of the flood control project. The referendum was widely publicized in the press and information packets were provided by the city at places of work and through the mail. As a result, individual voter's flood control benefits and tax costs if the referendum passed were clear.

A telephone survey was administered to a sample of registered voters in Roanoke one week following the bond referendum election (Shabman and Stephenson 1992). The voting behavior and motivations among voters and nonvoters discovered by the survey suggest that willingness to pay for flood control extends beyond individual landowners in the flood plain. People directly affected by flooding either at their residence or place of work did vote for the referendum. However, only about 10% of the city's population lives or works in the flood plain.

Therefore, the referendum could pass only if those people living and working outside the flood plain supported the project, and 56% of all voters favored the tax increase to support the bond issue.[4] Direct questions to all voters indicated that altruistic concerns for flood victims and concerns about the vitality of the Roanoke community motivated the support for the project among those outside the flood plain.

The Comparison for Identical Properties

The purpose of this paper is to compare the results from the different benefit estimation methods and to explain any differences among the estimates.[5] The PDA and hedonic price estimates of the benefits from flood protection fell the farther the property was from the river. Table 1 reports the CVM bids by "flood zone,"[6] excluding protest bids (always zero), which are shown in the last column. The expectation would be that the bids would fall with distance from the river, but that pattern does not hold. Also, the number of protest bids is greatest the farthest away from the river. A separate statistical analysis of the protest bids found that people who owned or occupied the house and were not landlords were more likely to protest the CVM survey. Protest was less likely the higher the likelihood of flooding at the respondent's house and the less concerned the respondent was with community effects.

Mean benefit estimates by flood zone were computed and are presented in figure 1 for each of five flood zones. The land price and PDA estimates are greatest the closest to the river, but the mean CVM bid peaks in the third flood zone away from the river. This CVM result is the same as reported in table 1, even though protest bids are included as

---

[4]The election turnout of 20% of the registered voters is consistent with all municipal elections. Also, the survey did question nonvoters. The results of that survey indicated that the election results, and reasons for those results, would not have been changed if voter turnout had been higher.

[5]The estimates are for the *same* land parcels (owners). Thus the hypothetical choice methods were a PDA estimate for parcel X and a CVM bid by the owner of parcel X. For the revealed choice methods, a land price difference estimate for parcel X and the voting behavior of the owner of parcel X is considered.

[6] Flood zones are areas where the frequency of flooding is approximately the same. For example, all areas where the likelihood of flooding was between the river bank and the "twenty-year" flood (i.e., the flood with a 5% chance of occurrence in any year) formed one flood zone.

zero bids in the CVM results of figure 1. The other striking result shown in figure 1 is that the hedonic price estimates exceed the other two estimates for all flood zones, with the gap between estimates greatest for the most flood-prone areas. Recall, however, that the land price estimates reflect the effect of the devastating 1985 flood event and that land prices were unresponsive to smaller flood events.

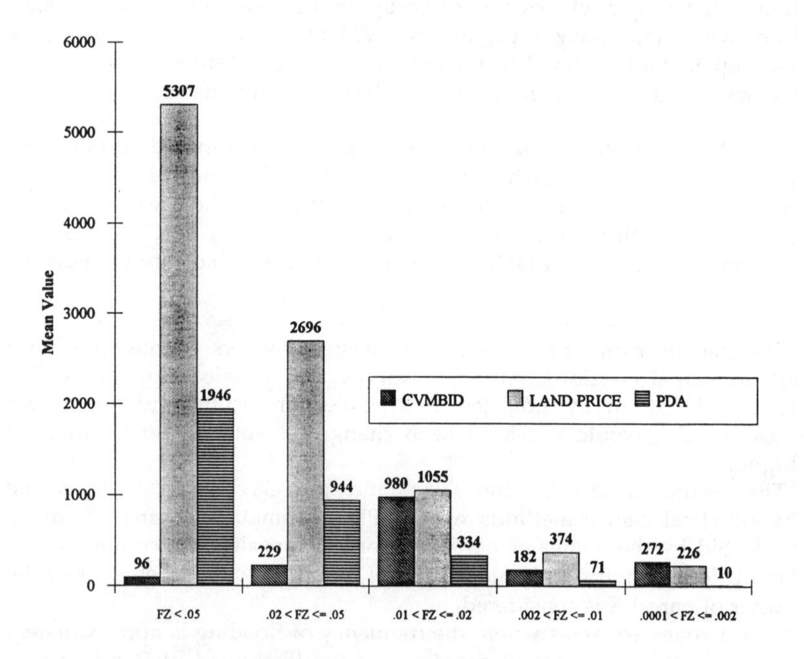

Figure 1. Mean Values of Estimates by Flood Zone

Flood Zone	CVM Bid Range		CVM Bid Mean	Number of Protest Bids
	Minimum	Maximum		
FZ < .05	0	600	178.57	5
.02 < FZ ≤ .05	0	2000	362.50	7
.01 < FZ < .02	0	5000	1225.00	2
.002 < FZ < .01	0	1500	322.22	4
.0001 < FZ < .002	0	2000	475.00	11

Table 1. CVM Bids by Flood Zone

The landowners had an opportunity in a well-publicized referendum to vote for the project and impose only a small tax on themselves. Table 2 reports on voting behavior in the referendum for the owners of the land parcels used in the CVM bid study. The voter information was taken from the records in the city's Office of Voter Registration. First, thirty-five people bid positive amounts (expressed a willingness to pay) in the CVM hypothetical survey. However, only twelve of these individuals bothered to vote in the election. Thus, those who bid the most were actually unlikely to vote when they had the opportunity to reveal their preferences. Those who bid zero were about as likely to vote as those who made positive bids. The protesters were the most likely to vote. The results shown in table 2 cast doubt on whether expressions of willingness-to-pay estimates in hypothetical CVM bids translate into actual choice behavior.

Class	CVM Bid > 0	CVM Bid = 0	Protest	Total
Registered and Voted	12	4	13	29
Registered and Did Not Vote	5	1	11	17
Not Registered	18	5	5	28
Total	35	10	29	74

Table 2. CVM Bids and Voting Behavior

Implications

The use of the PDA approach to benefit estimation is well established. However, as has been recognized for many years, PDA will not represent society's willingness to pay for flood risk reduction. Indeed, the results from the CVM and voting studies offer strong evidence that nonproperty benefits, which are ignored in the PDA approach, are significant in the support of flood risk reduction projects.

The hedonic price estimate may incorporate nonproperty concerns of land traders in the flood plain. Its advantage is that the estimates are derived from revealed market behavior. However, the necessary assumption for using this technique, that landowners have sound information on flood risk, is challenged by the empirical results from the Roanoke study. Land market traders appear unable to understand flood potential without some major flood to anchor their perceptions. Also, the large flood in 1985, although an infrequent occurrence, may have caused land market traders to overreact to the flood risk.

Taken together, the findings seem to support expanded use of the CVM technique, which appears to overcome both the information problem of the hedonic price method and the problem that PDA ignores nonproperty effects. However, the CVM bids did not follow a credible logic in relation to flood risk, and the CVM bidders did not carry their expressions of willingness to pay from the hypothetical setting of the survey into the voting booth.

The best that can be said from these comparisons is that the PDA and land price approaches did follow a credible logic since benefits fell with distance from the river. *However, the more fundamental question may be whether any measurement of benefits can be expected to be more than a rough guide to beginning and structuring the necessary collective choice decision on flood risk reduction.* With this perspective, no estimate is the "correct" measure of benefits and no estimate was expected to be "correct." Recall how benefit estimates have been used for guiding flood control investments in the past. The early applications of the land price method were expected to approximate only roughly willingness to pay. The benefit estimation effort was the basis for negotiations between landowners and flood control districts on mutually acceptable assessments to pay for project construction (Shabman 1989).

PDA benefit estimates since the late 1930s have been used because they flowed easily from the available hydrologic information, because they had an appealing "investment logic," and because they were adaptable to the political process that ranked projects. However, investment priorities were expected to recognize a variety of social objectives that might be served by flood risk reduction, including relief of human suffering and encouraging development of flood plain lands for agricultural and commercial uses. PDA benefit estimates have been an initial screen to determine project worth, but the benefit computations never have been, and were never intended to be, the final factor for determining project choice.

If this perspective on benefit estimation is adopted, then any suggestion to apply the CVM approach will need to be defended by demonstrating how CVM contributes to a better understanding of project effects and improves the reaching of political agreements. However, based on the evidence from this study, the CVM method does not yet offer reliable results that might be introduced into political negotiations. At present, there seems to be little reason to deviate from the PDA benefit estimation as long as federal funds continue to support flood control spending, and the limitations of the PDA benefit estimates are recognized when making flood control investment decisions.

Acknowledgments

Support for this project was provided by the U.S. Army Corps of Engineers, Institute for Water Resources, Fort Belvoir, Virginia.

References

Allee, D. J., B. T. Osgood, L. G. Antle, C. E. Simpkins, A. B. Motz, A. Van Der Slice, and W. F. Westbrook. 1980. *Human cost assessment—The impact of flooding and nonstructural solutions Tug Fork Valley, West Virginia and Kentucky*. U.S. Army Corps of Engineers, Institute for Water Resources. Research Report 85-R-4.

Blocker, T. J., and E. B. Rochford, Jr. 1986. *A critical evaluation of the human costs of flooding*. U.S. Army Corps of Engineers, Tulsa District. Contract No. DACW56-85-D-0160.

Donnelly, W. A. 1989. Hedonic price analysis of the effect of a floodplain on property values. *Water Resources Bulletin* 25 (3): 581–6.

Driscoll, P., B. Dietz, and J. Alwang. 1994. Welfare analysis when budget constraints are nonlinear: The case of flood hazard reduction. *Journal of Environmental Economics and Management,* April.

Federal Inter-Agency River Basin Committee, Subcommittee on Benefits and Costs. 1950. *Proposed practices for economical analysis of river basin projects*. Washington, D.C.

Maass, A. 1970. Public investment planning in the United States: Analysis and critique. *Public Policy* 27: 211–43.

Mitchell, R. C., and R. T. Carson. 1989. *Using surveys to value public goods.* Washington, D.C.: Resources for the Future.

Park, W. M., and W. L. Miller. 1982. Flood risk perceptions and overdevelopment in the floodplain. *Water Resources Bulletin* 18 (1): 89–94.

Roanoke Times and World-News. 1989. April 2, p. 1.

Shabman, L. 1973. *Decision making in the U.S. Army Corps of Engineers and the potential contribution of multiple objective analysis.* Technical report 24. Water Resources Research Center, Cornell University, Ithaca N.Y.

Shabman, L. 1989. The benefits and costs of flood control: Reflections on the Flood Control Act of 1936. In *The flood control challenge: Past, present, and future,* edited by H. Rosen and M. Reass. Chicago: Public Works Historical Society.

Shabman, L., and K. Stephenson. 1992. The possibility of community-wide flood control benefits: Evidence from voting behavior in a bond referendum. *Water Resources Research* 28: 959–64.

Shabman L., and E. Thunberg. 1991. Determinants of land owners willingness to pay for flood hazard reduction.*Water Resources Bulletin* 27: 657–65.

Soule, D. M., and C. M. Vaughan. 1973. Flood protection benefits as reflected in property value changes. *Water Resources Bulletin* 9 (5): 918–22.

Thompson, M. E., and H. H. Stoevener. 1983. Estimating residential flood control benefits using implicit price equations. *Water Resources Bulletin* 19 (6): 889–95.

U.S. Army Corps of Engineers. 1984. *Roanoke River Upper Basin, Virginia, final feasibility report and environmental impact statement for flood damage reduction.* Wilmington, N.C., January.

Water Resources Council. 1969. *Report to the Water Resources Council by the Special Task Force on Procedures for Evaluation of Water and Land Resource Projects.* Washington, D.C.

Water Resources Council. 1973. Establishment of principles and standards for planning water and related land resources. *Federal Register* 38 (Part III). Washington, D.C.

Water Resources Council. 1983. *Economic and environmental principles and guidelines for water and related land resource implementation studies*. Washington, D.C.: U.S. Government Printing Office.

White, G. F. 1936. The limit of economic justification for flood protection. *The Journal of Land and Public Utility Regulation* 5: 133–45.

Uncertainty and Time Preference in Shore Protection

Kevin O'Grady[1] and Leonard Shabman[2]

Abstract

From a review of economic, psychologic, and geographic literature, three points of criticism of the economic model of choice are 1) when faced with losses, individuals tend not to be averse to risk; 2) when faced with low-probability hazards, individuals tend to ignore the hazard altogether (truncate low probability); and 3) when faced with choices over time, individuals have different rates at which they trade off benefits now versus later (time preference rate). The main objective of the study was to determine whether these criticisms are supported and to draw conclusions regarding government policy for flooding and erosion hazards. The analysis is based on survey data from shoreline property owners on the Great Lakes.

Introduction

This research examines the standard economic model for assessing the benefits (damages avoided) of shore protection, that is, the Discounted Expected Net Benefit (DENB) model (O'Grady 1992). The ability of this model to assess the actual utility associated with shore protection activity has been criticized. Some criticism focuses on the model's assumptions regarding time preference rates, probability, and risk attitudes. Following a discussion of criticism, the study approach is explained. Subsequent sections discuss the empirical results and their implications.

---

[1]Senior Analyst, Planning and Management Consultants, Ltd., Carbondale, IL 62903
[2]Professor, Department of Agricultural and Applied Economics, Virginia Polytechnic Institute and State University, Blacksburg, VA 24061

The economic model of rational choice holds that, given the constraints of prices and income, an individual ranks all possible choices using one dimension of measurement, utility, then chooses the option that yields maximum utility. To account for behavior involving uncertain events distributed over time, the model has been embellished to yield expected utility theory (EUT) and discounted utility theory (DUT). Each theory weighs choices by two dimensions: utility and the probability of event occurrence in EUT, and utility and a time preference rate in DUT.

In practice, the abstractness of the concept of utility renders the discounted expected utility model directly inapplicable to choice problems; utility is difficult to measure. As such, the DENB model is used as a measure of utility where the utility of benefits and costs is measured in monetary terms. For the individual, the DENB model can be formulated as follows:

$$\text{DENB} = \sum_{t=0}^{T} \frac{\alpha\, c_{wot} - \delta\, c_{wt}}{(1+r)^t} - \sum_{t=0}^{T} \frac{K_t}{(1+r)^t} + \frac{V_T}{(1+r)^T},$$

where

DENB = discounted expected net benefit of hazard reduction

c = the cost of flooding/erosion damage in time period t, $t = (0, 1, 2, ..., T)$, with (w) and without (wo) mitigative action

T = the time period in which the property is sold

K = the cost of mitigation over time

V = the salvage value of protection works at the time of sale of the property

r = the time preference rate (discount rate)

α = the probability density function associated with the damage cost when no protective action is taken

δ = the probability density function associated with the damage cost when protective action is taken

Note that α and δ represent probability distributions of damage frequency. That is, there is a range of possible damage levels, and probabilities of damage can be associated with each damage level. These distributions may be based on erosion processes, water levels, and storm intensity. By taking protective action, a property owner shifts the probability of damage for a given level of water levels or storm intensity. It is assumed here, for the sake of simplicity, that the elements in the other terms are known with certainty.

Criticism of the Economic Model

With respect to EUT, much of the criticism arose in the late 1940s and early 1950s immediately after its axiomatic formalization. DUT has received less attention from the critics until recent years (Loewenstein and Thaler 1989). Much of the criticism, but by no means all, has come from psychologists concerned with the descriptive accuracy of axioms, individually and/or collectively. Cognitive psychological research, focusing on human limitations in decision making, suggests that people have difficulty perceiving and combining the three dimensions of EUT and DUT: utility, probability, and time preference. The following sections address the issues of risk attitude and the manner in which probability and the personal rate of time preference enter the net benefits model.

Time preference

In the evaluation of public projects, there has been considerable debate about the appropriate rate to employ for discounting (Mikesell 1977; Prest and Turvey 1965). The opportunity cost of capital and various measures of the market rate of interest are offered as appropriate. Alternatively, arguments are made for some rate lower than these rates to reflect long-term social planning. Despite this debate, the range of rates considered theoretically appropriate is relatively narrow, whereas it has been observed that the time preference rates implied by people's choices may be as high as several hundred percent (Hausman 1979; Gately 1980; Houston 1983). These authors drew their conclusions from observed purchases of electricity-using durable goods. Landsberger (1971) estimated that the time preference rate over a variety of consumer goods is about thirty percent. Furthermore, based on observed purchases, it has been argued that time preference rates vary considerably with the decision in question (Ruderman, Levine, and McMahon 1987). Research suggests that the time preference rate falls with an increase in the absolute value of the consequence of an event (Loewenstein and Thaler 1989). DUT assumes that the time preference rate is constant over time. However, research suggests it is a function of the delay time expected in experiencing an event. Benzion, Rapoport, and Yagil (1989) and Thaler (1981) found that the time preference rates implied by experimental choice decisions tended to fall the more temporally distant a future event.

Probability

There is evidence to suggest that natural hazards are perceived as less aversive than man-made hazards (Rowe 1977). This may arise because of a phenomenon outlined by Kunreuther (1985): individuals will focus on the catastrophic nature of technological hazards but will tend to "focus on the low probability aspect of a natural disaster, claiming 'it won't happen to me.'" The only available applied research on natural hazards suggests that low probabilities, at some point, are truncated to zero, ignoring altogether the objective probability of the loss (Slovic et al. 1977; Schoemaker and Kunreuther 1979).

Risk attitude

DENB models embody the assumption of risk-neutral behavior by the decision maker. (DUT and EUT make no assumption regarding risk attitude.) In the DENB model no utility (or disutility) is assigned to the benefits or costs on the basis of a decision maker's preferences regarding risk; benefits and costs are weighted only by their respective probabilities (Pratt 1964). However, there is evidence that individuals tend to be risk seeking when faced with a choice between losses, and risk averse when faced with a choice among gains (Kahneman and Tversky 1979). A field study of individuals living in flood and earthquake hazard zones supports the notion of risk-taking behavior in the domain of losses, although there is some conflicting evidence (Kunreuther et al. 1978; Brookshire et al. 1985).

Study Approach

A combination of high water levels and severe storms caused considerable property damage on the shores of the Great Lakes in the mid-1980s. This led to a public outcry for the U.S. Army Corps of Engineers, among others, to take some mitigative action. A variety of structural and nonstructural measures have been proposed (Shabman et al. 1989). Although all measures may be subject to some form of benefit-cost analysis, the high cost of structural water level regulation measures encourages an examination of methods of benefit assessment. Further-more, changes in federal policy in the 1980s have altered the cost-share rules for flood control projects. The Water Resources Development Act of 1986 will increase the cost to local beneficiaries such that they will be required to pay as much as 50% of project cost; the result is that there is increased pressure to examine methods for the assessment of benefits. The Great Lakes situation presented an opportunity for an applied examination of the benefits assessment model.

Hypotheses

From the research discussed above, three hypotheses regarding the DENB model were developed:

1. Property owners faced with hazards are not risk averse in their protection decisions.

2. Property owners vary in the time preference rates they apply to protection decisions.

3. Property owners tend to truncate the probability of rare damage events.

Much of the research discussed above was conducted in a controlled laboratory setting, where subjects were presented with hypothetical choice problems of little import to their well-being. This research effort is somewhat unique in that these hypotheses were tested in a field setting. The flooding and erosion hazards on the Great Lakes are salient and pressing issues to many shoreline property owners. These factors can be expected to reduce the hypothetical nature of the questions and responses.

Sample

The U.S. Army Corps of Engineers has divided the shoreline of the Great Lakes into a number of reaches. Based on information from the Corps' Chicago district, a sample of property owners was chosen from reaches where (a) there was an observed erosion problem, (b) at least 95% of the population lived in single-family dwellings, and (c) 40% to 60% of the properties had undergone some activity to protect against flooding or erosion. The types of action specifically considered were (i) raise, move, and/or reinforce residence, (ii) build or repair groin/jetty, and (iii) build or repair sea wall, breakwater, or dike. These criteria ensured that there would be considerable sample variation between those individuals who did and who did not take protective action against flooding and erosion.

A sample of property owners was surveyed by mail. The survey procedure involved three mailings. The first mailing introduced the property owners to the purpose of the survey and urged them to respond to the enclosed questionnaire. The second mailing consisted of a postcard urging those who had not yet responded to do so. The third mailing was sent to those property owners who had not returned a completed survey; it urged them to respond and it included another copy of the survey instrument. The mailings took place on July 9, July 23, and August 6, 1991.

Empirical model

The property owner's decision to take action to protect property against flooding and/or erosion (PROT) can be explained as a function of the following variables. These variables reflect the arguments in the general model.

TPR	=	landowner's personal time preference rate
PROB	=	probability of experiencing damage
RATT	=	the property owner's risk attitude
PRPR	=	shoreline protection taken by a previous owner
PVAL	=	the market value of the property
KSP	=	the perceived capital cost of shore protection
ERO	=	severity of erosion
FLO	=	existence of a flooding problem
INS	=	commercial insurance to compensate flood damage
RESID	=	existence of a residence on the property
AID	=	expectation of government disaster aid
CAP	=	the degree that protection expenditure is expected to be capitalized into the market price of the property
COA	=	membership in a property owners' coalition
INC	=	income

Operational definitions of these variables are presented below.

Protection Adoption (PROT). The dependent variable PROT in the protection adoption model was defined such that if property owners had not taken any action, then PROT was set to 0; if they had taken action, then PROT was set to 1. Action is taken to mean at least one of raising, moving, or reinforcing a residence, or building or repairing a groin/jetty, sea wall, breakwater, or dike.

Time Preference Rate (TPR). The time preference rate was determined from responses to a hypothetical question. The question was prefaced by describing a hypothetical shoreline property.

> You are the new owner of a waterfront lot. The lot has 100 feet of waterfront and is worth $40,000. The residence is set back 150 feet from the shore. After purchasing the lot you become aware that the lot has an erosion problem. An appraiser tells you that the erosion is reducing the market value of your property.
>
> Suppose you consult a contractor who describes two erosion control projects, which both cost the same amount:
> Project A will reduce the yearly loss of property value by $600. The project will be fully effective as soon as it is built.

Project B will reduce the yearly loss of property value by $1,000. However, this project will not become effective for 5 years.

Which project would you prefer? (Circle one number only.)

1 PROJECT A, WHICH REDUCES YEARLY EROSION DAMAGE BY $600 BEGINNING IMMEDIATELY
2 PROJECT B, WHICH REDUCES YEARLY EROSION DAMAGE BY $1,000 BEGINNING IN 5 YEARS
3 EITHER PROJECT WOULD BE ALL RIGHT
4 I AM NOT SURE

A choice indicating indifference between projects reflects a time preference rate of 10%. The time preference rate implied by the response to this question was coded as follows:

(i) response = "project B," then TPR = 0
(ii) response = "either project," then TPR = 1
(iii) response = "project A," then TPR = 2

Thus, TPR ranges from 0 to 2 corresponding to a rate lower than, equal to, or higher than 10%.

Probability Truncation (PROB). In the truncation question, the respondent was asked which property he or she preferred to live on, one with a low probability of high storm damage, the other with a high probability of low storm damage. This question presented point estimates of storm damage such that the expected values of both options were equal ($250). The question was worded as follows:

Suppose you had a choice between two pieces of shoreline property. The only differences between the pieces of property are the chance and level of storm damage:

	Annual Chance of Damage	Level of Damage for Each Storm
Property A	10% chance	$2,500
Property B	2% chance	$12,500

Which property would you prefer to live on? (Circle one number only.)

1 PROPERTY A, WITH A HIGH CHANCE OF LOW-DAMAGE STORMS
2 PROPERTY B, WITH A LOW CHANCE OF HIGH-DAMAGE STORMS
3 EITHER PROPERTY WOULD BE ALL RIGHT
4 I AM NOT SURE

The degree of probability truncation implied by the response to this question is coded as follows:

i) response = "property A," then PROB = 0
ii) response = "either property," then PROB = 1
iii) response = "property B," then PROB = 2

To the extent that an individual truncates the low probability of storm damage, he or she will be more likely to choose the property with the low probability of storm damage. Thus, as the degree of truncation increases, PROB increases.

Risk Attitude (RATT). An index of risk preference (RATT) was constructed from three questions taken together. These questions were worded as follows:

Suppose there was a 5% chance each year of a storm causing between $3,000 and $5,000 damage to the property described earlier.

(A) Would you buy an insurance policy for $200 per year if it covered damages only in the $3,000 to $5,000 range? (Circle one number only.)

1 YES ---> GO TO QUESTION B
2 NO ---> GO TO QUESTION C
3 I AM INDIFFERENT ---> GO TO QUESTION D
4 I AM NOT SURE ---> GO TO QUESTION D

(B) Would you buy the insurance policy if it cost $275 per year?

1 YES ---> GO TO QUESTION D
2 NO ---> GO TO QUESTION D

(C) Would you buy the insurance policy if it cost $125 per year?

1 YES
2 NO

Question A was constructed such that the expected storm damage would lie between $150 and $250. Respondents were given this range as opposed to a point estimate of expected damage to make the question more realistic, as insurance policies cover a range of possible damages. The price of the insurance policy was set at $200, the midpoint of the expected storm damage range. A negative response to A would suggest the respondent is risk seeking, an affirmative response suggesting risk-averse behavior. The prices assigned to the insurance policies in B and C ($275 and $125, respectively) were set above and below the range of expected storm damages ($150 to $250). Thus, an individual preferring one of these policies could be considered to have a relatively strong preference for or against risk. The respondent's risk attitude (RATT) was coded as follows:

i) response to A = Yes and response to B = Yes, then RATT = 5
ii) response to A = Yes and response to B = No, then RATT = 4
iii) response to A = Indifferent, then RATT = 3
iv) response to A = No and response to C = Yes, then RATT = 2
v) response to A = No and response to C = No, then RATT = 1

Thus, where an individual responds and is sure of his or her preference, RATT ranges from 1 to 5, where 5 corresponds to a risk-averse attitude, 3 corresponds to risk neutral, and 1 corresponds to a risk-seeking attitude. By responding "no" to both A and C (risk seeking) an individual shows a preference for taking the storm risk even though the cost of the insurance ($125) is less than the lowest level of expected storm damage ($150).

Previous Action (PRPR). Where a previous owner took no action, then PRPR was set to 0. If a previous owner did take action, then PRPR was set to 1. It is expected that this variable is indirectly related to PROT. Previous action may preclude the necessity of present action.

Property Value (PVAL). The value of the shoreline property (PVAL) was estimated in 1990 dollars. For those property owners taking action in the past, the property value was deflated based on the year they first reported taking action. It was not necessary to deflate the property values of those not taking action, as those values were reported in 1990 dollars. As property value increases it is expected that an owner is more likely to take protective action.

Cost of Protection (KSP). The cost of shore protection (KSP) was estimated in 1990 dollars. For those who took action, their costs were converted to 1990 dollars. For those not taking action, an estimate of protection cost was elicited. As KSP rises it is less likely that protective action will be taken.

Erosion Level (ERO). The property owner's perceived erosion problem was elicited with the following question:

> What is your best estimate of the average amount of your property eroded each year since you have owned this property?
>
> 1 I HAVE NO EROSION PROBLEM
> 2 LESS THAN 1 FOOT PER YEAR
> 3 1 TO 2 FEET PER YEAR
> 4 2 TO 3 FEET PER YEAR
> 5 OVER 3 FEET PER YEAR
> 6 I'M NOT SURE

ERO was coded as follows:

i) response = 1, then ERO = 1
ii) response = 2, then ERO = 2
iii) response = 3, then ERO = 3
iv) response = 4, then ERO = 4
v) response = 5, then ERO = 5

It is expected that ERO will be directly related to PROT.

Flooding Level (FLO). Where there was no flooding problem, FLO was set to 0; where there was a problem, FLO was set to 1. It is expected that FLO will be directly related to PROT.

Commercial Insurance (INS). Where a respondent replied that he or she held insurance, then INS was set to 1; INS was set to 0 if insurance was not held. Because insurance can be a substitute for shore protection, it is expected that INS is indirectly related to PROT.

Residence (RESID). Where there was no residence, RESID was set to 0; where a residence existed, RESID was set to 1. It is expected that RESID will be directly related to PROT.

Government Aid (AID). The variable AID was set to 1 if aid was expected and AID was set to 0 if it was not expected. It is expected that AID will be indirectly related to PROT, as it is a substitute for protection.

Capitalization (CAP). A question was designed to capture the degree to which a property owner expects that property improvements will be reflected in the market value of the property. The question was worded as follows:

> How do you think the market value of a shorefront property is affected by money spent on shore protection?
>
> 1 THE MARKET VALUE RISES BY MORE THAN THE AMOUNT SPENT
> 2 THE MARKET VALUE RISES BY THE SAME AS THE AMOUNT SPENT
> 3 THE MARKET VALUE RISES BY LESS THAN THE AMOUNT SPENT
> 4 THE MARKET VALUE IS UNAFFECTED BY THE AMOUNT SPENT

CAP was coded as follows:

i) response = 1, then CAP = 4
ii) response = 2, then CAP = 3
iii) response = 3, then CAP = 2
iv) response = 4, then CAP = 1

As CAP increases, it is expected that PROT will increase.

Coalition Membership (COA). Shoreline property damage incurred in the mid-1980s prompted some property owners to form organizations (or coalitions) to pressure governments to take action to reduce the chance of such damage in the future. However, it is difficult to determine, a priori, whether members would be more or less likely to take protective action. Coalition membership may make property owners more aware of the hazard, increasing the chance they will take action (Kunreuther et al. 1978). Additionally, to the extent the members see government action as a last resort, their individual actions having failed, there will be a positive relationship between membership and action. However, to the extent that any members perceive the responsibility of shore protection as lying with governments, they may be less likely to take action themselves. This variable will be maintained in the analysis as a control for the effects of the views of the coalitions. Where the respondents indicated they were members of a property owners' coalition, COA was set to 1; COA was set to 0 if they were not members.

Family Income (INC). The income variable (INC) was defined to reflect the income (in 1990 dollars) at the time of action. For those property owners taking action in the past, the income was deflated based on the year they first reported taking action. It was not necessary to deflate the incomes of those not taking action, as their income was reported in 1990 dollars. As INC rises, it is expected that PROT will increase.

Empirical Results

The survey response rate of over 52% (486 owners) suggests that the sample of respondents took an interest in the subject matter. Approximately 26% of respondents expected government aid in the event of flood and/or erosion damage. On average the respondents are over sixty years old and have owned their property almost twenty years. They are also a highly educated group, with an average education of 3.9 on a scale where 4.0 corresponds to the completion of a college degree. Even though a considerable number of these individuals reported an annual family income of less than $25,000, the mean income was approximately $68,000. On average the respondents appeared interested in the survey, were well educated, and had considerable experience as shorefront residents. Of the entire sample, approximately 39% took some action to protect their property from flooding and/or erosion; 61% did not.

A set of three questions on the survey was designed to elicit a person's attitude toward risk. The variable RATT was coded such that it ranged from 1 to 5, where the value of 1 indicated strong risk-seeking behavior and 5 indicated strong risk aversion; a value of 3 indicated risk neutrality. The average risk attitude score was 2.39, indicating people tended to be risk seeking (standard deviation of 1.53). Of the 371 people who responded to the set of questions, 246 indicated a score of 1 or 2. A minority of the respondents (125 of the 371) were risk neutral or risk averse. This result supports the findings of some laboratory studies that found that people tended to be risk seeking when faced with losses (Kunreuther et al. 1978; Schoemaker and Kunreuther 1979).

There was considerable dispersion among the responses in terms of the implied time preference rate (TPR): 40 responses implied a low rate, 31 implied a medium rate (10%), and 249 responses implied a high rate. There was also considerable dispersion in the responses to the probability truncation question: 87 responses indicated a low level of truncation, 32 indicated a moderate level, and 225 indicated a high level.

Probit analysis was used to estimate the functional relationship set out in a previous section. Probit analysis is appropriate when the dependent variable of the relationship is dichotomous, for example a binary variable taking only the value of 0 or 1. Such is the case in estimating the adoption protection model, where 1 indicates that action was taken and 0 indicates that it was not. The basis of Probit models is the normal cumulative density function. This function restricts the estimates of probabilities to lie between 0 and 1.

The initial estimation contained a large number of explanatory variables. Final estimates were obtained after a process of removing some of the variables that were of low statistical significance. The final estimation is presented in table 1. Additionally, simple correlation coefficients among the independent variables did not point out any potential sources of multicollinearity. Unfortunately, no formal procedures exist for diagnosing multicollinearity in Probit or Logit models. The variables that were removed were those that were highly statistically insignificant. Although some relatively insignificant variables remain in the final model, these variables have theoretical appeal.

The Chi-squared test indicates a statistically significant relationship. The model correctly categorizes 121 of 152 observations (79.6%), although there is a tendency to predict that a property owner did take action when in fact he or she did not (21 cases out of 31 mis-predicted). Both TPR and PROB display a strong statistical significance. The sign on PROB is negative as expected, that is, as a property owner ignores the chance of a storm (truncates the probability) that owner is less likely to take protective action. The sign on TPR is positive, which is opposite of what was initially hypothesized. However, the initial hypothesis assumed that the costs of shore protection would fall in the present while benefits would accrue over a longer time; thus, a high time preference rate would reduce benefits more than costs and would tend to reduce the likelihood that protective measures would be undertaken. The framing of the time preference rate question presented a situation where some individuals saw the erosion of property as something that must be stopped immediately; that is, the benefits were immediate and presumably costs could be put off until later. Thus, a high time preference rate would tend to be positively related to the likelihood that action was taken.

The coefficients on the variables ERO, FLO, RATT, INC, AID, COA, and RESID were of expected sign except for RATT. As the level of flooding and erosion increases (FLO and ERO), a person is more likely to take protective action; properties with a home (RESID) are more likely to be protected; where government aid is expected (AID), properties are less likely to be protected; coalition members are more likely to have taken action; and those with higher incomes are more likely to act.

Independent Variable	Coefficient	Standard Error	Prob-value
Intercept	-3.2075	0.8568	0.0002
ERO	0.4733	0.1030	0.0000
FLO	0.4117	0.3988	0.3019
RATT	-0.1332	0.0897	0.1375
INC	0.7150E-05	0.3420E-05	0.0366
AID	-0.5156	0.3105	0.0969
TPR	0.3628	0.2024	0.0730
PROB	-0.2531	0.1489	0.0892
COA	0.4263	0.3085	0.1670
RESID	1.1373	0.6913	0.1000

* * * * * * *

Chi-Squared = 60.494, Prob-value for Chi-Squared = 0.3089E-11

* * * * * * *

Frequencies of Predicted and Actual Outcomes

		Predicted		Total
		0	1	
Actual	0	95	10	105
Actual	1	21	26	47
Total		116	36	152

Table 1. Probit Analysis of Protection Adoption Model

RATT, with a negative sign, appears to be an anomaly. It was expected that those individuals who were more risk averse would tend to adopt protective measures. However, risk-averse individuals may have initially purchased properties that were not subject to flooding or erosion hazards; as such, they would be less likely to take protective action. Many respondents expressed an aversion to hazard insurance, indicating that it could not replace lost property. Because of some respondents' expressed aversion to insurance, the variable RATT may be reflecting an attitude toward insurance rather than risk attitude. To the extent that insurance and physical protection measures are substitutes, it can be expected that those persons preferring insurance may be less likely to adopt physical protection measures, hence the negative sign on RATT.

Summary and Implications

In recent years shoreline residential property owners along the Great Lakes have petitioned government agencies to take action to mitigate the effects of flooding and erosion on the lakes. This direct public pressure, particularly at a time of government spending restraint, has focused attention on the rationale and techniques used by public planners in assessing the benefits of various measures that might be taken and encouraging behavioral changes on the part of the public to reduce potential damage.

Benefit assessment

In assessing the benefits of hazard reduction, a public planner may rely on the concept of willingness to pay (WTP) as a reflection of those benefits (Herfindahl and Kneese 1974). That is, the benefits of a proposed hazard reduction measure are equal to what each individual, rational and fully informed, would be willing to pay for the hazard reduction (summed over all individuals). The WTP reflects the utility associated with the hazard reduction measure in question. In practice, as utility is difficult to measure, planners have come to rely on the damages-avoided notion, that is, the property damages that can be avoided by undertaking a particular measure are taken to represent WTP for that measure. The DENB model is a measure of damages avoided.

Care must be taken in using damages avoided as a measure of WTP for shore protection. That is, several assumptions must be made for DENB to be an appropriate proxy. Two of these assumptions follow:

1. All shoreline property owners are risk neutral. To the extent that a property owner is risk averse (seeking), the use of the DENB model will underestimate (overestimate) that person's utility derived from shore protection. From this study, it appears that the majority of property owners are risk seeking when making decisions about flooding and/or erosion hazards they may face. Similar findings appear in the literature (Schoemaker and Kunreuther 1979; Kunreuther et al. 1978).

2. It is assumed that the public planner and all the shoreline property owners in question have the same information about the probability and severity of hazard events and that they all apply the same time preference rate to intertemporal choice decisions regarding hazard reduction. To the extent that these assumptions do not hold, the level and direction of inaccuracy of DENB as a measure of WTP is not readily obvious. This study found that individuals varied considerably in the time preference rates they applied to their decisions and in the degree to which they truncated low probabilities. The time preference rate issue may offer more of a problem than the truncation issue. By ignoring truncation in a damages-avoided analysis, any estimation error lies in one direction: the analyst will overestimate the expected damage relative to the property owner. However, for any time preference rate chosen by an analyst and applied to all individuals, it is not clear whether that rate will over- or underestimate the individual's rate.

In accounting for variations in time preference rates and probability truncation, the model developed here is more descriptive than the typical benefit-cost model that holds these factors constant. However, despite the potential biases discussed above, the applied benefit-cost model remains a useful tool. Basic elements of the model were important in explaining the likelihood that individuals would adopt shore protection. To improve the descriptive capacity of the model, the analyst will need to direct attention toward the time preference rates specific to the individuals in question and to their understanding of the probability of hazard events.

With respect to probability truncation, it may be possible to develop a measure useful in assessing benefits. No doubt obtaining such a measure would be difficult. Here, information programs may be useful; the more a person understands the probability of potential hazards, the more accurate the benefits assessment model. The need for a measure of probability truncation disappears.

Behavioral implications

Although the DENB model is a method of accounting for the benefits individuals place on hazard reduction, it is also a behavioral model. This is reflected in the arguments of the model and in the assumptions. For example, any variation between the public planner and property owners in their time preference rates suggests the planner's model will not be able to account for some behavior.

One measure that might be considered to influence the behavior of property owners is that of hazard insurance. To the extent an individual is insured against a hazard, that person will be less likely to desire public protection from any hazards. This study suggests that property owners tend to be risk seeking and tend to truncate the low likelihood of hazard damage. Both of these forces will tend to reduce the demand for hazard insurance. Indeed, many people may choose not to buy insurance even at highly subsidized prices. This was the finding of Kunreuther et al. (1978) in a study of people living in earthquake and flood zones. This is not to say that hazard insurance is not an option open to policymakers. Perhaps two basic types of policies may be developed that are attractive to two types of property owners: policies with a low ceiling, designed to cover the high-probability, low-damage events; and policies with a high deductible to cover only the low-probability, high-damage events.

One method of indirectly influencing people to take protective action is to provide subsidized financing for them to undertake such action. However, the findings of this study suggest this measure would meet with limited success. Those individuals who are experiencing flooding and erosion tend to display a high time preference rate, above 10%. For some, this time preference rate may be well above the market finance rate. Thus, a finance rate below the time preference rate may not encourage borrowing to undertake protective action, although, no doubt, there may be some very low finance rate that might encourage such borrowing. This suggests that the finance rate is not preventing individuals from protecting their property. It is more likely that, for those experiencing flooding and erosion, the total cost of protection may be the prohibiting factor.

With respect to information programs, this study provides evidence that some individuals expect government aid in mitigating flood and erosion losses (26% in this sample) and that this expectation reduces the chance that protective action will be taken. This suggests that any information programs should make clear the role the government will take concerning such issues. The extent and type of any aid must be made clear.

The empirical results of this study suggest that there is a tendency for individuals to truncate the likelihood of low-probability hazard events; that is, they ignore low-probability events. Furthermore, this tendency reduces the likelihood that individuals will take action to protect themselves. As such, in designing information programs, it may be more appropriate to express hazard probabilities in periods of longer than an annual recurrence. Using a ten- or twenty-five-year period as the basis of presenting probability information may provide property owners with a better understanding of the risks they will face during their ownership of the property.

Acknowledgments

This research was funded by the Institute for Water Resources, U.S. Army Corps of Engineers. The Chicago district of the Corps supplied the initial mailing list and provided support in the sample selection process.

References

Benzion, U., A. Rapoport, and J. Yagil. 1989. Discount rates inferred from decisions: An experimental study. *Management Science* 35: 270–84.

Brookshire, D. S., M. A. Thayer, J. Tschirhart, and W. D. Schultze. 1985. A test of the expected utility model: Evidence from earthquake risks. *Journal of Political Economy* 93: 369–89.

Gately, D. 1980. Individual discount rates and the purchase and utilization of energy-using durables: Comment. *Bell Journal of Economics* 11: 373–4.

Hausman, J. A. 1979. Individual discount rates and the purchase and utilization of energy-using durables. *Bell Journal of Economics* 10: 33–54.

Herfindahl, O. C., and A. V. Kneese. 1974. *Economic theory of natural resources*. Columbus: Charles E. Merrill.

Houston, D. A. 1983. Implicit discount rates and the purchase of untried, energy-saving durable goods. *Journal of Consumer Research* 10: 23646.

Kahneman, D., and A. Tversky. 1979. Prospect theory: An analysis of decision under risk. *Econometrica* 47: 263–91.

Kunreuther, H. 1985. Natural and technical hazards: Similarities and differences. In *Environmental impact assessment, technology assessment, and risk analysis*, edited by V. T. Covello. Berlin: Springer-Verlag.

Kunreuther, H., et al. 1978. *Disaster insurance protection: Public policy lessons*. New York: John Wiley and Sons.

Landsberger, M. 1971. Consumer discount rate and the horizon: New evidence. *Journal of Political Economy* 79: 1346–59.

Loewenstein, G., and R. H. Thaler. 1989. Intertemporal choice. *Journal of Economic Perspectives* 3: 181–93.

Mikesell, R. F. 1977. *The rate of discount for evaluating public projects.* Washington: American Enterprise Institute.

O'Grady, K. L. 1992. Facing natural hazards: Uncertain and intertemporal elements of choosing shore protection along the Great Lakes. Ph.D. diss., Virginia Polytechnic Institute and State University, Blacksburg, Virginia.

Pratt, J. W. 1964. Risk aversion in the small and in the large. *Econometrica* 32: 122–36.

Prest, A., and R. Turvey. 1965. Cost-benefit analysis: A survey. *Economic Journal* 75: 683–735.

Rowe, W. D. 1977. *An anatomy of risk.* New York: John Wiley and Sons.

Ruderman, H., M. Levine, and J. McMahon. 1987. Energy-efficiency choice in the purchase of residential appliances. In *Energy efficiency: Perspectives on individual behavior,* edited by W. Kempton and M. Neiman. Washington: American Council for an Energy-Efficient Economy.

Schoemaker, P. J. H., and H. C. Kunreuther. 1979. An experimental study of insurance decisions. *Journal of Risk and Insurance* 46: 603–18.

Shabman, L., et al. 1989. *Living with the Lakes: Opportunities and challenges. Annex C: Interests, policies and decision making: Prospects for managing the water levels issue in the Great Lakes—St. Lawrence River Basin.* Washington: International Joint Commission Water Levels Reference Study, June.

Slovic, P., et al. 1977. Preference for insuring against probable small losses: Insurance implications. *Journal of Risk and Insurance* 44: 237–58.

Thaler, R. 1981. Some empirical evidence on dynamic inconsistency. *Economics Letters* 8: 201–7.

Public Perceptions of Fresh Water Issues

Robert E. O'Connor,[1] Richard J. Bord,[2] and Ann Fisher[3]

Abstract

Risk management takes place in a context of public perceptions of fresh water quality, quantity, and availability. Regardless of their consistency or inconsistency with expert opinions, public perceptions can, and often do, influence policy decisions. This paper reports our review of more than 500 studies of what Americans believe about fresh water quality, quantity, and availability. Most people view quality problems as serious and getting worse, regardless of the source (surface water or ground water) or use (such as for drinking, agriculture, or recreation). Although people in the East only care about water quantity during drought periods, Westerners are frequently worried. Research on perceptions of risk to fresh water quality, quantity, and availability should be placed in the context of comparative risk perceptions, should measure salience, and would benefit from attention to policy options.

Introduction

Fresh water is essential to human life. Yet most Americans, with occasional dramatic exceptions involving floods or droughts, have treated fresh water quite cavalierly. And why not? Before World War II, a sparse population coupled with a generally temperate climate and abundant rainfall in most of the country relegated water to the taken-for-granted category. That lack of concern is being replaced by nationwide apprehension toward water quality and, in some places, serious

[1]Associate Professor of Political Science, Pennsylvania State University, University Park, PA 16802
[2]Associate Professor of Sociology, Pennsylvania State University, University Park, PA 16802
[3]Senior Research Associate, Dept. of Agricultural Economics and Rural Sociology, Pennsylvania State University, University Park, PA 16802

questions about quantity and availability (Dunlap, Gallup, and Gallup 1992). Post-World War II population growth and migration into arid regions, industrial-based prosperity, and the increased use of chemicals in agriculture have placed serious burdens on water quality everywhere and, particularly in the West, on quantity (Ingram 1990). The modern environmental movement, sparked by dramatic pollution incidents such as Love Canal, has raised public consciousness of risks to water quality and, to a lesser extent, quantity and availability.

This chapter summarizes what Americans think about fresh water quality, quantity, and availability in an era of publicity about water pollution, potential shortages, and higher water bills. We know that many Americans no longer take safe fresh water for granted, but just how central is this issue in the realm of everyday concerns? How much do people worry about water problems and what threats do they perceive to water quality, quantity, and availability? Do they perceive water as posing risks to their own well-being?

Public risk perceptions are linked directly to the salience of fresh water problems. Public risk perceptions affect water use directly as people decide whether their lake is fit for swimming, their tap water is safe to drink, or their time in the shower should be shortened. In addition, risk perceptions are important because they influence policy formation. Risk perceptions have set limits on options considered by decision makers, have influenced the outcomes of elections and referenda, and have structured the composition and strength of interest groups concerned with water policy.

An accurate understanding of public perceptions is important for water professionals for two reasons. First, it can result in more effective public education programs so that people have a better understanding of water problems and options to mitigate them. Second, it can help governments respond effectively to public concerns.

There are hundreds of studies tapping peoples' beliefs and preferences relating to water issues, but none squarely aim at establishing perceived risk in the way experts use the concept. As a result, we cannot answer questions about the public's risk perceptions as definitively as we would like. Experts often express risk as the magnitude of a threat to health or welfare, weighted by the likelihood that it will occur. In contrast, lay judgments of risk typically incorporate many dimensions and are made relative to other risks (Slovic 1987). Studies abound that ask people to judge the relative risks of smoking, skiing, hazardous chemicals, nuclear power plants, and other potential threats to health and welfare, but we know of no studies that include water in these evaluations. This forces us to infer perceived risk from several kinds of less-than-perfect data.

Measuring Public Attitudes and Inferring Perceived Risk

To measure public attitudes, we begin with Gilbert White's (1966) five approaches to discerning public opinion. We add a sixth approach that examines historical descriptions and policy studies. These narratives tend to be indirect and unsystematic, yet they often include rich detail about environmental perspectives. Each of these six approaches provides specific insights not available from the others.

Approach	Features
Narratives	Examine historical descriptions and policy studies
Textual Analysis	Systematically examine literature, paintings, speeches, and articles that show attitudes
Surveys	Ask people their opinions
Consumer Choices	Infer attitudes from behavior, including recreation uses and residential decisions
Ethnography	Study cases of how people make decisions related to fresh water
Considered Judgment	Provide information to groups that then discuss and take positions on water issues

Table 1. Approaches to Measuring Public Attitudes

The approaches complement each other. Textual analysis identifies dominant or consistent value themes in rhetoric, literature, and other informational media during different historical periods. Surveys ask a sample of the population questions posed by researchers. The consumer choices approach involves inferring values from actual behavior. Consumer choices, such as the number of recreational users at a lake, money paid for irrigation water, or the additional property value of riverfront homes, reflect attitudes toward fresh water. Ethnographies involve intensive observations of individuals or groups as they go about their daily lives. Finally, White (1966, 130) wrote that "subjecting people to experimental situations in which they are asked to voice opinions after being exposed to a variety of information and persuasion and to interaction with peers seeking answers to the same problem" would be a revealing way to assess attitudes. This approach resembles the Public Agenda Foundation's notion of "considered judgment" (1986). It begins with the traditional focus group procedure of learning from group discussion, but then provides the group with scientific information as well as the positions of different concerned organizations. Advancing beyond their original reactions, the members of the group provide their considered judgment on the issue.

We used these six categories to sort available studies that shed light on the public's attitudes about fresh water. The data forming the basis for this report are extensive, but less than ideal. Despite reviewing hundreds of articles, books, and reports written in the 1970s, 1980s, and 1990s, we still lack important information. We would have preferred more studies designed to estimate specific perceived water risks; more longitudinal studies, those using the same items across populations and times; more studies incorporating adequate numbers of minorities and working-class people; and more studies juxtaposing water issues with other social/environmental problems.

In this paper, we concentrate on the American public's overall perceptions of fresh water. Our criterion for inference is consistency. If the same result surfaces in many studies (especially in several of the categories listed in table 1), across time and place, we consider it valid. The references reflect only examples from the hundreds of studies examined for this paper.

People's perceptions of water quality, quantity, and availability are an integral part of environmental attitudes. Therefore, we first report changes that have occurred in people's assumptions about the natural world. Next, we discuss public perceptions of fresh water quality. Then, we turn to the data on the public's perceptions of water quantity and availability. Finally, we summarize the results, draw implications for the limits to our knowledge of the public's risk perceptions, and suggest new directions toward a better understanding of how the public views fresh water.

Changing World Views and Their Implications for Water Issues

What people pay attention to and consider important depends on their beliefs about the nature of the world and the values accompanying those beliefs. As time passes, different world views become dominant and form a relatively unquestioned set of assumptions that guide perception, discussion, and decision making (Kuhn 1970).

During the last thirty years, many social commentators have identified the emergence of a new set of beliefs and values pertaining to the natural environment and have offered varying, but greatly overlapping, interpretations of that new world view and its implications (Cotgrove 1982; Dunlap and Van Liere 1978; Harman 1979; Milbrath 1984; Olsen, Lodwick, and Dunlap 1992). Their basic argument is that many Americans have rejected a technological world view and have adopted a more ecological one. The technological world view centers on the notion of people controlling nature, using natural resources to enhance their quality of life by fostering economic growth (McPhee 1989).

Throughout most of this century, discourse on fresh water reflected the technological world view (Field Institute 1990). People viewed fresh water as something to manipulate to produce energy, to enhance development, to provide recreation areas, to irrigate crops, or even to eliminate if it stood in the way of more development (table 2). In contrast, the ecological world view emphasizes people living in harmony with nature, with preservation taking precedence over development. In the last three decades, water policy debates, the formation of new environmental organizations and the enormous growth of membership in the older ones (Ingram and Mann 1989), the presentation of

	World View	
	Technological	Ecological
Humanity and nature	Humanity needs to control nature, which is essentially incompetent	Humanity must live in harmony with nature
Quality of life	Depends on controlling nature	Is linked to preserving natural environments
Priority value	Economic growth	Aesthetic and ecological values
Role of water resources	Instrumental, use for economic development (e.g., energy, irrigation)	Given the history of use, preserve remaining wilderness

Table 2. Core Beliefs of World Views

ecological positions in the popular media, the initiation of "green" marketing schemes, the rise of eco-tourism, and the large increases in ecological sentiments in surveys (Dunlap and Scarce 1991) demonstrate the emergence of an ecological world view. The emergence of the ecological world view is not simply a reaction to a degraded environment, for many aspects of environmental quality actually have been improving (Portney 1990). Instead, the new world view is a product of interest groups, media attention, and numerous sources of disillusionment with the dominant culture.

Perceptions of water-borne risks must be viewed in the context of this changing world view. Increased sensitivity to environmental problems carries over to perceptions and decisions about water quality, quantity, and availability. The large percentage of the American people reporting themselves to be extremely concerned about the environment (Dunlap 1987) suggests they also are concerned about the essential natural resource, water.

Attitudes Toward Fresh Water Quality

Studies of public perceptions of water quality produce quite consistent results, regardless of whether the questions mention drinking water, surface water, underground water supplies, or just fresh water in general. When people think about water, they often do not make distinctions that experts consider critical. For example, experts typically distinguish between sources, such as surface and ground water, and among uses, such as drinking, irrigation, recreation, or habitat preservation. Results related to water quality parallel survey findings on general environmental issues. We reduced hundreds of studies to eleven basic conclusions about American public opinion on water quality.

- Most people believe that our country has serious water quality problems and that the situation is getting worse.

A majority of Americans express great concern for water quality problems and think that conditions are not improving (Dunlap and Scarce 1991; Gallup Organization 1972–91; Louis Harris and Associates 1972–91). Figure 1 shows results from four questions that national samples have answered over several years. The outcome: Americans are worried about their water and are increasingly pessimistic.

- Water quality ranks high among people's *environmental* concerns, but lags behind concerns about the economy, crime, and drugs.

Placed in the context of environmental problems, water quality is a most significant concern. Placed in the broader context of social and economic problems, however, water quality is less important (Dunlap 1989). This raises the issue of salience: To what degree do water quality problems dominate thinking or discussion? The evidence implies relatively low salience. People want clean water, but sometimes they want other things from government even more.

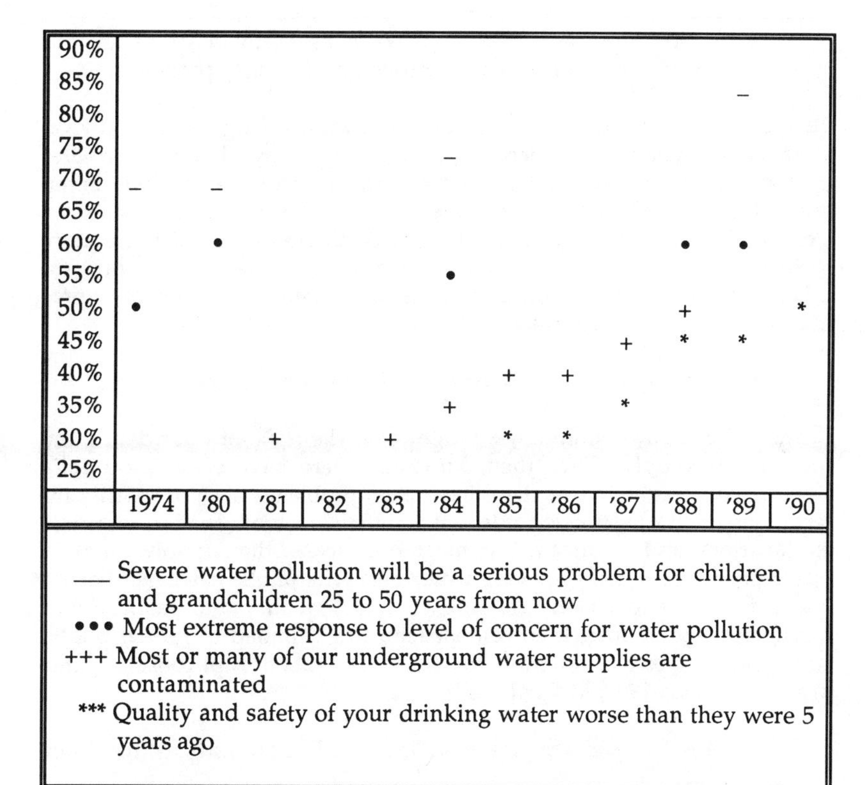

Figure 1. Agreement with Negative Judgments

- Water quality concerns are linked closely to the perceived health risks of hazardous chemical wastes.

Improperly handled hazardous chemical waste (along with radioactive waste) engenders the highest levels of environmental concern (Opinion Research Surveys 1972–91). Seven studies done between 1982 and 1992 indicate that most Americans view toxic wastes as the most important environmental problem (Louis Harris and Associates 1972–91; Roper Center 1982–91). Other threats to water quality, such as acid rain, coal-burning power plants, depletion of wetlands, agricultural runoff, mining wastes, drought, aging and leaking urban water systems, and competition for water among users are viewed as serious by a bare majority or, usually, fewer. Questions that link chemical pollution and water, either surface or ground, generate levels of concern comparable to that engendered by chemical wastes alone.

- Most people rate their own water as adequate or better, but perceive that others have serious water quality problems.

People generally judge the quality of their own drinking water, and local streams and lakes, to be acceptable, although a trend toward greater pessimism has shrunk the number to a slight majority in polls taken since 1989 (Louis Harris and Associates 1972–91; Opinion Research Corporation 1972–91). Water quality problems are perceived to be more serious as the reference area shifts from local, to state, to country, to other countries; people tend to assume that others' problems are worse than their own (Dunlap 1987).

- Water quality simply is not an issue in most communities.

In the 1950s many communities fought over the issue of fluoridating the local water supply. Since then, however, there have been few public battles about water. When there have been problems, such as giardia infestations or shortages due to droughts, officials have turned to technical explanations and solutions. For most Americans, their involvement in water policy is limited to paying water bills. The public has never been a major player in decisions relating to water quality and institutional mechanisms promoting involvement tend to be limited to formal public comments on environmental impact statements for proposed water projects (Ingram 1990; McCool 1987).

- Most people are not well informed about threats to water quality.

Traditionally having enjoyed ample supplies of good water and rarely having been involved in water decisions, the public cannot be expected to know much about what affects water quality (Opinion Research Corporation 1972–91; Roper Center 1972–91). A 1990 survey encompassing Ohio and West Virginia, for example, found 21 percent to 30 percent of the respondents could not identify a single important water problem in their state (Survey Sampling 1990). Maryland polls focusing on Chesapeake Bay pollution indicate low awareness of nonpoint sources (Maryland State Polls 1988–90).

Most people know that their water comes from a water authority that pipes it to their homes, but have little information about its origin or ultimate destination. Most people understand pollution from factories that dump into waterways, but fewer citizens associate pollution with runoff from a municipal landfill or agricultural chemicals used hundreds of miles away. Media images of plant effluent flowing into a stream or rusting fifty-five gallon drums are vivid and memorable. Nonpoint pollution is difficult to dramatize visually (Lovrich et al. 1986).

There is evidence, however, that public education programs have been effective in some areas. Faced with serious groundwater pollution and a highly publicized toxic waste disposal program, Minnesotans seem to have become aware of nonpoint pollution (Minnesota Center for Social Research 1985, 1987, 1989). California (*Los Angeles Times* 1990, 1991) and some towns in Massachusetts (Berdan 1983) have run successful public education programs regarding water conservation. In Illinois, a public education program was successful in enhancing public support for wetlands restoration (Smardon 1989).

- People are reluctant to acknowledge that their own activities could threaten water quality.

Not surprisingly, the different "publics" that affect water quality look elsewhere for the causes of water pollution. Most farmers understand nonpoint pollution and realize that farmers play an important role in generating such pollution, but assert that their farm is not to blame (Christensen and Norris 1983). Boaters want improved water quality, but oppose limitations on recreational boating that threaten water quality and wildlife (Florestano and Rathbun 1981).

- Industry receives disproportionate blame for water quality problems.

People overestimate the contribution of manufacturing, especially the chemical and oil industries, to water pollution while underestimating the share coming from nonpoint sources (Opinion Research Corporation 1972–91; Roper Center 1972–91). People view industry as making significant progress in dealing with pollution problems, but at an inadequate pace (Roper Center 1972–91).

- Almost everyone wants the federal government to establish strict water quality standards, but the public is less certain about which level of government would best enforce those standards.

A substantial majority of the American people hold the federal government responsible for establishing and maintaining strict water quality standards (Louis Harris and Associates 1972–91; Opinion Research Corporation 1972–91). Opinion polls over the last twenty years have found as much as 96 percent of the public assigning responsibility for clean water to the federal government (Dunlap and Scarce 1991). The public consistently expresses support for tough water quality standards and enforcement.

The public is less united, however, on the issue of which level of government should be responsible for enforcing federally established standards (Dunlap and Scarce 1991; Welch 1981). While approximately half of the public supports enforcement by the national government, the other half is divided between state and local governments. This finding is consistent with research that indicates many people have a proprietary view of water and want control mechanisms located closer to home.

- Americans express a willingness to pay for water quality.

Majorities consistently say they are willing to spend more and to pay higher taxes for a cleaner environment in general, and to protect water quality and to clean up contaminated waterways specifically (Cotgrove 1982; Dunlap and Scarce 1991; Feldman 1991; *Los Angeles Times* 1990, 1991; Mitchell and Carson 1989; Opinion Research Corporation 1972–91). Willingness to pay, however, varies considerably with how the question is asked. It is also possible that willingness-to-pay questions elicit socially desirable responses. Having already identified water quality as a serious problem, the respondent might find it incongruous to refuse to pay to ameliorate that problem.

On the other hand, residents of some states have supported the strict enforcement of water pollution controls even at the expense of losing job-intensive industries (Keeter 1990). Although there appears to be substantial support for water quality, the public's actual willingness to pay is difficult to determine. Survey results could be a function of social desirability and the survey format, a lack of understanding of what can be accomplished for each dollar, a real commitment to sacrifice for clean water, or some combination of the three.

- Demographic differences in attitudes are small.

There are no substantial differences in water quality judgments along educational, income, racial, or regional lines (Asch and Seneca 1980; Dunlap 1989; Field Institute 1990, 1992; Olsen and Highstreet 1987; Roper Center 1972–91). More women than men consistently express high levels of concern for water quality problems, but these gender differences seldom exceed even five percent. Some studies show African Americans, the elderly, and those with less than a high school education to be somewhat less concerned. In many cases, however, the observed differences are a result of these respondents being more willing to choose a "don't know" response. Although there are some regional differences in how water quality influences recreation behavior, in their other attitudes toward water quality Americans are not sharply divided along demographic lines.

In summary, the American people may not know much about what affects water quality and who is really responsible for pollution, but they value clean water highly.

Attitudes Toward Fresh Water Quantity and Availability

Water quantity and availability are separate concerns. Large quantities of available but contaminated water are not useful, nor are large quantities of clean water that are difficult to transport to areas of shortages. Nevertheless, the public tends to assume that quantity equals availability. We will treat the two issues as one in the subsequent discussion. There are seven noteworthy conclusions.

- Except during drought episodes, people east of the 100th meridian rarely think about water quantity problems.

In the United States, continuing discussions of water shortages are generally limited to the continental states west of the 100th meridian, excluding the Pacific Northwest. Elsewhere, water quantity concerns ebb and flow depending on whether there is a drought. As a result, national surveys have seldom included quantity questions, and our conclusions must rely primarily on the views of Westerners.

- Water quantity concerns are endemic to the West and Southwest, and reached intense levels in California during the drought years of the late 1980s.

As the population grows in the West and water supplies dwindle, public concern increases (Hundley 1992; Ingram 1990). A 1990 California survey (Field Institute 1990) found that 91 percent believed there was a water shortage, 85 percent said it was serious, and 40 percent viewed the problem as extremely serious.

- Westerners have a proprietary attitude about water.

While water agency and government officials tend to view water from a large regional perspective, the public thinks of water in terms of their own state or immediate region (Public Agenda Foundation 1986). Concern over maintaining water availability often has lead to "turf politics" (Shelley and Wijawawickrema 1984; Yankelovich 1991). People in water-rich areas of the West are willing to share their water, but are worried about having it stolen.

- Like other Americans, Westerners are not well informed about water problems.

The realities concerning Western water include the following:
- irrigation consumes most water in the dry Western states (although the growth in use now is mostly residential and industrial)
- many ground water sources are being depleted
- the federal government has heavily subsidized the development of Western water supplies
- although not as strong as it once was, the water rights doctrine provides "use it or lose it" incentives for farmers

Most Westerners, however, believe that residential and industrial users take most of the water. People have almost no understanding of the role agriculture plays in water quantity conflicts and little realization of the threat to water quality posed by agricultural practices. The public tends to assume that farmers are good stewards of this precious resource. Most people consider water to be an entirely renewable resource. In addition, few people are familiar with technical concepts (for example, water rights, market incentives, fair market value, and transferring water) that must be understood for informed discussion of water policy in the West (Public Agenda Foundation 1986). Nevertheless, there are indications that Westerners are becoming better informed as a result of increased media coverage of water issues (Field Institute 1992). The Omnibus Water Act of 1992, which relies more heavily on market incentives that will shift water to those uses people are most willing to pay for, may also stimulate public interest in water issues.

- Voluntary water conservation is acceptable, but enforced conservation generally is not.

During periods of drought, voluntary water conservation has been remarkably successful (Bruvold 1979). People adopt a number of water-saving strategies when the need is clear and the option is suggested, not demanded. Although there is some evidence that people are willing to accept mandatory constraints during periods of prolonged drought and perceived crisis, there is concern that everyone bear the same burden (Hundley 1992). Polls consistently demonstrate a reluctance to accept increased fees as incentives for water conservation (Flack and Greenberg 1987; Lord, Chase, and Winterfield 1983).

- The public supports reclaimed water for selected uses.

Large majorities support the use of reclaimed wastewater for toilet flushing, irrigation, and golf course maintenance. The public is less supportive of uses involving personal contact, such as bathing or food preparation (Bauman and Kasperson 1974; Bruvold 1981, 1988; Bruvold, Olson, and Rigby 1981; Sims and Bauman 1974).

Although water managers have been reluctant to promote the use of reclaimed water because of expectations of negative public reactions, there is evidence that public education programs can substantially increase public acceptance. For example, a majority in Denver initially opposed a proposal to use reclaimed water for drinking purposes. A public education program changed this to 63 percent not objecting, 25 percent opposed, and 12 percent undecided (Hadeed 1977).

- Support for preservation of wild and scenic areas is growing and reflects increased questions about the value of development.

Polls in Western states consistently demonstrate agreement that additional development should be discouraged and that areas still in a wild state be left that way (Field Institute 1990, 1992). There is a willingness to allow areas with high water usage to continue their consumption rate, but a reluctance to open water wilderness areas to recreation or other use (*Los Angeles Times* 1990, 1991).

Conclusions

Available data unambiguously indicate substantial, and increasing, levels of concern for most environmental issues, including water quality. Most people view industry, especially the chemical and petrochemical industries, as the major polluter. Fears of toxic chemical waste—the number one environmental concern—along with fear of nuclear power plants and nuclear waste are highly correlated with water pollution concerns. Regarding water quantity, chronic and periodic drought stimulate concerns. What are the implications of these findings for assessing public perceptions as they relate to water quality and quantity issues?

General risk perception studies indicate that most people do not make risk judgments the same way that technical experts do. Experts rely on a "probabilistic risk assessment" technique that uses a "fault-tree" approach to trace sequences of events that could lead to problems or undesired outcomes (Fischhoff, Slovic, and Lichtenstein 1978). The model produces estimates of, for example, the likelihood of an extra death within a specified group over a defined period of time. The general public, however, includes a number of "qualitative" factors in forming risk judgments; risks tend to be perceived as more serious to the extent that they are involuntary, delayed, poorly understood, potentially catastrophic, dreaded, and fatal (Slovic 1987). For example, nuclear power and toxic chemicals exhibit these characteristics and are perceived as high risks.

Little research focuses specifically on the perceived risk of water quality and quantity, so we are forced to extrapolate from what we know about risk judgments to what we expect about water concerns (Scherer 1990). To the extent that water concerns are linked to fears of toxic chemical pollution, perceived risk should be substantial. Toxic chemical exposure is viewed as long-term and involuntary (usually imposed by industry), characterized by dreaded and potentially fatal threats to health (cancer), and is poorly understood. We know that people judge toxic chemicals to be the most prominent threat to water quality. If studies were done, we would expect to find that the majority who say they are greatly concerned about water quality would also report high levels of perceived risk.

However, a majority of Americans judge their own drinking water to be of acceptable quality, even though they report great concern about water quality in general. This suggests that, to many Americans, water quality is not salient—an immediate and compelling problem. Natural hazard assessments have tended to neglect salience. It may be that water quality concerns generate health risk perceptions significant enough to produce vigilance, but not compelling enough to result in public demand for political action, except in local crisis situations. Clean water laws, after all, were a result of concerted efforts by environmental groups, not mass movements (Yeager 1991).

Other water quality problems, such as algae blooms, giardia contamination, discolored water, or even acid rain, are unlikely to be perceived as particularly risky. They tend not to generate the anxiety and dread characteristic of contamination by toxic chemicals. Periodic "natural" water quality threats are more likely to be perceived as nuisances to be dealt with by technology while undramatic, invisible problems like acid rain are not perceived as personal threats.

Water quantity issues are also not likely to induce high perceived risk. Even natural hazards in disaster areas, such as flood plains, are not salient enough to produce workable natural hazard policies (Wright and

Rossi 1981). The mere fact that most residents return to flood plains after floods, as do most residents of areas experiencing recurrent, devastating fires, indicates that the risks, at whatever perceptual levels, are tolerable. Droughts generate substantial concern and effort to deal with the negative outcomes, but the problem is generally viewed as manageable and not life threatening.

We propose three directions to guide research on risk perceptions about water issues. First, perceptions of risks involving water quality, quantity, and availability make more sense in the context of *comparative risk perceptions*. Studies of environmental attitudes, including perceptions of threats *to* and *from* water quality and quantity, must go beyond the simple assessment of perceived seriousness or levels of concern. Researchers must cast water quality and quantity studies directly in the light of comparative risk perception research. These studies must examine how people judge water quality and quantity risks compared to other known risks.

Second, just as crucial, environmental perception studies must measure *salience*. In many studies we examined, it was too easy to indicate concern for a problem defined by the researcher. What is important is the extent to which that issue dominates thinking, attention, and plans for action. We know virtually nothing about the salience dimension from available environmental research (Whyte 1986). We expect that salience would vary by the type of use being threatened, with risks to drinking water being more salient than threats to recreation, habitat, or biodiversity. The evidence on this is sparse, partly because available studies often do not distinguish among uses. To claim that Americans are shifting world views from a technological to an ecological perspective may be premature in the absence of a better understanding of the salience of environmental issues. Local contamination of water supplies by toxic chemicals certainly can result in high perceived risk. It is much more difficult to make statements about the public's risk perceptions of other water problems.

Third, research on perceived risks to and from fresh water would benefit from attention to *policy options*. How the public perceives fresh water affects both their behavior and the policy options available to governments. As both state and national governments are reassessing water policy in the 1990s, research can help inform this reassessment by reporting what people expect, fear, find acceptable, and even demand. This research direction focuses on how beliefs influence risk perceptions and policy choices. We need to know more about how the public views real options that policymakers face. As the probability increases that good water will become substantially more expensive, people are more likely to take an interest in—and even become involved in—policy discussions and decisions.

Acknowledgments

The authors appreciate the financial support of the National Geographic Society (Grant #4782-92); the work of research assistants Kelly Finley, Sharon Gripp, Susan Mann, and William Wheeler; and the advice of Elizabeth David, William Hansen, Charles Howe, Helen Ingram, Mary Jo Kealy, John Munro, and William Weyrick.

References

Asch, P., and J. J. Seneca. 1980. The incidence of water quality: A county level analysis. *Water Resources Research* 16 (2): 319–24.

Bauman, D. D., and R. E. Kasperson. 1974. Public acceptance of renovated waste water: Myth and reality. *Water Resources Research* 10 (4): 667–74.

Berdan, M. R. 1983. Wellesley's water program planning: Demand strategy. *Journal of New England Water Works Association* 97 (1): 82–92.

Bruvold, W. H. 1979. Residential response to drought in central California. *Water Resources Research* 15 (6): 1297–1304.

Bruvold, W. H. 1981. Community evaluation of adopted uses of reclaimed water. *Water Resources Research* 17 (3): 487–90.

Bruvold, W. H. 1988. Public opinion on water reuse options. *Journal of the Water Pollution Control Federation* 60 (1): 45–9.

Bruvold, W. H., B. H. Olson, and M. Rigby. 1981. Public policy for the use of reclaimed water. *Environmental Management* 5 (2): 95–107.

Christensen, L. A., and P. E. Norris. 1983. Soil conservation and water quality improvement: What farmers think. *Journal of Soil and Water Conservation* 38 (1): 15–20.

Cotgrove, S. 1982. *Catastrophe or cornucopia: The environment, politics and the future.* New York: John Wiley and Sons.

Dunlap, R. E. 1987. Public opinion on the environment in the Reagan era. *Environment* 29 (6): 7–37.

Dunlap, R. E. 1989. Public opinion and environmental policy. In *Environmental politics and policy: Theories and evidence,* edited by J. P. Lester. Durham, N.C.: Duke University Press.

Dunlap, R. E., G. H. Gallup, and A. M. Gallup. 1992. *The health of the planet survey*. Princeton: The George H. Gallup International Institute.

Dunlap, R. E., and R. Scarce. 1991. The polls—poll trends: Environmental problems and protection. *Public Opinion Quarterly* 55 (4): 651–72.

Dunlap, R. E., and K. D. Van Liere. 1978. The new environmental paradigm: A proposed measuring instrument and preliminary results. *Journal of Environmental Education* 9 (Summer): 10–9.

Feldman, D. L. 1991. *Water resources management: In search of an environmental ethic*. Baltimore: Johns Hopkins University Press.

Field Institute. 1990. *Californians' views on water*. Sacramento: Association of California Water Agencies.

Field Institute. 1992. *An update on Californians' views on water*. Sacramento: Association of California Water Agencies.

Fischhoff, B., P. Slovic, and S. Lichtenstein. 1978. Fault trees: Sensitivity of estimated failure probabilities to problem representation. *Journal of Experimental Psychology: Human Perception and Performance* 4 (1): 330–44.

Flack, J. E., and J. Greenberg. 1987. Public attitudes toward water conservation. *Journal of the American Water Works Association* 79 (1): 46–51.

Florestano, P. S., and P. A. Rathbun. 1981. Public opinion and interest group positions on Chesapeake Bay issues: Implications for resource management. *Coastal Zone Management Journal* 9 (1) :19–39.

Gallup Organization. 1972–1991. *Gallup Polls*. Princeton: Gallup Organization.

Hadeed, S. J. 1977. Potable water from wastewater—Denver's program. *Water Pollution Control Federation Journal* 49 (8): 1757–8.

Harman, W. W. 1979. *An incomplete guide to the future*. New York: W. W. Norton Co.

Hundley, N., Jr. 1992. *The great thirst: Californians and water, 1770s–1990s*. Berkeley: University of California Press.

Ingram, H. M. 1990. *Water politics: Continuity and change*. Albuquerque: University of New Mexico Press.

Ingram, H., and D. E. Mann. 1989. Interest groups and environmental policy. In *Environmental politics and policy: Theories and evidence,* edited by J. P. Lester. Durham, N.C.: Duke University Press.

Keeter, S. 1990. *The commonwealth poll*. Richmond: Survey Research Laboratory, Virginia Commonwealth University.

Kuhn, T. S. 1970. *The structure of scientific revolutions*. 2nd ed. Chicago: University of Chicago Press.

Lord, W. B., J. A. Chase, and L. A. Winterfield. 1983. Choosing the optimal water conservation policy. *Journal of the American Water Works Association* 75 (7): 324–9.

Los Angeles Times. 1990. Poll.

Los Angeles Times. 1991. Poll.

Louis Harris and Associates. 1972–91. *Harris Surveys*. Chapel Hill, N.C.: Institute for Research on the Social Sciences.

Lovrich, N. P., J. C. Pierce, T. Tsurutani, and A. Takematsu. 1986. Policy relevant information and public attitudes: Is public ignorance a barrier to nonpoint pollution management. *Water Resources Bulletin* 22 (2): 229–35.

Maryland State Polls. 1988–90. College Park: University of Maryland.

McCool, D. 1987. *Command of the waters*. Berkeley: University of California Press.

McPhee, J. 1989. *The control of nature*. New York: Farrar Straus Giroux.

Milbrath, L. W. 1984. *Environmentalists: Vanguard for a new society.* Albany: State University of New York Press.

Minnesota Center for Social Research. 1985. *1985 Minnesota fall survey.* Minneapolis: University of Minnesota.

Minnesota Center for Social Research. 1987. *1987 Minnesota state survey.* Minneapolis: University of Minnesota.

Minnesota Center for Social Research. 1989. *1989 results and technical report—Minnesota water quality survey.* Minneapolis: University of Minnesota.

Mitchell, R. C., and R. T. Carson. 1989. *Using surveys to value public goods: The contingent valuation method.* Washington, D.C.: Resources for the Future.

Olsen, D., and A. L. Highstreet. 1987. Socioeconomic factors affecting water conservation in southern Texas. *Journal of the American Water Works Association* 79(1): 59–68.

Olsen, M. E., D. G. Lodwick, and R. E. Dunlap. 1992. *Viewing the world ecologically.* San Francisco: Westview Press.

Opinion Research Corporation. 1972–91. *Opinion research corporation surveys.* Chicago: Opinion Research Corporation.

Portney, P. R. 1990. Overall assessment and future directions. In *Public policies for environmental protection,* edited by P. R. Portney. Washington, D.C.: Resources for the Future.

Public Agenda Foundation. 1986. *Water efficiency in the West: The public's view.* New York: Public Agenda Foundation.

Roper Center. 1972–91. *Roper reports.* Storrs, Conn.: Roper Center.

Scherer, C. W. 1990. Communicating water quality risk. *Journal of Soil and Water Conservation* 45 (2): 200.

Shelley, F. M., and C. Wijawawickrema. 1984. Local opposition to the transfer of water supplies: An Oklahoma City case study. *Water Resources Bulletin* 20 (5): 721–7.

Sims, J. H., and D. D. Bauman. 1974. Renovated waste water: The question of public acceptance. *Water Resources Research* 10 (4): 659–65.

Slovic, P. 1987. Perception of risk. *Science* 236 (4799): 280–5.

Smardon, R. C. 1989. Human perception of utilization of wetlands for waste assimilation, or How do you make a silk purse out of a sow's ear? *Constructed wetlands for wastewater treatment: Municipal, industrial, and agricultural.* Chelsea, Mich.: Lewis Publishers.

Survey Sampling, Inc. 1990. *West Virginia opinion survey and Ohio public opinion survey.* Fairfield, Conn.: Survey Sampling.

Welch, S. 1981. *Public attitudes of Nebraskans toward water policy.* Washington, D.C.: Office of Water Research and Technology, U.S. Department of the Interior.

White, G. F. 1966. Formation and role of public attitudes. In *Geography, resources, and environment,* Vol. I, edited by R. W. Kates and I. Burton. Chicago: University of Chicago Press.

Whyte, A. V. T. 1986. From hazard perception to human ecology. In *Geography, resources, and environment,* Vol. II, edited by R. W. Kates and I. Burton. Chicago: University of Chicago Press.

Wright, J. D., and P. M. Rossi, eds. 1981. *Social science and natural hazards.* Cambridge: ABT Books.

Yankelovich, D. 1991. *Coming to public judgment: Making democracy work in a complex world.* Syracuse, N.Y.: Syracuse University Press.

Yeager, P. C. 1991. *The limits of law: The public regulation of private pollution.* Cambridge: Cambridge University Press.

Subjective De-Biasing of Data Sets: A Bayesian Approach

M. Elisabeth Paté-Cornell[1]

Abstract

In this paper, we examine the relevance of data sets (for instance, of past incidents) for risk management decisions when there are reasons to believe that all types of incidents have not been reported at the same rate. Our objective is to infer from the data reports what actually happened in order to assess the potential benefits of different safety measures. We use a simple Bayesian model to correct ("de-bias") the data sets given the nonreport rates, which are assessed (subjectively) by experts and encoded as the probabilities of reports given different characteristics of the events of interest. We compute a probability distribution for the past number of events given the past number of reports. We illustrate the method by the cases of two data sets: incidents in anesthesia in Australia, and oil spills in the Gulf of Mexico. In the first case, the "de-biasing" allows correcting for the fact that some types of incidents, such as technical malfunctions, are more likely to be reported when they occur than anesthetist mistakes. In the second case, we have to account for the fact that the rates of oil spill reports in different incident categories have increased over the years, perhaps at the same time as the rates of incidents themselves.

Report Biases in Databases

Some databases come from direct surveys (e.g., marketing surveys), others are voluntary reports of events (e.g., reports of incidents in airline operations). In both cases, there is almost always a certain rate of "nonresponse" (or "nonreport"). Nonreport is likely to occur, for example, when people are asked to record their own mistakes and are

[1]Professor, Department of Industrial Engineering and Engineering Management, Stanford University, Stanford, CA 94305

fearful of the consequences. The probability of nonreport can thus depend on some characteristics of the events of interest. Therefore, if there are reasons to believe that all types of incidents are not reported at the same rate, the databases involving voluntary incident reports cannot be used directly to compute future probabilities as equal to past frequencies of incidents reports.

The problem is twofold: how best to utilize existing data and how to improve data collection in the future. This paper outlines a Bayesian method for updating the results obtained from databases involving report biases to permit better probabilistic risk estimates. This method is basically the same as the one developed by Gaba and Winkler (1992) and Gaba (1993) but addresses a slightly different problem: they used a Bayesian model to study the implications of errors in survey data. The question addressed here is, given the rate at which a particular type of incident was reported, what really happened? The approach is based on expert opinions about the past probabilities of reports of incidents conditional on the event for different types of incidents. These probabilities may vary over time.

In our recent research, we encountered this problem in two specific instances: anesthesia incidents and oil spills due to pipeline leaks in the Gulf of Mexico. In our study of anesthesia, we assess the patient risks and the benefits of some organizational improvements (Paté-Cornell 1992a, 1992b). We use the data provided by the Australian Incident Monitoring Study (AIMS). This database, gathered in Australia and New Zealand in the last four years, contains about 2,000 cases of incidents, of which 67 resulted in patient death or severe brain damage (Webb et al. 1993). The corresponding number of operations was estimated at 1.12 million. In our current study, we first assess the patient risk of death or brain damage by categories of "initiating events." We then evaluate the benefits of different management improvements that should decrease either the probabilities of the initiating events or the probabilities of the patient's death conditional on the initiating events. It soon became clear that the rates of reporting, although apparently fairly stable over time, varied greatly according to three main factors: the types of facilities (rural hospitals and clinics versus urban facilities), the experience of the anesthetist (established practitioners versus residents), and the nature of the initiating incident (mechanical problems versus syringe swaps and drug errors). These factors, of course, are dependent: there are more residents in large university hospitals, and according to our experts, they are more likely to experience nonventilation problems than their more experienced colleagues because, for example, they may not yet be fully familiar with the ventilation machine. Therefore, we need to "clean up" this data set to account for report biases and dependences among bias factors.

The rates of reports are, by definition, unrecorded. Our approach is to encode and use expert opinions to obtain estimates of the probabilities of reports given events. We can then proceed to a first updating of the data (i.e., of the frequency of reports) to obtain an initial estimate of the frequencies of the initiating events per operation. This first-order correction gives us an adjustment of the ranking of the different incident types and suffices to change the order of priorities of some of the considered improvements (e.g., improving the supervision of the residents or reducing the time on duty of anesthesiologists in state-run hospitals). Actual Bayesian updating of these probabilities of initiating events then allows us to derive a probability distribution for the number of past initiating events given the number of reports, and provides a first estimate of the uncertainties. Further treatment of uncertainties allows us to account for the fact that the experts may be uncertain and vague about the probabilities of reports given different types of initiating events and different types of anesthetists. Finally, once the incidents have been divided among categories defined by the three considered parameters (types of initiating events, types of facilities, and types of anesthetists), the sample sizes in some categories may be small. This, in turn, introduces uncertainties in the evaluation of the probabilities of some categories of incident reports in the future given their frequencies in the past. All these uncertainties may affect the risk assessment results in each category and the risk reduction estimates for different management improvements.

The second case with similar problems is that of pipeline leaks in the Gulf of Mexico. It came up in a study about ways of reducing the risks to human life and the environment of unburied and leaky marine pipelines in the Gulf. The record shows that pipelines (about 18,000 miles) leak on average every four days (Woodson 1991). Therefore, although most of the spills are very small, the problem cannot a priori be considered trivial. The existing database also shows that the numbers of reports vary across pipe diameter ranges (which obviously affects the amount of oil spilled). Furthermore, the numbers of spill reports have increased significantly over time. The rules of enforcement have become stricter in the last twenty years, the fines stiffer, and the general awareness of environmental damage has provided more incentives for the oil companies and the regulatory agencies to record incidents and fix the defective pipes. It could also be that the pipes are aging and corroding and simply leak more often. Because the costs of inspecting and fixing marine pipelines are high, it was proposed to perform a risk analysis based on zonation for the Gulf in order to set priorities among the different zones. The existing database is a starting point, but it needs to be "debiased" for reporting errors before significant figures can be extracted from it.

These kinds of problems are extremely common in detection studies. The airlines have set anonymous data incident report systems to learn from their employees' experiences. The Atlanta-based Center for Disease Control regularly encounters these types of questions. Its approach is to encode educated guesses, often as a lower bound of the probabilities of reporting given the events, and to quote the number of reports with these guesses. The result is thus often something like, "Phenomenon X has been reported in y percent of cases, but the real incidence could be as high as z percent."

We start here with a first-order correction of past frequencies of events using point estimates of the probabilities of report given events. These figures are encoded as expert opinions, accounting for the different factors that can affect the rate of reporting and the possible dependencies among these factors. We then proceed to an actual Bayesian treatment of uncertainties (Raiffa and Schlaiffer 1961) about the future frequencies of events. Finally, past report rates may have varied over time and across event categories. The treatment of biased data sets can then be improved in many different ways: by gathering larger samples, by better assessment of the past reporting rates, and by improvement of the reporting process and forms.

"De-Biasing" the Database: Bayesian Treatment

First-order correction: Point estimate

Event E is the event of interest (e.g., an initiating incident in anesthesia that may or may not lead to severe consequences), characterized by different factors, for example, in our study of anesthesia: X (type of incident, e.g., nonventilation of the patient), Y (experience of the anesthetist, e.g., unsupervised resident), or Z (type of facility, e.g., urban teaching hospital). The partition of events by type X is indexed in i and defines a set of initiating events Ei (e.g., occurrence of a tube-disconnect incident).

Report R is a record (report) of an occurrence of event E. It is assumed that all records correspond to an event (no false positive). Similarly, it is assumed that the type of event (indexed in i) is indicated by the record, that it is part of the report form, and has been identified (event Ri, e.g., report of a drug overdose incident).

The corrective factor is the probability $p(R \mid E)$ that, given that an incident occurred, it was reported in the database. It can be encoded from experts and used directly for simple updating. (Note that all marginal probabilities are expressed here per operation.) By definition of conditional probability

$$p(E,R) = p(R) \times p(E \mid R) = p(E) \times p(R \mid E)\ . \tag{1}$$

Assuming that there is no report without an event ($p(E \mid R) = 1$),

$$p(E) = \frac{p(R) \times p(E \mid R)}{p(R \mid E)} = \frac{p(R)}{p(R \mid E)}\ . \tag{2}$$

Most of the time, the probability of report conditional on the event depends on the values of the conditioning factors X, Y, Z (=> probability $p(\mathrm{i} \mid E,X,Y,Z, \ldots))$. By definition, these conditional probabilities cannot be obtained from statistical records and are encoded as expert opinions (using means, in case of additional uncertainties, to obtain a single probability estimate for each case). Simple use of the Bayes formula and expansion of the joint distribution involving X, Y, Z, etc., then allows accounting for dependencies among the factors X, Y, and Z to obtain the conditional probabilities of report $p(Ri \mid Ei)$ of each type of incident. The use of a belief network—influence diagram without decision nodes—can provide an easy graphic representation of the dependencies and a way of computing their effects on the results. Figure 1 shows such a network for the anesthesia case.

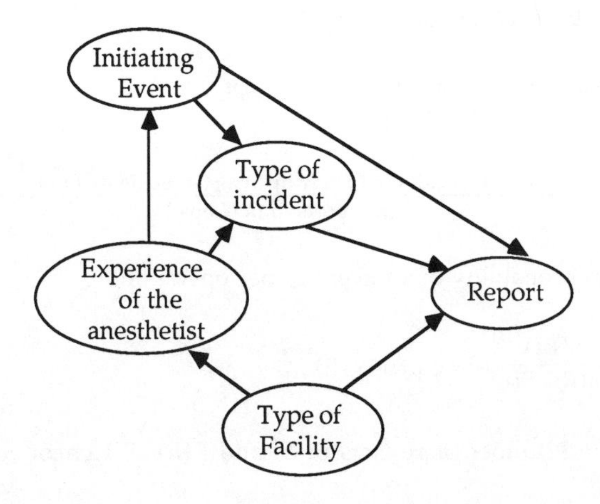

Figure 1. Dependencies among Events and Reports

If it is true that incidents such as drug errors (e.g., injecting a patient with the wrong drug) are more likely to be caused by residents than by experienced anesthetists, that there are more residents in urban hospitals than in rural clinics, and that residents' errors are less likely to be reported than errors of more experienced people, then the data must be updated to allow proper computation of the actual probability of drug error (per operation). The belief network then yields a probability of report of drug error conditional on the different factors and (by inference) an updated probability distribution of the error types *Ei* given an event *E*. The frequencies of reports φ_{Ri} of the different types of events *Ei* (wrong drug, ventilation problems, etc.) are updated by conditional distributions depending on the values of the conditioning factors (*X, Y, Z*) to obtain the probability of event *Ei* per operation.

Bayesian treatment of the number of past incidents

The problem is now the following: for each type of initiating incident *Ei* (e.g., esophageal intubation), we do not know how many occurred, but we have had a number *ji* of reports in the past *N* operations ["ops." below]. We want to derive from the experts' rates of reports *ri* a probability distribution for the number of the events *Ei* that occurred. Bayes' theorem is then used to derive the *probability distribution* of the number of incidents *ki*, given the number of reports *ji* and the number of operations *N*:

$$p(ki \text{ incidents } Ei \mid ji \text{ reports in } N \text{ ops.}) =$$

$$\frac{p(ki \text{ incidents } Ei \text{ and } ji \text{ reports in } N \text{ ops.})}{p(i \text{ reports } Ri \text{ in } N \text{ ops.})} =$$

$$\frac{p(ji \text{ reports} \mid ki \text{ incidents } Ei \text{ in } N \text{ ops.}) \times p(ki \text{ incidents } Ei \text{ in } N \text{ ops.})}{p(ji \text{ reports } Ri \text{ in } N \text{ ops.})} . \quad (3)$$

Let *pi* be the probability of an event *Ei* per operation,

$$pi = \frac{p(Ri)}{p(Ri \mid Ei)} \quad (\text{<= equation 2}) ,$$

with $p(Ri)$ = Number of reports of *Ei* and $P(Ri \mid Ei)$ encoded as expert opinion.

The probability *ri* of report *Ri* given an incident *Ei* is $ri = p(Ri \mid Ei)$.

The probability pri of report Ri (and an incident Ei, since there is no report without incident) is

$$pri = p(Ei, Ri) = p(Ri) \times p(Ei \mid Ri) = p(Ei) \times p(Ri \mid Ei) . \quad (4)$$

The database yields the probability $p(Ri)$ as the frequency of reports. The conditional probability ri is assessed by expert opinions, and the probabilities pi and pri are computed by simple multiplication. One can then compute the different distributions of equation (3):

1. The distribution of the number of reports given the number of incidents in N operations is binomial (Bernoulli trials) with parameters ri and ki:

   => $p(ji$ reports $\mid ki$ incidents in N ops.$)$ =

   $$\binom{ki}{ji} \; ri^{ji} \times (1 - ri)^{ki-ji} . \quad (5)$$

2. The distribution of the number of incidents in N operations is binomial with parameters pi and N:

   => $p(ki$ incidents in N ops.$)$ =

   $$\binom{N}{ki} \; pi^{ki} \times (1 - pi)^{N-ki} . \quad (6)$$

3. The distribution of the number of reports in N operations is binomial with parameters pri and N:

   => $p(ji$ reports in N ops.$)$ =

   $$\binom{N}{ji} \; pri^{ji} \times (1 - pri)^{N-ji} . \quad (7)$$

Equation (3) then allows computing a probability distribution for the number of incidents that occurred, given the number of reports.

Additional uncertainties occur if the experts are unsure about the conditional probabilities of reports given incidents, and if the sample size, given the number of incidents in some categories, is very small. The first type of uncertainties (about expert opinions) can be treated by considering the probability $p(Ri \mid Ei)$ as a random variable. This variable can be described (by the experts), for example, by a two-parameter Beta distribution. One can then proceed to a Bayesian updating to obtain the corresponding distribution for $p(Ei)$. Treatment of the second type (small sample) can also start with the assessment of a Beta distribution for $p(Ri \mid Ei)$, this time as a function of the sample size (see, for example, Howard 1970). Both issues are beyond the scope of this paper.

Effect of the variation of the report rates over time

It has been assumed above that both the rate of incidents and the rate of reports given incidents were stable over time. This is clearly not true in the case of many reporting systems that have improved over time while the rate of incidents may have increased, such as the reports of oil spills from pipelines in the Gulf of Mexico over the last twenty years (Woodson 1991). In that case, it seems that the reporting systems have improved as the means of incident detection have improved and the incentives for reports have increased. It is thus logical to think that both effects (increase in the rate of incidents and increase of the conditional probability of report given the events) are combined in the time variations of the data. To assess the actual rates of pipeline leakage (in the past) based on the report data, one must consider the possibility that both the density of pipelines and the number of corrosion holes in the pipes have increased. The probabilities of report given the incidents at different dates in the past are assessed from experts in the oil industry (preferably people who were directly involved in the reporting process).

The treatment sequence for this kind of uncertainty is then the following:

1. Update the data per year and per pipe diameter using expert opinions.

2. Decide whether an assumption of stability of the underlying process (frequency of the events) is justified once the data have been "cleaned up."

3. If the frequency of events appears stable over time, proceed to an analysis of the frequency of spills per pipe diameter based on the full database (combining all years after correction for the increase in the rate of reports).

4. If the frequency of events seems to have increased over time, assess the rate of increase after correction of the data to compensate for report biases (e.g., by regression analysis).

This process first can be performed by first-order point estimates of the probabilities involved (mean future frequencies), then with uncertainty analysis if the data and the decisions for which they will be used justify the additional computations.

Improving the Procedure of Data Collection

Obviously, an analysis of biases and uncertainties is justified only to the extent that it may affect the decisions that it is intended to support. In the case of our anesthesia study, the main objective is to set priorities among risk reduction measures based on management improvements. There may be, for instance, an underreporting of residents' errors (and consequently of the types of incidents that they are most likely to face). If unaccounted for, this underreporting may lead to an underestimation of the benefits of improving their supervision. A single point probability analysis of report biases is thus a helpful first step. It provides an initial basis for the decision to proceed with further uncertainty analysis given the repetition of operations and the desire of the decision makers to consider the full range of uncertainties.

The Bayesian analysis outlined above suggests several ways of improving the process of reporting and the treatment of the data once they have been gathered. One can address in sequence the following questions:

1. Given the existing data, how can the probability of future events of different categories be best estimated (do we need an uncertainty analysis to reflect the vagueness of expert opinions and small sample sizes in the data set)?

2. How can we find more about the probability of reports given the events in the existing data set?

In the method described above, the choice of the experts asked to assess these figures is critical: they can be the supervisors of residents, anesthesiologists, and authorities in the field who understand best the match between the reports and actual performance. They can be the practitioners themselves and the residents who can explain why they feel that they do or don't have to report problems (or why they should not). They can be the operating room nurses who have observed the process for a long time (they have been, so far, one of our best sources of information). The key issue is to examine the incentives to fill out voluntarily an anonymous report as part of an effort to improve collective learning. There may be a strong incentive for the conscientious practitioner if he believes that by reporting what is often his own mistake, he may prevent an accident in the future. Several elements seem to be critical to the success of data gathering:

- a feeling of safety (that the report is not going to be traced back for reprisals)
- a positive attitude of the hierarchy towards the reporting process
- the conviction that the information will be used in a positive way in the future (feedback about the actual use of the information is critical in this respect)

In the case of reports of oil leaks in the Gulf of Mexico, the reports may or may not be anonymous, and there is an additional punitive motivation: the fines that are levied on the oil companies by federal and state authorities. At this time, the oil companies have an incentive to concern themselves with their environmental image, which may make them more diligent toward cleaning large, trackable sheens and less inclined to acknowledge responsibility for small ones. For the risk analyst, the questions are as follows:

1. Do we need a larger data set to compensate for the small sample sizes that can result from a fine partition of the set of events (accounting for relevant factors)?

   This may be desirable if the sample size causes large uncertainties in the probabilities of future reports and if the choice among future policies requires reducing these uncertainties. For example, the analysis of uncertainties proposed above may suggest that the same data should also be gathered in the United States (it is generally assumed that the rates of incidents are similar in both countries).

2. How can we improve the data-gathering process to decrease the probability of nonreports and allow better risk assessment in each category?

   It can be that after a first round of computation, some incidents are shown to contribute little to the overall risk; therefore, requesting their reporting is considered a waste of time and eventually decreases the overall incentive to report. On the other hand, if the distinction between serious and less serious incidents is unclear, it may be useful to keep the complete data set.

Conclusions

Cleaning up a database that is biased by incomplete reports may be a complex task because it requires first an appropriate partition of the set of events of interest (understanding the factors that influence the rates of reports) and because it necessarily involves expert opinions. The experts may introduce their own personal biases in the analysis, and there are often uncertainties about the likelihood functions that they supply. This kind of updating, however, is necessary and unavoidable if one is to set priorities among several policies based on a collection of incidents in which the report rates can vary from type to type. In the two cases that are described here, it seems clear that updating the data is needed to improve the available information and to make cost-effective choices among possible risk reduction policies.

References

Gaba, A. 1993. Inferences with an unknown noise level in a Bernoulli process. *Management Science* 39 (October).

Gaba, A., and R. L. Winkler. 1992. Implications of errors in survey data: A Bayesian model. *Management Science* 38 (July).

Howard, R. A. 1970. Decision analysis: Perspectives on inference, decision, and experimentation. *Proceedings of the IEEE* 58 (May).

Paté-Cornell, M. E. 1992a. Management of patient safety in anesthesia: Risk analysis model and critical organizational factors. Research report to the Anesthesia Patient Safety Foundation, June.

Paté-Cornell, M. E. 1992b. Management of patient safety in anesthesia. *Anesthesia Patient Safety Newsletter,* Summer.

Raiffa, H., and R. Schlaiffer. 1961. *Applied statistical decision theory.* Cambridge, Mass.: Harvard University Press.

Webb, R. K., et al. 1993. The Australia incident monitoring study: An analysis of 2,000 incident reports *Anesthesia and Intensive Care* 21 (5): 520–8.

Woodson, R. D. 1991. A critical review of offshore pipeline failures. Report to the National Research Council, Washington, D.C.

Bayes' Theorem and Quantitative Risk Assessment

Stan Kaplan[1]

Abstract

This paper argues that for a quantitative risk analysis (QRA) to be useful for public and private decisionmaking, and for rallying the support necessary to implement those decisions, it is necessary that the QRA results be "trustable." *Trustable* means that the results are based solidly and logically on all the relevant evidence available. This, in turn, means that the quantitative results must be derived from the evidence using Bayes' theorem. Thus, it argues that we should strive to make our QRAs more clearly and explicitly Bayesian, and in this way make them more "evidence dependent" than "personality dependent."

Purpose

The purpose of this short paper is to argue that an explicit Bayesian foundation is not only desirable but necessary if a quantitative risk assessment (QRA) is to fulfill the purposes for which it is done.

What are the purposes for which QRAs are done? There are many purposes and reasons for doing a QRA, but it seems that they can all be subsumed in the following two major categories:

- To assist in decisionmaking
- To assist in decision implementation

With these in mind, our line of argument follows.

[1]Vice President and Technical Director; PLG, Inc.; 4590 MacArthur Blvd., Suite 400; Newport Beach, CA 92660-2027

Line of Argument

1. All decisions are made under uncertainty.

2. It is logically impossible, therefore, to make a valid decision unless we have quantified all the uncertainties involved in the decision.

3. Similarly, unless the uncertainties have been quantified and made explicit, it is difficult, and perhaps impossible, to gain the understanding and support of the people who are necessary to get the decision properly implemented.

4. The role of QRA is, in fact, to quantify all the uncertainties, i.e., the uncertainties in the costs, benefits, and risks for each branch of the decision tree, as shown in figure 1.

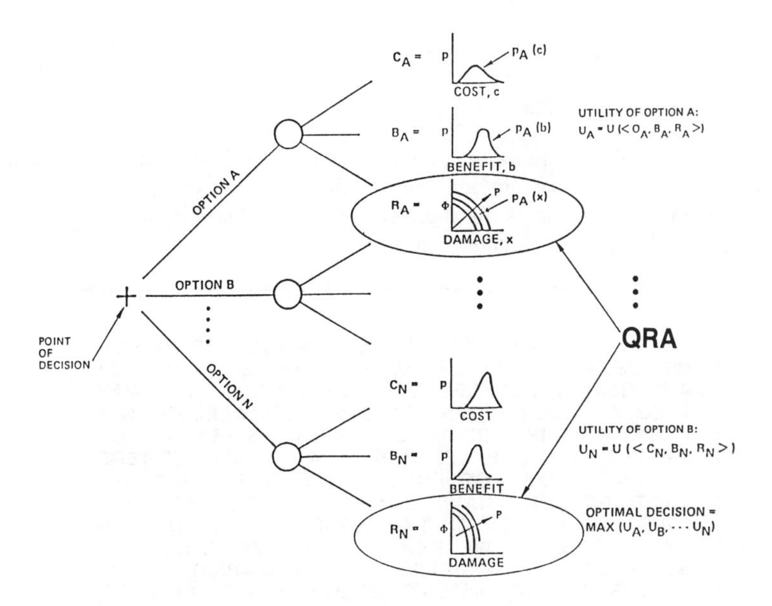

Figure 1. Anatomy of a Decision—the Role of QRA

5. The appropriate language for quantifying uncertainty is the language of probability curves, with "probability" understood in the subjective/objective sense of LaPlace, Bayes, and Jaynes.

6. For these probability curves to be useful in the decisionmaking process, and for getting the decision implemented, they must be "trustworthy." We must be able to trust that the curves reflect and express the total body of relevant information and evidence available. They should not express, or be influenced by, anyone's position, politics, personality, persuasiveness, opinion, or wishful thinking. In particular, we must guard that they not be influenced by our human tendency to dismiss, to not look at, or to not register unpleasant and undesirable things.

   In short, the curves must be "Evidence dependent, not personality dependent."

7. The only way to achieve this degree of trustworthiness is through a rigorous and disciplined use of Bayes' theorem.

8. That is why Bayes' theorem is the necessary foundation of QRA.

Probability and Credibility

To use Bayes' theorem properly, we must first understand what the word "probability" means in the Bayesian sense. One very good expression of this sense is the definition given by DeMorgan in figure 2. Another is that of E. T. Jaynes in figure 3.

WE HAVE LOWER GRADES OF KNOWLEDGE, WHICH WE USUALLY CALL DEGREES OF BELIEF, BUT THEY ARE REALLY DEGREES OF KNOWLEDGE . . . IT MAY SEEM A STRANGE THING TO TREAT KNOWLEDGE AS A MAGNITUDE, IN THE SAME MANNER AS LENGTH, OR WEIGHT, OR SURFACE. THIS IS WHAT ALL WRITERS DO WHO TREAT OF PROBABILITY, AND WHAT ALL THEIR READERS HAVE DONE, LONG BEFORE THEY EVER SAW A BOOK ON THE SUBJECT . . . BY DEGREE OF PROBABILITY WE REALLY MEAN, OR OUGHT TO MEAN, DEGREE OF BELIEF . . . PROBABILITY THEN, REFERS TO AND IMPLIES BELIEF, MORE OR LESS, AND BELIEF IS BUT ANOTHER NAME FOR IMPERFECT KNOWLEDGE, OR IT MAY BE, EXPRESSES THE MIND IN A STATE OF IMPERFECT KNOWLEDGE.

Figure 2. DeMorgan's Definition of Probability (DeMorgan 1847)

Jaynes' definition goes a notch deeper, and in particular, explains the dual subjective/objective sense of probability. Probability is "subjective" in that it refers to a quantity, a degree of plausibility, that

PROBABILITY THEORY IS AN EXTENSION OF LOGIC, WHICH DESCRIBES THE INDUCTIVE REASONING OF AN IDEALIZED BEING WHO REPRESENTS DEGREES OF PLAUSIBILITY BY REAL NUMBERS. THE NUMERICAL VALUE OF ANY PROBABILITY (A/B) WILL IN GENERAL DEPEND NOT ONLY ON A AND B, BUT ALSO ON THE ENTIRE BACKGROUND OF OTHER PROPOSITIONS THAT THIS BEING IS TAKING INTO ACCOUNT. A PROBABILITY ASSIGNMENT IS 'SUBJECTIVE' IN THE SENSE THAT IT DESCRIBES A STATE OF KNOWLEDGE RATHER THAN ANY PROPERTY OF THE 'REAL' WORLD; BUT IS COMPLETELY 'OBJECTIVE' IN THE SENSE THAT IT IS INDEPENDENT OF THE PERSONALITY OF THE USER; TWO BEINGS FACED WITH THE SAME TOTAL BACKGROUND OF KNOWLEDGE MUST ASSIGN THE SAME PROBABILITIES.

Figure 3. Jaynes' Definition of Probability (Howson 1989)

exists not in the real world but only in our minds. On the other hand, it is completely "objective" in that it depends only on the body of evidence available and not on the personality or mood of the user.

Lack of understanding of this duality has been a source of intense, sometimes violent, controversy for hundreds of years in the history of probability and statistics, and it continues today within the field of risk assessment, although, thank goodness, slowly and steadily, clarity is spreading.

Clarity would spread much faster if it were not for the semantic problem that the word "probability" has been interpreted in so many different senses throughout these hundreds of years. Also, if it were not for the fact that we humans get attached to our particular interpretations and fight like the devil against any change.

After confronting this situation within risk assessment for many years, the author was moved to formulate two theorems on human communication, given in figure 4. After fighting so hard for the Bayes/LaPlace interpretation, the author is now wondering if it would be best simply to start using a new word. Jaynes suggests the word "plausibility," which is a very good candidate. In the same spirit, the author would like to suggest the word "credibility" (Howson 1989). If we then speak of the "credibility of proposition *A*," it is clearer that we are talking about a property of *A* itself, not of us. The objective aspect is thus emphasized, not the subjective.

THEOREM 1:	**50% of the problems in the world result from people using the same words with different meanings.**
THEOREM 2:	**The other 50% come from people using different words with the same meanings.**

Figure 4. Kaplan's Communication Theorems

The credibility of *A*, then, given evidence *E*, can be thought of as a mathematical mapping from the space of evidence packages, *E*, to the space of real numbers between 0 and 1. It is a direct mapping, so to speak; it does not pass through our heads. Thus, it is objective, not subjective.

What is the mechanics of this mapping function? That is Bayes' theorem.

Derivation of Bayes' Theorem

The derivation of Bayes' theorem is given in figure 5. It begins by defining a scale of credibility, here denoted by *p*. For convenience, and for historical reasons, we choose this scale to lie between 0 and 1, and then calibrate the scale in a "Bureau of Standards" sense, using a lottery basket or similar device.

The scale then turns out to have a number of interesting properties, two of which are shown just below the scale in figure 5. We recognize these as two of the axioms of probability theory, except that here they appear not as axioms but as consequences of the way we calibrated the scale.

The second property, involving what we would now call the conditional credibility, is symmetric in *A* and *E*. Using the symmetry, and dividing through by *p(E)*, we obtain Bayes' theorem, as shown.

WHAT IS BAYES THEOREM?

0 0.5 1.0

$$p(A) + p(\bar{A}) = 1.0$$

$$p(A \,\&\, E) = p(A)\, p(E|A) = p(E)\, p(A|E)$$

therefore: $$p(E)\, p(A|E) = p(A)\, p(E|A)$$

BAYES' THEOREM $$p(A|E) = p(A) \left[\frac{p(E|A)}{p(E)} \right]$$

"POSTERIOR" "PRIOR" CORRECTION FACTOR

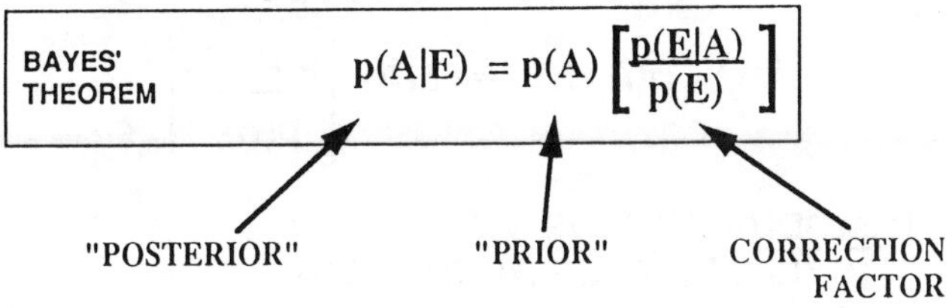

Figure 5. Bayesian Reliability Assessment Methodology

The Meaning of Bayes' Theorem

The theorem tells us how the credibility of proposition *A* changes when a new piece of evidence, *E*, is added. The credibility after the evidence, *p(A/E)*, is equal to the credibility prior to the evidence, *p(A)*, times a correction factor, as shown.

The key term in the correction factor is the numerator *p(E/A)*, known as the "likelihood function."

Observe that the theorem is derived directly from the calibration of the credibility scale and is thus a fundamental property of the scale, a fundamental property of what we have chosen to mean by "credibility."

Much controversy has arisen, as we said, about this theorem. To an orthodox statistician, it is formally true, like any other theorem in probability theory, but not much good for anything. To a Bayesian, it is not just another theorem; it is the fundamental logical law governing the process of evidence evaluation. To an extremist Bayesian, like the present author, it is not just a fundamental logical principle; it is actually the definition of logic. It is what we mean by rational thinking.

In support of this extremist interpretation, we show in figure 6 that the fundamental principles of logic, the Modus Ponens and Modus Tollens, can be seen as special cases of Bayes' theorem.

THE FUNDAMENTAL PRINCIPLES OF LOGIC SEEN AS SPECIAL CASES OF BAYES' THEOREM

$$p(A \mid B) = p(A)\left[\frac{p(B \mid A)}{p(B)}\right]$$

1: MODUS PONENS (SYLLOGISM OF ARISTOTLE)

IF A THEN B	$p(B \mid A) = 1.0$
A	$p(A) = 1 = p(A \mid B)$
$\Rightarrow$ B	$\Rightarrow p(B) = 1$

2: MODUS TOLLENS (REDUCTIO AD ABSURDUM)

IF A THEN B	$p(B \mid A) = 1$
NOT B	$p(B) = 0$
$\Rightarrow$ NOT A	$\Rightarrow p(A) = 0$

3: PLAUSIBLE REASONING

IF A THEN B	$p(B \mid A) = 1$
B	
$\Rightarrow$ A MORE LIKELY	$\Rightarrow p(A \mid B) \geqslant p(A)$

B IS UNLIKELY	p(B) IS SMALL
EXCEPT WHEN A IS TRUE	$p(B \mid A)$ IS SIZEABLE
B IS TRUE	
$\Rightarrow$ A IS MUCH MORE LIKELY	$\Rightarrow p(A \mid B) \gg p(A)$

Figure 6. The Fundamental Principles of Logic

Conclusion

We have thus seen that, as the basic principle for the evaluation of evidence, Bayes' theorem is necessary for QRA. In addition, we have shown that the theorem itself is "bedrock" in terms of it being the definition of logical thinking. It thus constitutes a solid and necessary foundation that gives to a QRA the trustworthiness it needs in order to have the proper impact on decision and action.

References

DeMorgan, A. 1847. *Formal logic*. London: Taylor and Walton.

Howson, C., and P. Urbach. 1989. *Scientific reasoning: The Bayesian approach*. LaSalle, Ill.: Open Court Publishing.

Risk-Based Analysis for Flood Damage Reduction

Darryl W. Davis[1] and Michael W. Burnham[2]

Abstract

The traditional approach of the U.S. Army Corps of Engineers to flood damage reduction studies is compared to risk-based analysis, a method of performing studies in which uncertainty in technical data is explicitly taken into account. Use of the risk-based method results in explicit consideration of uncertainty in estimated flow-flood frequency, flood stage, and flood damage, and hence in project flood damage reduction benefits. Several examples are cited.

Introduction and Definitions

Risk-based analysis is a method of performing studies in which uncertainty in technical data is explicitly taken into account. With such analysis, trade-offs between alternatives, risk, and consequences are made highly visible and quantified. As used herein, risk refers to the possibility or chance of harm, loss, or injury, as in the risk of flooding. Uncertainty refers to the quantity or state of being uncertain, as in uncertainty in stage for a stated flood.

Risk-based analysis applied to flood damage reduction studies results in explicit consideration of uncertainty in estimated flow-flood frequency, flood stage, and flood damage, and hence in project flood damage reduction benefits. Further, the performance reliability (in a

---

[1]Director, Hydrologic Engineering Center, 609 Second Street, Davis, CA 95616

[2]Chief, Planning Analysis Division, Hydrologic Engineering Center, 609 Second Street, Davis, CA 95616

stage or damage-reduction sense) is quantified for consideration in subsequent trade-off analysis.

Traditional Corps Approach

We historically used our best estimates of flow frequency, stage-flow rating (water surface profiles), and stage damage to formulate and evaluate project alternatives. The selected project is one that reasonably maximizes net flood damage reduction benefits subject to acceptable performance. Performance is usually characterized by a degree of protection. Uncertainty is considered by application of professional judgment, by conducting sensitivity analysis, and in the case of levee/ flood wall projects, by the addition of freeboard to ensure performance for the design flood. Risk is considered by nominating several sizes of a project, each with acceptable performance, and selecting the preferred alternative.

Flow frequency is developed by applying adopted federal interagency guidelines, Bulletin 17B (U.S. Department of Interior 1982), when gaged data are available, and rainfall-runoff models, such as HEC-1 (Hydrologic Engineering Center 1990), when watershed modeling is appropriate. Uncertainty is considered by making an adjustment (expected probability) to the frequency curve. This adjustment is made to correct the curve for bias resulting from basing it on a short length of record.

Stage-flow ratings are developed for most situations by water surface profile computations using, for example, HEC-2 (Hydrologic Engineering Center 1991). When flow is complex or circumstances unusual, unsteady flow or two-dimensional model computations are needed. Models are adjusted based on observed high-water marks, available rating curves, and published guidelines. Uncertainty is sometimes considered by performing sensitivity analysis through evaluating the results of reasonable adjustments of model variables. The outcome of sensitivity analysis may result in the adoption of model coefficients to ensure that computed water surface profiles are on the conservative side. Freeboard serves the basic purpose of accommodating uncertainty in stage resulting from intractable hydraulic (conveyance) factors for the adopted design flow.

The stage-damage curve provides a summary statement of damage as a function of river stage or elevation. Damage is sensitive to a variety of factors that are frequently recognized as important in understanding variation in damage across structures but rarely empirically verified. Uncertainty is sometimes considered by performing sensitivity analysis.

Risk-Based Approach

The risk-based approach is similar to traditional practice in that the basic data are the same; the significant departure is that uncertainty is now explicitly quantified. Best estimates are made of flow frequency, stage flow, and stage damage. Project alternatives are formulated and evaluated, and the selected project is that which reasonably maximizes expected net economic benefits subject to acceptable performance. The difference is that uncertainty in technical data is quantified and explicitly included in evaluating project performance and benefits. Because of the risk-based approach, performance can now be stated in terms of reliability of achieving stated performance goals. Also, adjustments/additions of features to accommodate uncertainty, such as adding freeboard for levee/flood walls, is not done.

The methods recommended in Bulletin 17B, *Guidelines for Determining Flood Flow Frequency*, can be used to quantify flow-frequency uncertainty as needed for risk-based analysis. For locations and conditions without a valid gage record, or where stage-frequency is estimated directly, uncertainty is quantified by the method of "order statistics" (U.S. Army Corps of Engineers 1993). An equivalent record length is adopted for the nonanalytic frequency curve.

Uncertainty in water surface profiles (stage-flow ratings) can be described by associating a distribution of error with the rating. A standard deviation of stage errors taken as normally distributed, for example, often characterizes uncertainty adequately. At a gaged location, study of field measurements compared to the adopted rating curve and to stages computed with a calibrated profile computation model can provide the basis for quantifying the uncertainty. The more common circumstance encountered in studies is no gage at a site and some high-water marks recorded for recent flood events. Study of nearby gaged ratings, sensitivity of stage to calibrated model coefficients, and professional judgment form the bases for quantifying uncertainty in this case.

Uncertainty in flood damage is estimated by performing analysis of the uncertainty in component elements, such as structure elevation, structure value, content value, and depth-damage relationship. The result is a stage-damage curve with an associated standard deviation of error in damage for each stage in the tabulated curve.

Some Examples

Chester Creek (68 sq. mi.) is a tributary of the Schuylkill River near Philadelphia. A replacement levee/flood wall (overtopped in a major flood) is proposed for the lower several miles of stream. A stream gage exists within the proposed protected area and high-water marks are available for the major overtopping flood.

Flow frequency is developed from the gaged record, following Bulletin 17B guidelines. Uncertainty in flow is quantified using confidence-limit methods documented in Bulletin 17B for a record length of sixty-five years (the available record length). An HEC-2 model was developed and calibrated for the overtopping flood. Comparison of the gaged rating with field measurements, rating computed with the HEC-2 model, and selected sensitivity studies suggests uncertainty in stage can be represented with a normal probability distribution and a standard deviation of 1.0 foot. The uncertainty in damage is characterized by a standard deviation of 10% of the damage estimate.

Several sizes of levee/flood wall are formulated and evaluated. The method of evaluation often used for risk-based analysis is Monte Carlo simulation. This method is analogous to evaluating the performance and benefits of a very large number (several thousands) of possible future flood events. A project with a levee height of 27.0 feet (no freeboard added) is the alternative that maximizes net flood damage reduction benefits. The project will pass the 100-year flood with 97.5% reliability, should it occur. This is a high reliability. For comparison, the project formulated by traditional Corps of Engineers methods results in selecting the levee/flood wall alternative that provides 100-year protection. The levee/flood wall, with 3 feet of freeboard added, is 26.8 feet in height. No information on performance reliability is developed in traditional analysis; presumably the performance is assumed (incorrectly) to be completely assured.

Jackson, Mississippi, is partially protected from Pearl River flooding by an existing Corps levee project. The city has grown in recent years. Studies are under way to raise the existing levees and extend them both upstream and downstream to protect the new development. A stream gage with ninety-one years of record exists within the study area. High-water marks are available throughout the study area for a recent large flood.

Flow and stage uncertainty are developed similar to the Chester Creek study. Uncertainty in flow is quantified using Bulletin 17B criteria. Because of the flatness of the stream and excellent calibration results, the stage uncertainty likely can be represented by a normal probability distribution with a standard deviation of between .5 feet and .8 feet. Several additional study reaches are needed to analyze flooding adequately within the area. The flow-frequency curve at the gage is representative of the flooding frequency throughout the study area; the gage stage-flow rating is not, however. Ratings for the other several reaches are developed from multiple-profile HEC-2 runs. Rating uncertainty is based on high-water mark calibration and sensitivity results for the respective reaches, and professional judgment by the hydraulic engineer. Damage uncertainty is represented similar to tnat at Chester Creek.

Preliminary risk-based analysis levee-sizing studies indicate that a levee height corresponding passing the 100-year flood stage with 90% reliability is economically justified.

Castro Valley is a small watershed east of San Francisco Bay that drains about fifteen square miles of a highly urban area. Castro Valley is real, but the scenario that follows has been developed to enable discussion for a setting that differs from the two described previously. Past studies have examined alternative solutions to flooding problems. The watershed is gaged near the stream outlet but not for any significant continuous period. A few major floods have been recorded. No systematically measured high-water marks are available. Project alternatives are channel improvements, detention storage, and out-of-basin diversions.

Flow frequency is developed by application of a catchment model (e.g., HEC-1) for several hypothetical floods. The model is developed and calibrated based on previous regional studies and the several gaged floods. The hypothetical storms are developed from NOAA Atlas 2, verified with locally published storm data. To compute flow uncertainty, the order statistics method is applied with an equivalent record length of twenty years. This length is adopted based on the quality of model calibration, comparison with nearby longer record gages, and professional judgment. An HEC-2 model is developed using cross-sectional geometry derived from aerial spot elevation mapping technology, field reconnaissance and published guidelines for stream and overbank roughness coefficients, and a few local resident-noted flood marks from recent events. Stage-flow ratings are developed from multiple profile runs of the calibrated HEC-2 model. Uncertainty in stage rating is based on recommended minimums (U.S. Army Corps of Engineers 1993), sensitivity studies, and professional judgment. The standard deviation of stage uncertainty for the several reaches ranges from 1.5 feet to 2.5 feet. Damage estimates use results of recent conventional field surveys based on aerial spot elevation mapping technology, local assessor records, and field verification. The standard deviation of error in damage from the adopted curve is estimated to be 20%.

Project alternatives are formulated and evaluated; the alternative with maximum expected net benefits is a channel project. The cross section of the channel is varied by reach and shaped and treated for appearance in urban reaches, and to provide quality wildlife habitat for several secluded reaches. The project will pass the fifty-year flood within banks (no freeboard) with 50% reliability, and the twenty-year flood with 80% reliability.

Observations

Adopting the risk-based approach brings flood damage reduction project formulation/evaluation in line with methods now adopted within the Corps for major rehabilitation projects, and soon to be adopted methods for geotechnical and structural analysis. It makes explicit the effects of uncertainty in critical technical inputs. It provides the basis for rational comparison of increased/decreased project size/cost/performance and trade-offs. It responds to OMB, congressional, local sponsor, and other critics that imply that traditional Corps methods result in project features that are not explicitly evaluated based on risk nor necessarily economically justified.

The technical studies needed to perform risk-based analysis are not markedly different from studies performed presently; only the explicit quantification of the critical uncertainties is new. Studies have been under way only a short time, so experience is lacking. Work to date suggests that study costs, once the learning stage is successfully accomplished, will not increase and in some instances, could decrease. For an initial major study, an increase over present effort of a person-month each of study manager, hydraulic engineer, and economist may be required. Later, no increase should occur.

References

Hydrologic Engineering Center. 1990. *HEC-1, flood hydrograph package user's manual*. September. Davis, Calif.

Hydrologic Engineering Center. 1991. *HEC-2, water surface profiles user's manual*. February. Davis, Calif.

U.S. Army Corps of Engineers. 1993. *Risk analysis framework for evaluation of hydrology/hydraulics and economics in flood damage reduction studies*. November. EC 1105-2-250.

U.S. Army Corps of Engineers. 1993. *Uncertainty estimates for non-analytic frequency curves*. Draft, November. ETL 1110-2-XXXX.

U.S. Department of Interior. Geological Survey. 1982. *Guidelines for determining flood flow frequency*. Bulletin #17B. March.

Quantifying Flood Damage Uncertainty

David A. Moser[1]

Abstract

The U.S. Army Corps of Engineers (Corps) has adopted a risk-based analysis procedure to analyze the effect of uncertainty on flood damages and the benefits of flood damage reduction projects. The purpose is to use the results of the analysis in project selection. These procedures combine the uncertainties in discharge-frequency, stage-discharge, and stage-damage relationships. This paper describes the quantification of the uncertainty in stage-damage. This procedure accounts for the uncertainties in the parameters that underlie estimated flood damages.

Introduction

The application of a risk-analysis framework to flood damage reduction studies requires the identification, quantification, and evaluation of risk and uncertainty from various sources. Estimates of economic damages from flooding are frequently considered subject to significant errors. These errors then lead to errors in the estimation of flood damage reduction benefits. The purpose of this paper is to describe procedures for quantifying the error in flood damage estimates. In addition, this paper will describe an approach to the development of summary measures of uncertainty that can be combined with hydrologic and hydraulic sources of uncertainty. These will be used to derive a probability distribution for expected annual flood damages and flood damage reduction benefits.

---

[1]U.S. Army Corps of Engineers, Institute for Water Resources, Casey Building, Fort Belvoir, VA

The views expressed in this paper are solely those of the author and do not necessarily reflect the policy of the U.S. Army Corps of Engineers.

For expository purposes and completeness the discussion first presents the identification, quantification, and assessment of risk and uncertainty in a stage-damage relationship. It is assumed for this initial section that data collection and analysis have been designed with the goal of preserving the inherent uncertainty in the economic data.

The Risk-Based Analysis Framework for the Stage-Damage Curve

The stage-damage curve is a summary statement of the direct economic cost of floodwater inundation for a specified river segment or reach. It is a summary expression in the sense that it combines the data that are used to estimate flood damage. The procedures for developing a stage-damage curve for a particular river reach are well known (Corps 1988). For residential areas, the following information is synthesized in the stage-damage curve:

- number of structures by building and construction types
- structure values by building type and usage
- values by building type and usage
- first-floor elevation of structure
- damage as a percent of value for structure and contents for various flood depths
- flood depths at damage location corresponding to river stages at a reference location

Similar types of information on value, depth-damage, and flood depth at a site are necessary to develop the stage-damage curve for some other land-use types. The sources of errors and uncertainties in the underlying components of the stage-damage curve can best be examined by looking at each component individually.

Table 1 shows the major components required for the development of a residential stage-damage curve and identifies some common sources of error. The emphasis is on residential damages, but the general approach could be applied to other flood damage categories. A comprehensive risk-based analysis framework quantifies the risk and uncertainty in each of these components for the location being studied and develops an estimate of the resulting total error in the stage-damage curve.

Damage Category	Data Category	Source of Errors and Uncertainty
Residential Structure	Number of structures by structure and construction type	Enumerating structures Classifying structures
	Elevation of the first floor for each structure	Surveying Topographic map Contour line interpolation
	Depreciated replacement value of each structure	Real estate appraisal Replacement cost estimation —effective age Depreciation estimation Market value estimation
	Structure damage	Postflood structure damage survey Depth-damage curve —floodwater velocity —duration of flood —sediment load —building material —internal construction —condition —flood warning
Residential Content	Depreciated replacement value of contents	Content inventory survey Content-to-structure value ratio
	Elevation of the first floor for each structure	Surveying Topographic map Contour line interpolation
	Content damage	Postflood content damage survey Depth-damage curve —floodwater velocity —duration of flood —sediment load —condition —content location —flood warning

SOURCE: U.S. Army Corps of Engineers (1988)

Table 1. Residential Flood Damage Categories and Sources of Uncertainty

Often, data either has not been or cannot be developed to quantify all the errors. The approach to developing the stage-damage curve for use in the risk-based analysis framework follows the approach that is currently used in many urban flood damage studies. This involves the following steps:

1. Enumerate and classify each structure in the study area by building and construction type.

2. Establish the elevation of the first floor for each structure using topographic maps, aerial photographs, surveys, and hand levels.

3. Estimate the value of each structure using real estate appraisals, recent sales prices, property tax assessments, replacement cost estimates, or surveys.

4. Estimate the value of the contents using the structure value combined with an estimate of the content-to-structure value ratio for each particular structure type.

5. Estimate the structure damage at various water depths for each building by combining the value of structure with a depth-percent damage curve for that building and construction type.

6. Estimate the content damage at various water depths for each building by combining the value of contents with a content depth-percent damage curve for that building type.

7. Determine the depth of water at each location by combining the elevation of a reference flood at the location with the elevation of the first floor of the structure.

8. Aggregate estimated damages at all locations for the reference flood and repeat for all other floods.

Figure 1 shows a schematic representation of the sources of risk and uncertainty in the components of the stage-damage curve. This figure also shows that these would be represented as uncertainty bounds for the stage-damage curve. The basic procedure recommended here is to combine all the underlying sources of uncertainty about the economic component of flood damage into the stage-damage curve. For some land-use types, particularly industrial, the stage-damage relationship is directly estimated using surveys and interviews. Approaches to quantifying uncertainty where interview methods are used will be described later in this paper.

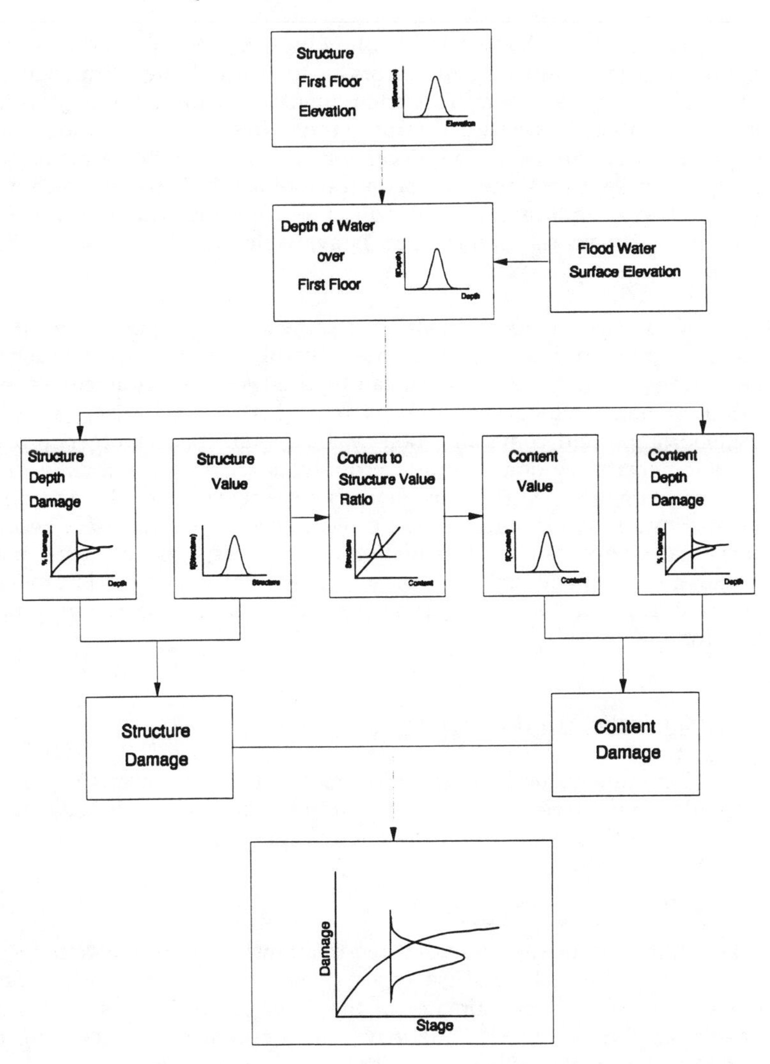

Figure 1. Schematic of Development of Stage-Damage Curve with Uncertainty

The sections immediately following will discuss methods and approaches to quantifying the errors in the underlying components of the stage-damage curve. The components considered are structure values, content-to-structure value ratios, first-floor elevations, and depth-damage curves.[2,3] The discussion will then provide a method for carrying these errors and uncertainties forward into the stage-damage curve. The concluding section will offer recommendations for proceeding when no risk information is available, such as for an existing stage-damage curve.

This paper only addresses the uncertainty in quantifying inundation flood damage and flood control benefits to residential structures.[4,5] The basic approach can be used to quantify uncertainty in other damage categories, such as damages to businesses, utilities, transportation and communication systems, and flood emergency costs. Similar uncertainty quantification procedures are required for the "with project" damages necessary to quantify project benefits. The additional intensification and location benefit categories are not addressed. Because these categories require speculation on the response of flood plain occupants to a flood control alternative, sensitivity analysis is probably most appropriate for investigating the effects of uncertainty on project benefits.

Uncertainties in Structure Values

Structure value is a crucial source of risk and uncertainty in the stage-damage curve. It is especially influential since it directly determines the estimate of structure damages and indirectly influences the

---

[2]The estimation of the stage-damage relationship for some commercial and most industrial land-use types does not rely on depth-damage curves. Instead, direct estimates of the damages at various depths are developed through on-site interviews. An approach to quantifying the uncertainty in these values is described later in this paper.
[3]For this paper, it is assumed that residential structures in each category and construction type are enumerated accurately. This source of measurement error could readily be incorporated in the analysis, however.
[4]See U.S. Water Resources Council (1983) for a complete description of all benefit categories.
[5]See U.S. Army Corps of Engineers (1992b, 1992c) for discussion of applying risk-based analysis to other water resources areas.

estimate of content damages. The approach followed here uses the depreciated replacement value as the appropriate measure of structure value.[6]

One of several approaches may be used to estimate the depreciated replacement value of a structure. There are a variety of methods used to estimate structure value. These include replacement cost estimating as well as methods that require adjustments based on real estate assessment data and sales prices. Both of these last methods require subtracting out the land component of the assessed value or sales price. The presumption is that the remainder is the depreciated replacement value of the structure. There are obviously many opportunities for error in the adjustments that must be made. In addition, there is no general method for quantifying the amount of error in the resulting structure values. One approach to quantifying the error is to compare the results from a sample of individual structures to the results using the replacement cost estimating method. An easy, yet useful, approach to quantifying the range of errors in structure values using real estate assessments is to query local real estate experts and appraisers.

Whatever the method used to estimate the depreciated replacement value of the structure, sampling and sample statistical approaches can be used effectively to minimize the data collection and extensive structure-by-structure analysis. Once the uncertainties in structure values are quantified, this information needs to be incorporated into the analysis and carried forward to the estimation of damages.

Structure Category Type	Mean	Standard Deviation
One Story—No Basement	65,206	74,261
One Story—With Basement	57,598	41,951
Two Story—No Basement	97,817	88,333
Two Story—With Basement	76,315	76,724
Split Level—No Basement	95,666	84,357
Split Level—With Basement	84,854	65,755
Mobile Home	15,550	8,998

Table 2. Structure Value Statistics from FIA Claims Data

Table 2 shows the means and standard deviations of depreciated replacement cost of structures calculated from the Federal Insurance

[6]This is based on employing the "rational planner" model and the willingness-to-pay principle.

Administration (FIA) claims data. The analysis also shows that structure values are not normally distributed, but that a lognormal distribution provides a close fit.

Content-to-Structure Value Ratio

A common method for estimating the content value of residences is to apply a ratio that relates content value to structure value. This approach mimics that typically used by residential casualty insurers in setting rates and content coverage for homeowners insurance. The value of contents found in any structure is highly variable, however. It may reflect the wealth and income of the occupants, their personal tastes and lifestyles, and a variety of other factors. Insurance companies typically assume that contents are 50% to 75% of the structure value. These percentages reflect the ratio of *replacement* cost of contents to the *replacement* cost of the structure. For flood damage studies, the structure value should be the depreciated replacement cost, while contents should similarly reflect the cash value of the contents given their current age and condition. Therefore, the casualty insurance experience and values are not entirely transferable to National Economic Development flood damage analysis.

Structure Category Type	Number of Cases	Mean	Standard Deviation	Minimum	Maximum
One Story—No Basement	71,629	0.434	0.250	0.100	2.497
One Story—With Basement	8,094	0.435	0.217	0.100	2.457
Two Story—No Basement	16,056	0.402	0.259	0.100	2.492
Two Story—With Basement	21,723	0.441	0.248	0.100	2.500
Split Level—No Basement	1,005	0.421	0.286	0.105	2.493
Split Level—With Basement	1,807	0.435	0.230	0.102	2.463
Mobile Home	2,283	0.636	0.378	0.102	2.474
All Categories	122,597	0.435	0.253	0.100	2.500

Table 3. Content-to-Structure Value Ratios from FIA Claims Data

Figure 2 shows the frequency distribution of the content-to-structure value ratio for a two-story, no-basement residence. Note that it is not normally distributed but skewed to the right. Table 3 contains the means and standard deviations of content-to-structure value ratios for several residential structure categories calculated from FIA flood insurance claims records.

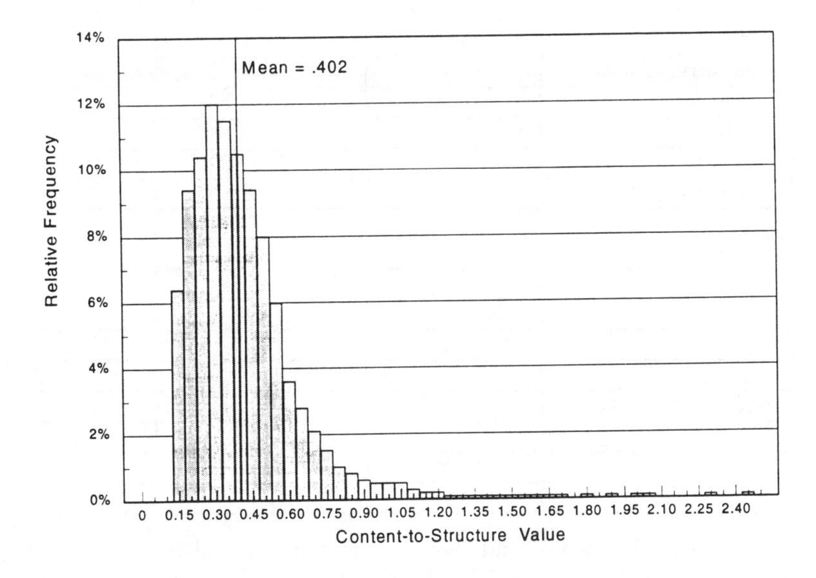

Figure 2. Distribution of Content-to-Structure Value Ratio from FIA Claims Data—Two Story, No Basement

In some instances, content values may have been developed by using survey or inventory methods. Where content values are directly measured there will still be uncertainty in the actual content value due to errors in inventories, pricing, and age. It is difficult to judge the overall effect of these potential sources of uncertainty. One easily implemented method is to request that the individual completing the survey or the inventory provide an estimate of the accuracy of the provided information.

First-Floor Elevation

The estimation of flood damage using depth-percent damage relationships requires the elevation of the first floor of the structure for both structure and content damages. One of the following methods may be used to establish this elevation for each structure in the study area:

- field surveys
- aerial surveys
- topographic maps

Method	Contour Interval (feet)	Error[1] (feet)	Standard Deviation[2] (feet)
Field survey			
Hand level	NA	± 0.2 @ 50'	0.10
Stadia	NA	± 0.4 @ 500'	0.20
Conventional level	NA	± 0.05 @ 800'	0.03
Automatic level	NA	± 0.03 @ 800'	0.02
Aerial survey			
	2'	± 0.59	0.30
	5'	± 1.18	0.60
	10'	± 2.94	1.50
Topographic map			
	2'	± 1.18	0.60
	5'	± 2.94	1.50
	10'	± 5.88	3.00

[1]Errors for aerial survey and topographic maps are calculated at a 99% confidence level, assuming the deviations from the true elevation are normally distributed with zero mean and indicated standard deviations.
[2]Standard deviation for field survey assumes that error represents a 99% confidence interval and assuming normal distribution.
SOURCE: U.S. Army Corps of Engineers (1986, calculated from tables 5.1 and 5.3)

Table 4: Estimates of Errors and Standard Deviations of First-Floor Elevations

The use of these methods and the elevation errors are described in *Accuracy of Computed Water Surface Profiles* (Corps 1986). Table 4 summarizes the quantification of elevation uncertainties for each of these methods.

The basic approach to incorporating these first-floor elevation uncertainties in quantifying the uncertainty in the stage-damage curve is to rely on the error in the method used to measure these elevations. When using one of the methods shown in Table 4, the indicated standard deviation can be used as one parameter in a normal distribution representing the first-floor elevation for each structure in the residential inventory. The mean value in this distribution would be estimated as the measured elevation.

Depth-Damage Curves

The final elements in developing the stage-damage curve are the structure and content depth-damage functions. Each of these functions relates flood damage as a percentage of the value of the structure or contents at various depths of flooding above the first-floor elevation. A useful reference for understanding the approaches to the development and use of depth-damage curves can be found in *Catalog of Residential Depth-Damage Functions Used by the Corps of Engineers in Flood Damage Estimation* (Corps 1992a). As is noted in this report, there are a variety of methods used by the Corps. In addition, the report displays many examples of functions used by Corps districts. All these curves, however, are deterministic in that there is no uncertainty in the percent damage. It is well known that the percent damage at a particular depth of flooding is highly variable across structures, even of the same type, as well as locations. Table 5 shows some of the variables that can influence a depth-damage function. Generally, the effect of any particular variable on a depth-damage curve is not statistically verified. It may be assumed that the influence of all the variables shown in Table 5 is revealed in the postflood damages, but the specific reason for a high or low percent damage is unknown. In flood damage studies where effects such as sediment or velocity are believed to be important, the typical approach is to shift the depth-damage function arbitrarily to account for these influences.

There has not been a thorough analysis of the existing depth-damage curves used by the Corps nor those developed by the FIA based on the federal flood insurance claims records.[7]

---

[7]The complete FIA claims records are currently being statistically analyzed, and measures of the dispersion in percent damage are being calculated.

VARIABLES	EFFECTS
Hydrologic Variables	
velocity	Velocity is a major factor aggravating structure and content damage. Additional force creates greater danger of foundation collapse and forceful destruction of contents.
duration	Duration may be the most significant factor in the destruction of building fabric.
sediment	Sediment can be particularly damaging to the workings of mechanical equipment and can create cleanup problems.
frequency	Repeated saturations can have a cumulative effect on the deterioration of building fabric and the working of mechanical equipment.
Structural Variables	
building material	Steel frame and brick buildings tend to be more durable in withstanding inundation and less susceptible to collapse than other material.
inside construction	Styrofoam and similar types of insulation are less susceptible to damage than fiberglass and wool fiber insulation. Most drywall and any plaster will crumble under prolonged inundation.
condition	Even the best building materials can collapse under stress if the construction is poor or is in deteriorated condition.
age	Age may not be a highly significant factor in itself, except that it may serve as an indicator of condition and building material.
content location	Small variations in interior location of contents can result in wide variation in damage.
Institutional Factors	
flood warning	Both content and structural loss can be reduced through flood fighting and evacuation activities when there is adequate warning.

SOURCE: U.S. Army Corps of Engineers (1988, V-37, V-38)

Table 5. Variables that Influence the Depth-Damage Relationship

Deriving Stage-Damage Curve with Uncertainty Using Monte Carlo Simulation

The various components of the Figure 1 diagram showing the development of the stage-damage curve with uncertainty have been described. The remaining problem is to combine the various sources of risk and uncertainty to derive the overall risk and uncertainty associated with the stage-damage curve. Analytically determining the joint risk or joint uncertainty from the underlying components is extremely difficult in many cases. An alternative approach is to use Monte Carlo simulation to derive a numerical approximation of the analytical solution. This involves developing a risk-based flood damage model where the various parameters are described by probability distributions rather than as deterministic, single values. At each river stage these distributions are "sampled" and the resulting values of damages recorded. Multiple iterations allow the estimation of the distribution of damage at any stage. By rerunning the model with multiple stages, a complete stage-damage curve with uncertainty can be developed.

The stage-damage curve derivation is easily adaptable to a spreadsheet format. There are several Monte Carlo simulation "add-ins" for spreadsheets.[8] Table 6 shows an example of a spreadsheet, used with @RISK, to derive a stage-damage curve with uncertainty. For this example, structure value (column 6) is assumed to be normally distributed with a mean value equal to the depreciated replacement cost estimated using expert appraisers (column 4). The standard deviation of structure value (column 5) is based on an error of plus or minus 20% of the mean value. The standard deviation is approximated by dividing the error by 3.28, because the error range in this example is presumed to contain the 90% confidence interval. The values for the error range and level of confidence are assumed for this example. In a study these would vary by the methodology used to determine structure replacement value. The distributions of the content-to-structure value ratios (column 11) are assumed to be truncated lognormal with means (column 9) and standard deviations (column 10) determined from Table 3. The truncations are .10 and 2.50 and correspond to the minimum and maximum shown in Table 3. The values in column 6 and column 11 are combined to derive the content value with uncertainty shown in column 12. The uncertain structure values and content values are combined with the appropriate depth-damage relationship to determine structure, content, and total damages for each structure and for the entire reach.

---

[8]Two such "add-ins" are @RISK (Palisade Corp.) for LOTUS 1-2-3 2.X and 3.X (Lotus Development Corp.) and Crystal Ball (Decisioneering Inc.) for EXCEL (Microsoft).

For each stage on the stage-damage curve with uncertainty, the river stage is accurately known: assumed to be a reference flood stage plus 0 feet for the values shown in Table 6. The depth of flooding at each of the structures is uncertain due to the uncertainty in first-floor elevations. Column 1 shows the depth of flooding at each structure, calculated from the measured elevation of the first floor and the water surface elevation of a reference flood plus 0 feet[9]. In this example, the first-floor elevations were determined by using topographic maps with 2-foot contour intervals. Referring to Table 4, a minimum standard deviation of 0.6 feet was used for the first-floor elevation with error. Combining the water surface elevation with the first-floor elevation with error produces the depth of flooding at each structure with uncertainty shown in column 2. For each depth of flooding shown in column 2, a percent damage is determined from the depth-damage curves with uncertainty for each structure and its contents.

Combining the appropriate percent damage with the structure and content values provides an estimate of the structure damage (column 8) and content damage (column 14) from the specified flood: the reference flood. The total flood damages for each structure for a single iteration are shown in column 15. A complete simulation for a specific flood requires multiple iterations of the model to derive an accurate distribution of damages for that flood event.

The number of simulations required to achieve the desired level of accuracy is influenced by a number of factors. The number of iterations increases with 1) the variance and skew of the variable of interest; 2) reductions in the probability of contributory variables; and 3) the number of contributing variables. There is no definitive criterion on the number of iterations other than that they must be sufficient to achieve the desired level of accuracy. Each Monte Carlo simulation problem must be evaluated considering time and budget. Figure 3 shows the stage-damage curve with uncertainty developed with the spreadsheet described above, using 1000 iterations for each simulation.

One final aspect of the uncertainty in the stage-damage curve should be noted. Figure 4 shows the sensitivity of the mean stage-damage curve to the error in the first-floor elevation. Notice that a larger error does not change the damage curve at the greatest stages. Rather, as the error increases, the stage-damage curve is "pulled" to the left so that larger damages occur at lower stages. In addition, the error band at lower

---

[9]Note that the uncertainty about the water surface elevation is *not* included in the stage-damage with uncertainty calculations. Instead, this would be incorporated when all the sources of uncertainty are combined.

Depth of Flooding at Structure (feet) (1)	Depth of Flooding with Uncertainty (2)	Structure Type (3)	Structure Total Value ($1000) (4)	Std. Dev. Structure Value ($1000) (5)	Uncertain Structure Value ($1000) (6)	Structure Depth-Damage (7)	Structure Damage with Uncertainty ($1000) (8)	Mean Content-to-Structure Value Ratio (9)	Std. Dev. Content to Structure Value Ratio (10)	Uncertain Content-to-Structure Value Ratio (11)	Uncertain Content Value ($1000) (12)	Content Depth-Damage (13)	Content Damage With Uncertainty ($1000) (14)	Total Damage With Uncertainty ($1000) (15)
7.0	7.9	6	141	17	145.6	44.0%	64.1	0.441	0.248	0.213	30.9	49.0%	15.2	79.3
6.9	6.8	1	74	9	66.8	42.4%	28.3	0.434	0.250	0.355	23.7	49.0%	11.6	39.9
6.2	5.0	5	103	13	97.1	22.0%	21.4	0.402	0.259	0.548	53.1	33.0%	17.5	38.9
6.1	5.6	1	37	5	35.8	36.0%	12.9	0.434	0.250	0.674	24.1	43.4%	10.5	23.4
6.0	5.2	1	106	13	114.3	32.0%	36.6	0.434	0.250	0.364	41.5	41.8%	17.4	54.0
5.8	5.8	2	75	9	80.6	37.0%	29.8	0.435	0.217	0.461	37.2	43.0%	16.0	45.8
5.6	4.7	2	55	7	55.6	31.5%	17.5	0.435	0.217	0.558	31.1	37.2%	11.6	29.1
5.6	4.7	5	89	11	96.0	21.4%	20.5	0.402	0.259	0.378	36.3	31.5%	11.4	31.9
5.4	5.3	5	90	11	72.9	22.6%	16.5	0.402	0.259	0.599	43.7	34.8%	15.2	31.7
5.2	4.9	5	102	12	88.2	21.8%	19.2	0.402	0.259	0.567	50.0	32.5%	16.3	35.5
5.1	4.1	2	117	14	116.4	28.5%	33.2	0.435	0.217	0.713	82.9	33.6%	27.9	61.1
5.0	4.7	5	82	10	81.7	21.4%	17.5	0.402	0.259	0.710	57.9	31.5%	18.2	35.7
4.8	5.2	6	109	13	102.6	34.0%	34.9	0.441	0.248	0.193	19.9	40.0%	7.9	42.8
4.8	4.2	2	84	10	67.5	29.0%	19.6	0.435	0.217	0.473	31.9	34.2%	10.9	30.5
4.7	5.5	5	68	8	71.2	23.0%	16.4	0.402	0.259	0.488	34.7	36.0%	12.5	28.9
4.6	3.7	6	46	6	41.9	26.5%	11.1	0.441	0.248	0.163	6.8	31.5%	2.2	13.3
4.5	3.6	2	66	8	71.0	26.0%	18.4	0.435	0.217	0.817	57.9	31.0%	18.0	36.4
4.5	5.5	5	87	11	99.8	23.0%	22.9	0.402	0.259	0.143	14.2	36.0%	5.1	28.0
4.3	5.3	6	73	9	80.5	34.5%	27.8	0.441	0.248	0.205	16.5	40.5%	6.7	34.5
4.1	4.2	1	67	8	53.7	29.2%	15.7	0.434	0.250	0.456	24.5	37.8%	9.3	25.0

Table 6. Example Spreadsheet for Monte Carlo Simulation

stages also increases the greater the error in first-floor elevation. Understanding the impact of uncertainty on the stage-damage curve can be useful in detailing study tasks and spending so as to reduce the overall amount of uncertainty.

Stage-Damage Uncertainty Using the Opinions of Experts

The approach described in Figure 1 may not reflect the methodology typically used in estimating damages; for industrial property, for instance. As noted earlier, the stage-damage curve for these types of property may typically be developed individually during postflood surveys or personal interviews with plant managers, plant engineers, or other knowledgeable individuals and experts. Instead of combining structure and content values with depth-damage curves, damages at various water-surface elevations are directly approximated. In these situations, the respondents should be asked to provide a range of damages, as well as the most likely damage, at the elevations used. Additionally, the respondent could be asked about her/his confidence that the range contains the actual damage value that would occur: what is the chance that damages might be higher or lower? These opinions of experts on the

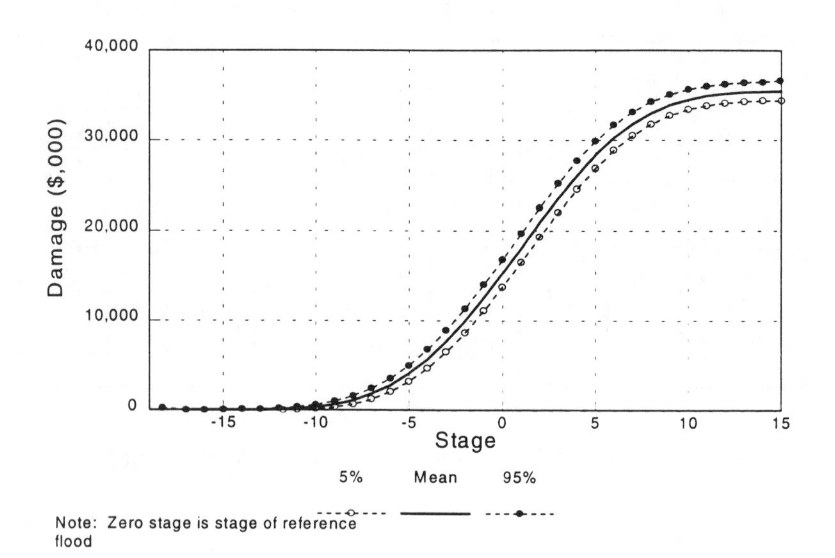

Figure 3. Stage-Damage Curve with Uncertainty

range and confidence can be used to calculate the parameters of a distribution to represent damage at each stage for the plant. If all the individual is willing to provide is the range, the analyst can use the midpoint of the range as the mean and one-fourth of the range as the standard deviation.[10]

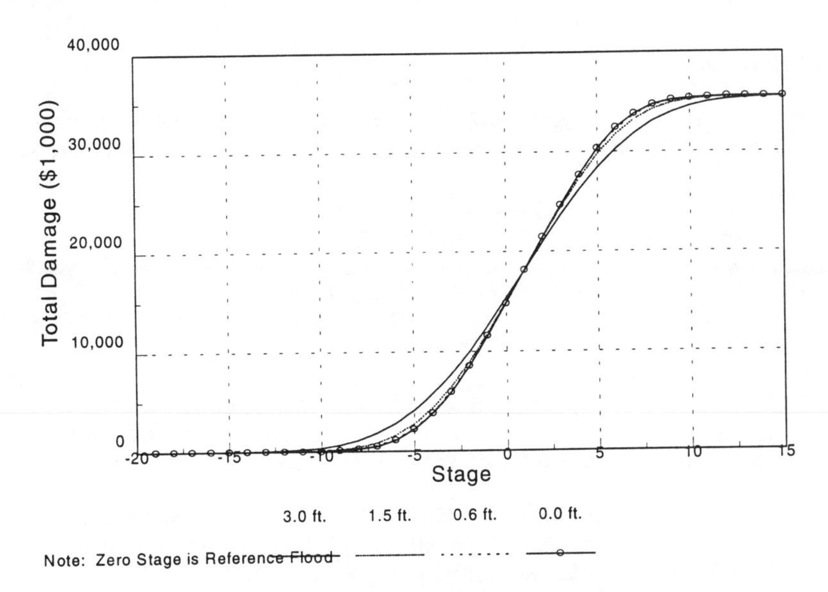

Figure 4. Sensitivity of Mean Stage-Damage Curve to Error in First-Floor Elevation

Conclusions

This paper describes an approach to quantifying the uncertainty in flood damages based on quantifying the underlying uncertainties in the components of the damage calculation. It shows that Monte Carlo simulation can be successfully used to combine the uncertainties. This analysis incorporates the uncertainties in the economic variables that influence the economic performance of alternative flood damage reduction projects. The resulting stage-damage curve with uncertainty can then be combined with the other major components in the flood damage reduction evaluation calculation, (the frequency-discharge relationship

---

[10]For a review of expert judgment under uncertainty see Kahneman, Slovic, and Tversky (1982).

with uncertainty and the stage-discharge relationship with uncertainty), to determine the distribution of flood damages and project benefits. This analysis provides information on both the engineering performance (reliability) and the economic performance to improve decisions on flood damage reduction projects.

References

Kahneman, D., P. Slovic, and A. Tversky, eds. 1982. *Judgment under uncertainty: Heuristics and biases*. New York: Cambridge University Press.

U.S. Army Corps of Engineers. 1986. *Accuracy of computed water Surface profiles*. Hydrologic Engineering Center.

U.S. Army Corps of Engineers. 1988. *National economic development procedures manual—Urban flood damage*. Institute for Water Resources, IWR Report 88-R-2.

U.S. Army Corps of Engineers. 1992a. *Catalog of residential depth-damage functions used by the Corps of Engineers in flood damage estimation*. Draft. Institute for Water Resources.

U.S. Army Corps of Engineers. 1992b. *Guidelines for risk and uncertainty analysis in water resources planning*. Vol. I, *Principles*. Institute for Water Resources, IWR Report 92-R-1.

U.S. Army Corps of Engineers. 1992c. *Guidelines for risk and uncertainty analysis in water resources planning*. Volume II, *Examples*. Institute for Water Resources, IWR Report 92-R-2.

U.S. Water Resources Council. 1983. *Economic and environmental guidelines for water and related land resources implementation studies*. U.S. Government Printing Office.

Flood Loss Assessment with Integrated Measures

Yacov Y. Haimes, Duan Li, Vijay Tulsiani,[1]
Eugene Z. Stakhiv, and David Moser[2]

Abstract

In most cases, the maximum flood loss reduction can be achieved only by an optimal combination of both structural and nonstructural measures. An integrative approach that combines the calculation of flood loss reduction through flood warning systems with the calculation of flood loss for a given flood control structure has been developed in this paper, thus facilitating the evaluation of combined structural measures and flood warning systems in reducing flood loss. The conventional expected value of flood loss, the conditional expected value of extreme flood loss, and the cost associated with the structural measures and flood warning systems are traded off and analyzed in a multiobjective framework.

Introduction

In most cases, the maximum flood loss reduction can be achieved only through an optimal combination of both structural and nonstructural flood control measures, because the adoption of integrated measures enlarges the feasible region of flood control measures as compared to the situations where structural or nonstructural measures alone are considered. Structural measures include the construction of reservoirs, levees, and flood walls. Nonstructural measures include flood plain land use planning, flood insurance, flood warning systems, and

---

[1]Director, Associate Director, and Research Assistant, respectively, Center for Risk Management of Engineering Systems, University of Virginia, Charlottesville, VA 22903
[2]Chief of Policy and Economist, respectively, Institute for Water Resources, U.S. Army Corps of Engineers, Casey Building, Fort Belvoir, VA 22060-5586

flood proofing. Various flood control measures prevent inundation of the flood plain in different ways and have different impacts on the flood damage-frequency relationship. A structural measure, such as an increase in reservoir height, affects the frequency—discharge relationship. Levees and flood walls confine the discharge within certain channels, thus changing the relationship between discharge and elevation. Most nonstructural measures, such as flood warning systems, modify the stage-damage relationship.

The idea to combine both structural and nonstructural measures in flood control is not new. Various research results have been reported that combine structural measures with other nonstructural measures, such as zoning, flood proofing, and flood insurance. Readers can refer to Thampapillai and Musgrave (1985), which provides a comprehensive survey in reviewing integrated structural and nonstructural measures in flood damage mitigation. Recently, research work on combining structural measures with flood warning systems has been reported (Haimes, Li, and Tulsiani 1992).

Issues of both design and operation are involved in structural and nonstructural measures. Building a reservoir is a structural measure in flood control. Determination of the height of the reservoir is a design issue, while determination of the amount of the release on a monthly or daily basis is an operational issue. Installing a flood warning system is a nonstructural measure in flood control. Determination of an acceptable reliability of a warning system is a design issue, with a consideration of the cost of the system, while determination of the flood warning threshold for various flood events is an operational issue. It is important to note that operational issues can be addressed only in a framework of dynamic optimization. For example, different levels of flood warning thresholds cause different probabilities of missed forecasts and false alarms, thus affecting the fraction of the community's future response (Haimes, Li, and Stakhiv 1990; Li et al. 1992). In this paper, only design options for both structural measures and flood warning systems are considered. The existing methodology for computing flood loss for a given structural measure, developed by the U.S. Army Hydrologic Engineering Center (1989), is extended to incorporate flood warning systems.

To compute flood loss for a given flood control structural measure, a widely used procedure developed by the U.S. Army Hydrologic Engineering Center (1989) investigates the relationships between discharge and frequency, discharge and elevation, and damage and elevation, so that a damage-frequency curve can be generated for an average annual flood loss. An integrative approach that combines the calculation of flood loss reduction through flood warning systems with the calculation of flood loss for a given flood control structure is developed in this paper, thus facilitating the evaluation of combined structural measures and flood warning systems in reducing flood loss.

Description of the Integrated Approach

In the procedure for computing flood damage that has been developed by the U.S. Army Hydrologic Engineering Center (1989), four functional relationships (or curves) (figure 1) are adopted to quantify each alternative of structural measures completely:

- curve of frequency versus discharge
- curve of elevation versus discharge
- curve of elevation versus damage
- curve of frequency versus damage

Note that the fourth curve can be derived if the other three are known. In general, the relationships of frequency versus discharge, elevation versus discharge, and damage versus elevation are constructed from real data so the curve of frequency versus damage can be derived to compare the expected flood damage for structural measures.

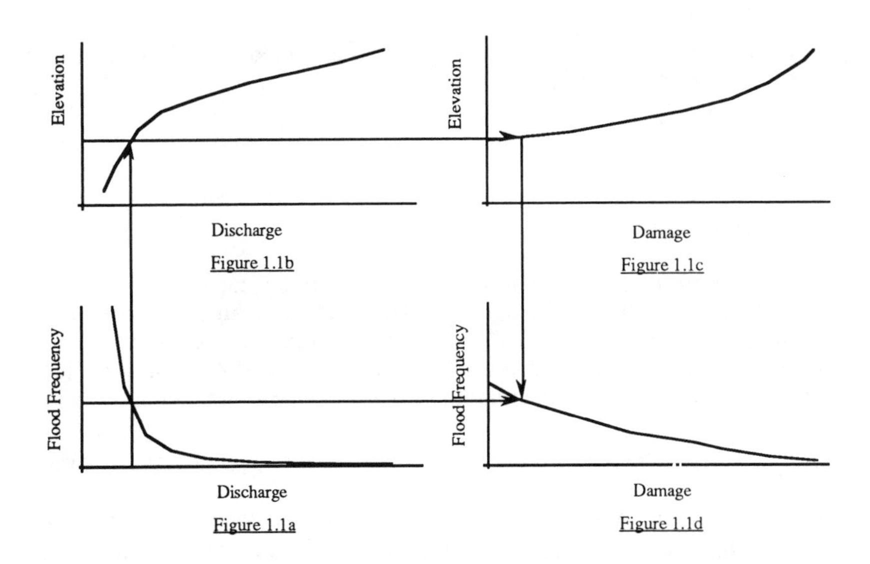

Figure 1. Curves for Quantifying Alternative Measures

The approach of discrete enumeration of all possible combinations of structural and flood warning systems is adopted in the development of the integrated approach. Assume that there are N feasible alternatives of structural measures and M feasible designs of flood warning systems. Therefore, there are $(N + 1)(M + 1)$ combinations of flood control alternatives, which includes one "do nothing" option, N options involving only structural measures, M options involving only flood warning systems, and NM options involving a combination of a structural measure and a flood warning system.

This paper is based on the premise that the introduction of a flood warning system will not affect the relationships between the frequency and discharge and between the elevation and discharge. It will, however, alter the curve of elevation versus damage, thus changing the relationship between frequency and damage.

To evaluate the flood loss reduction by installing a flood warning system, the concept of category-unit loss function by Krzysztofowicz and Davis (1983) is adopted. The main modification is in the use of the notation θ, which was originally used in Krzysztofowicz and Davis (1983) as the response degree of an individual and is adopted in this paper to represent the fraction of people in a community who respond to flood warnings.

The cost function of evacuation, C_E, is assumed to be a linear function of the response fraction

$$C_E = MC\,\theta\,, \tag{1}$$

where MC is the maximum evacuation cost for the community when a full response is present.

Assume that the elevation of the flood plain zone under consideration is Y and the flood stage is E. The flood loss function without a warning system is essentially given by the curve of elevation versus damage (figure 1c) for each given structural measure. Alternatively, the flood loss function without a warning system can be expressed by

$$L_{wo} = MD\,\delta(E - Y)\,, \tag{2}$$

where MD is the maximum possible damage to the community due to a flood of the highest magnitude and $\delta(E - Y)$ is the unit damage function specifying the fraction of MD that occurs when the depth of flooding is $(E - Y)$.

The flood loss with a warning system is assumed to be of the following form:

$$L_w = MC\,\theta + MD[1 - \theta\,MR(E - Y)]\delta(E - Y)\,, \tag{3}$$

where $MR(E - Y)$ is the unit reduction function specifying the reduction of the maximum flood loss, MD, when the depth of flooding is $(E - Y)$ and full response of the community is made, i.e., $\theta = 1$.

In summary, the flood loss reduction, L_{RD}, when a flood warning system is installed can be expressed as the difference between L_{wo} and L_w,

$$L_{RD} = \theta\,MD\,MR(E - Y)\delta(E - Y) - MC\,\theta\,. \tag{4}$$

The damage value of the maximum flood loss for a community, MD, and the functional form of the unit damage function can be obtained from the curve of elevation versus damage when structural measures are evaluated. The only additional information required to calculate the relationship of elevation and flood loss reduction with a flood warning system is the value of maximum evacuation cost, MC; the value of response fraction in the community, θ; and the unit reduction function, $MR(E - Y)$. A recent report by Jack Faucett Associates (1990) offers procedures for calculating the costs and benefits of flood warning systems, which are useful to determine the unit reduction function, $MR(E - Y)$, and to evaluate the trade-off between cost and flood loss. The resulting curve of elevation versus flood loss reduction can be viewed as a function parameterized by the response fraction θ. Note, however, that the flood loss reduction, L_{RD}, is a linear function of response fraction θ.

Reducing the value of damage in the curve of damage versus elevation for each structural measure by $L_{RD}(E;\theta)$ for each given value of elevation, E, yields the new relationship between elevation and damage when a flood warning system is introduced. Setting θ equal to one yields a maximum achievement of flood loss reduction. Combining this new curve of elevation versus damage with the curves of frequency versus discharge and elevation versus discharge provides a new relationship between frequency and damage for a combined structural measure and a flood warning system.

It is important to emphasize that the above analysis estimates the maximum flood loss reduction through combination of a structural measure and a warning system. A flood warning system may fail in its function to protect the community from incoming floods due to missed forecasts or a short lead time for evacuation. A more realistic flood loss associated with a warning system can be achieved by calculating an expected measure, for example, by $[pL_w + (1 - p)L_{wo}]$, where p is the probability that the flood warning system correctly issues an evacuation order given the occurrence of a flood event, or by $[\sum_{\lambda} p(\lambda)L_w(\lambda) + (1 - p)L_{wo}]$, where λ represents the lead time that is assumed to take discrete values; $p(\lambda)$ is the probability that the flood warning system correctly issues an evacuation order with lead time λ, given occurrence of a flood event; and $L_w(\lambda)$ is the flood loss with a warning system, given lead time λ. Certainly, data availability will always pose difficulties in using those more complicated calculations. Thus, in this paper, only equations (3) and (4) will be used to calculate the maximum flood loss reduction offered by a warning system.

Although the relationship of damage versus frequency provides a complete evaluation for each flood control alternative, it is necessary to compress information to generate a risk measure when various flood control alternatives are compared. The most commonly used risk measure is the expected value of the flood loss. Although the expected-value approach indicates the central tendency of flood damage of each flood control alternative, it fails to separate the extreme catastrophic flood events from the rest. The partitioned multiobjective risk method (PMRM) (Asbeck and Haimes 1984) adopts the concept of conditional expectation, which enables the analyst to isolate, quantify, and evaluate the impact of each flood control alternative on the consequences of extreme catastrophic flood events.

Multiobjective analysis is performed in this paper to evaluate the various flood control alternatives. Three objective functions are evaluated in this multiobjective framework. Consistent with the notations used in the PMRM (Asbeck and Haimes 1984) and for each flood control alternative, f_1 is used to denote the cost of a combined structural measure and flood warning system, f_5 the expected flood damage, and $f_4(\alpha)$ the conditional expectation of extreme flood damage whose return periods are greater than $\frac{1}{1 - \alpha}$, where α is the partitioning point on the probability axis. Both f_5 and $f_4(\alpha)$ can be derived from the curve of flood damage, L, versus frequency for each flood control alternative:

$$f_5 = \text{mean of } \{L\} \tag{5}$$

and

$$f_4(\alpha) = \text{mean of } \{L | \text{ return period of } L \geq \frac{1}{1-\alpha}\} . \tag{6}$$

A flood control option may have different impacts on the expected flood loss and on the expected flood loss with floods whose return periods exceed certain threshold levels. The integrative methodological framework developed in this paper provides more decision aids for determining the optimal flood control strategy. The added trade-off information about cost and expected extreme flood loss explicitly addresses the public concern about catastrophic flood loss.

Example Problem

An illustrative example problem is investigated in this section to demonstrate the integrated approach developed in the previous section. This example problem is based on a study undertaken by the U.S. Army Corps of Engineers (1990) for the South Fork and South Branch Potomac Rivers at Moorefield, West Virginia, in 1990. Two structural measures of flood control are available. Plan 1 is the zero-cost plan, i.e., the without-project alternative. Plan 4 is a structural plan and includes levees and flood walls to protect residential areas, industrial plants, businesses, schools, and commercial areas in northern and southern Moorefield.

For plan 1, the relationships of discharge versus elevation, frequency versus discharge, and frequency versus damage are available from data. The original data are used to perform regression analysis to obtain the following functional relationship:

$$E = 1.28767 \left(\frac{D}{1000} - 10.7\right)^{0.41567} + 812.44577 , \tag{7}$$

$$D = 10^6 \left[0.00211 \left(\frac{1}{F}\right)^{0.62411} + 0.00582 \right] , \tag{8}$$

and

$$L = -56.65856 \, (F)^{0.59494} + 9.02988 , \tag{9}$$

where E is the flood elevation in feet, and D is the discharge in cfs, F is the flood frequency, and L is the flood damage in millions of U.S. dollars. The relationship of elevation versus damage is obtained through the use of the other three relationships in equations (7) through (9):

$$L = -56.65856\left[\frac{0.00211}{\frac{1}{1000}\left\{\left(\frac{E-812.44577}{1.28767}\right)^{1/0.41567}+10.7\right\}-0.00582}\right]^{0.95326}$$

$$+\ 9.02988\,,$$

$$\text{if } E \geq 815.746\,, \tag{10}$$

and L is equal to zero otherwise.

For plan 4, an alternative elevation versus discharge curve was provided for the elevation in the channel (U.S. Army Corps of Engineers 1990). The flood loss depends, however, on the relationship of discharge versus elevation in the flood plain. Therefore, the data for the discharge versus flood plain elevation are assumed on the basis of the data provided for the discharge versus elevation in the channel. The corresponding relationship between elevation and discharge is expressed as

$$E = 1.55092\left(\frac{D}{1000} - 10.7\right)^{0.31598} + 812.48596\,. \tag{11}$$

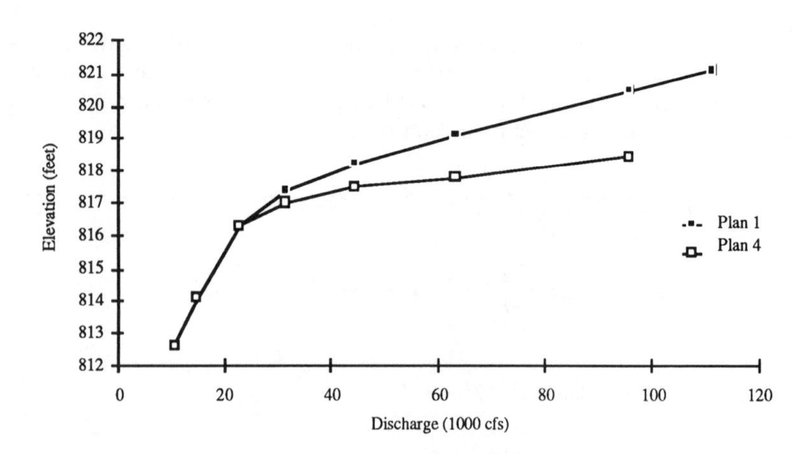

Figure 2. Comparative Discharge versus Elevation Curves for Plan 1 and Plan 4

Figure 2 depicts comparative plots of discharge versus elevation for plan 1 and plan 4. Since the elevation versus damage and the frequency versus discharge relationships do not change due to construction of levees, flood walls, and other structural measures that constitute plan 4, the frequency versus damage curve for plan 4 can be obtained by combining equations (8), (9), and (11),

$$L = \begin{cases} 521.70\,(0.02 - F) & \text{if } 0.02 \geq F \geq 0.0133 \\ -50.45478\,(F)^{0.53290} + 8.49407 & \text{if } F \leq 0.0133 \end{cases} \tag{12}$$

Figure 3 depicts the comparative plots of frequency versus damage for plan 1 and plan 4.

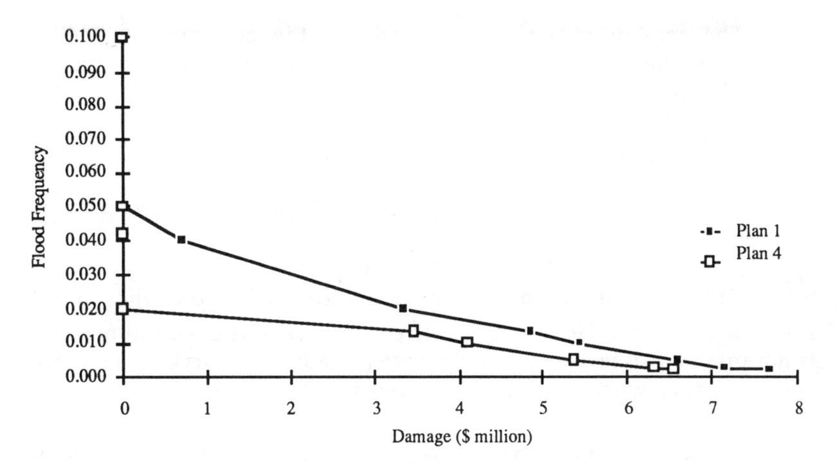

Figure 3. Comparative Frequency versus Damage Curves for Plan 1 and Plan 4

Assume that one design option for a flood warning system is available to be added to the existing structural measures. This added option of a flood warning system will provide two more integrated flood control measures, plan 1 + flood warning system and plan 4 + floodwarning system. Introduction of a flood warning system will change the relationship between elevation and damage.

The value of the maximum flood loss for a community, *MD*, and the functional form of the unit damage function can be obtained from equation (10). In this specific example, notice that the base elevation of the flood plain zone is 812.44577 feet; the maximum flood loss for the community, MD, is equal to \$9.02988 million; and the functional form of the unit damage function is

$$\delta(E - Y) =$$

$$1 - 6.27456\left[\frac{0.00211}{\frac{1}{1000}\left\{\left(\frac{E - 812.44577}{1.28767}\right)^{2.40576} + 10.7\right\} - 0.00582}\right]^{0.95326}$$

$$\text{if } E \geq 815.746\,, \tag{13}$$

and $\delta(E - Y)$ is equal to 0 otherwise.

The maximum evacuation cost for the community, MC, is assumed to be equal to \$50,000 and the unit reduction function is assumed to be

$$MR(E - 812.44577) = 0.25 + 0.04(E - 812.44577)$$

$$- 0.00333(E - 812.44577)^2\,. \tag{14}$$

Only the case of maximum flood loss reduction when a full response is present, i.e., $\theta = 1$, is studied. The cases where a full response is not present can be easily found by interpolating the curves of elevation versus damage with and without a flood warning system.

Using equation (4) with $\theta = 1$, the loss reduction when a flood warning system is introduced can be expressed as a function of flood elevation

$$L_{RD} =$$

\b\bc\{(9.02988 - 56.65856 \b\bc\[(\f(0.00211,\f(1,1000)

$$\left\{\left(\frac{E - 812.44577}{1.28767}\right)^{2.40576} + 10.7\right\} - 0.00582))^{0.95326}\;)$$

$$*\left\{0.25 + 0.04(E - 812.44577) - 0.00333(E - 812.44577)^2\right\} - 0.05\,,$$

$$\text{if } E \geq 815.746\,, \tag{15}$$

and L_{RD} is equal to 0 otherwise. Subtracting L_{RD} from L in equation (10) yields the curve of damage versus elevation when a flood warning system is introduced. Figure 4 depicts the comparative plots of damage versus elevation with and without a warning system.

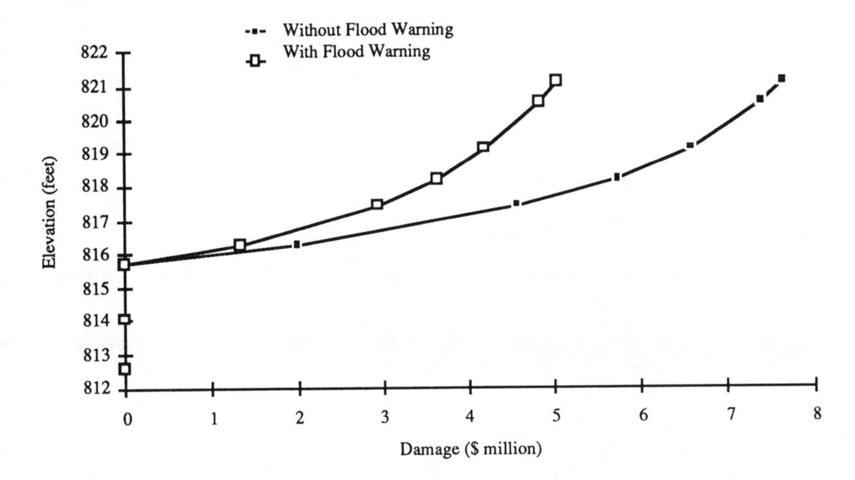

Figure 4. Comparative Plots of Elevation versus Damage for Plan 1 with and without Flood Warning System

Combining equations (7), (8), (10), and (15) yields the curve of damage versus frequency in figure 5 for plan 1 + flood warning system, with the following functional relationship:

$$L = -32.13928\,(F)^{0.54301} + 6.08851\,. \qquad (16)$$

Similarly, combining equations (11), (8), (10), and (15) yields the curve of damage versus frequency in figure 6 for the option of plan 4 + flood warning system, with the following functional relationship:

$$L = \begin{cases} 340.58\,(0.02 - F) & \text{if } 0.0133 \leq F \leq 0.02 \\ -29.39980\,(F)^{0.51246} + 5.46498 & \text{if } F \leq 0.0133 \end{cases} \qquad (17)$$

From the relationship of flood loss versus frequency, the expected flood loss and conditional expectation of extreme flood loss can be calculated using equations (5) and (6). Two partitioning points, $\alpha = 0.9$ and $\alpha = 0.99$, are selected in calculating the expected extreme flood loss. The results are summarized in table 1 for four flood control options.

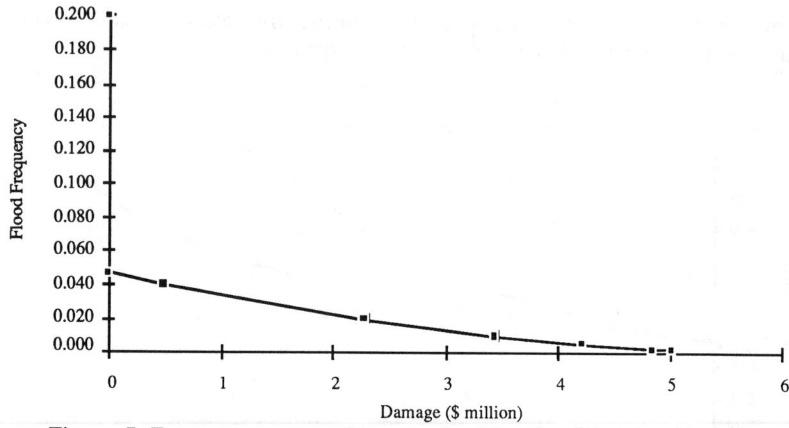

Figure 5. Frequency versus Damage for Plan 1 + Warning System

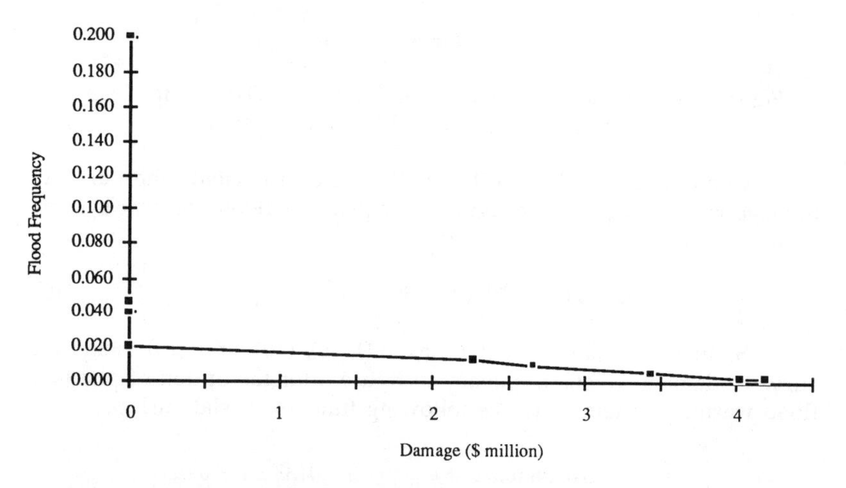

Figure 6. Frequency versus Damage for Plan 4 + Warning System

Since plan 1 is the option of doing nothing, it does not have an associated cost. The average annual cost for plan 4 is given as $0.865 million for a fifty-year level of protection (U.S. Army Corps of Engineers 1990). The average annual cost of the flood warning system is assumed to be $50,000.

Once the costs, the expected values, and the conditional expected values for the different plans are computed, a trade-off analysis can be performed in terms of costs and flood damages. Figure 7 depicts the resulting three trade-off curves of cost versus the expected value of flood loss, cost versus conditional expectation of extreme flood loss with α = 0.9, and cost versus conditional expectation of extreme flood loss with $\alpha = 0.99$.

	Average Annual Cost ($ million)	$f_5(L)$	$f_4(L \mid a = 0.9)$	$f_4(L \mid a = 0.99)$
Plan 1	0.000	0.377	3.428	6.657
Plan 1+W	0.050	0.159	1.586	4.086
Plan 4	0.865	0.204	2.515	5.705
Plan 4+W	0.915	0.084	1.412	3.348

Table 1. Costs and Flood Losses in Example Problem

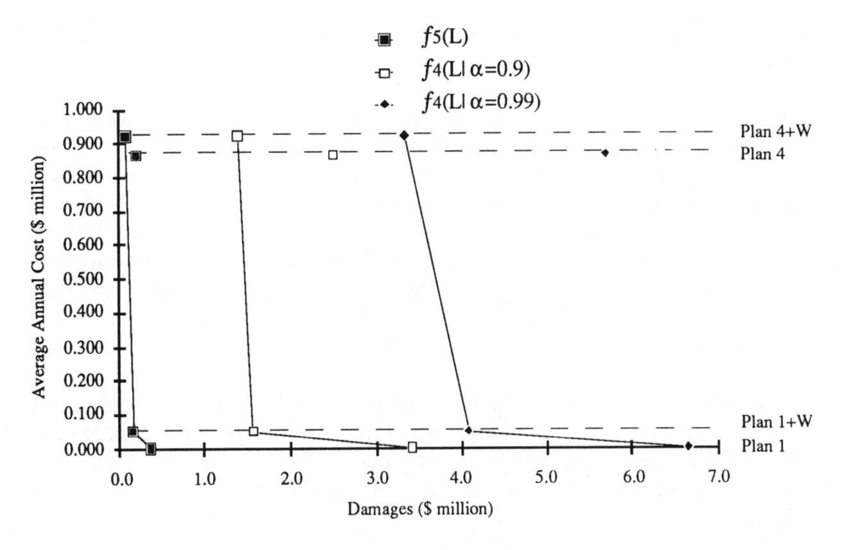

Figure 7. Trade-off between Cost and Flood Damage

Conclusion

This study has demonstrated that integrated flood control alternatives that use both structural measures and flood warning systems would reduce the vulnerability of a community to flood damage and add resiliency to its protection because of added redundancy. If one considers only structural measures, plan 4 in the case study represents a good solution to reduce the risk of flood losses. Plan 4, however, becomes inferior when the integrated flood control alternatives that use both structural measures and flood warning systems are considered. Since the integrated framework developed in this paper builds on the U.S. Army Corps of Engineers' previous work, it is very easy to adopt and implement. Given the analysis for each structural and flood warning system, the combined analysis is simple. The additional data requirement is minimal. The incorporation of the measure of risk of extreme events along with the expected flood loss in a multiobjective framework offers deeper insight in determining the best flood control strategy. Future investigation includes real case studies and incorporation of uncertainty analysis in the calculations.

References

Asbeck, A. H. S., and Y. Y. Haimes. 1984. The partitioned multiobjective risk method (PMRM). *Large Scale Systems* 6:13–38.

Haimes, Y. Y., D. Li, and E. Z. Stakhiv. 1990. Selection of optimal flood warning threshold. In *Risk-Based Decision Making in Water Resources*, edited by Y. Y. Haimes and E. Z. Stakhiv. New York: American Society of Civil Engineering.

Haimes, Y. Y., D. Li, and V. Tulsiani. 1992. Integration of structural measures and flood warning systems in flood damage reduction. A report to the Institute of Water Resources, U.S. Army Corps of Engineers. Charlottesville, Virginia: Environmental Systems Modeling, Inc.

Jack Faucett Associates. 1990. Procedures for calculating the costs and benefits of flood warning and preparedness systems. Final report submitted to U.S. Army Corps of Engineers, Water Resources Support Center, Institute for Water Resources, Fort Belvoir, Virginia 22060-5586.

Krzysztofowicz, R., and D. Davis. 1983. Category-unit functions for flood forecast-response system evaluation. *Water Resources Research* 19: 1476–80.

Li, D., Y. Y. Haimes, E. Z. Stakhiv, and D. Moser. 1992. Optimal flood warning threshold: A case study in Connesville, Pennsylvania. In *Risk-Based Decision Making in Water Resources V*, edited by Y. Y.

Haimes, D. A. Moser, and E. Z. Stakhiv. New York: American Society of Civil Engineering.

Thampapillai, D. J., and W. F. Musgrave. 1985. Flood damage mitigation: A review of structural and nonstructural measures and alternative decision frameworks. *Water Resources Research* 21: 411–24.

U.S. Army Corps of Engineers. 1990. *Integrated feasibility report and environmental impact statement: Local flood protection—Moorefield, West Virginia*. Fort Belvoir, Virginia.

U.S. Army Hydrologic Engineering Center. 1989. EAD: Expected annual flood damage computation. In User's Manual, CPD-30. Davis, California.

Risk-Based Budgeting for Maintenance Dredging

Michael R. Walsh and David A. Moser[1]

Abstract

The U.S. Army Corps of Engineers must estimate the budget required to conduct maintenance dredging at hundreds of sites across the country. The amount of funds needed to do the maintenance dredging is highly uncertain and there are risks associated with overestimates and underestimates. A risk-based approach to the budgeting process for maintenance dredging can help identify the uncertainty and assess and manage the associated risk. A concept and preliminary plan for a risk-based approach for developing budgets for maintenance dredging is presented.

Introduction

Each year, Corps of Engineers' Operations and Maintenance (O&M) managers are responsible for maintenance dredging at over five hundred dredging sites across the nation at an annual total cost ranging from about $300 million to $400 million. Figure 1 depicts the budgeted amounts, actual expenditures, and their difference for the direct costs of maintenance dredging and associated surveys during the five fiscal years (FY) 1988–92. Data are derived from the Corps of Engineers' Management Information System (COEMIS) and the Corps' Automated Budget System (ABS). Engineering and design (E&D) and supervisory and administrative (S&A) costs are not included in figure 1. The data shows that the Corps overbudgeted by about $96 million in FY 88, but steadily reduced overbudgeting until FY 91, when there was a small underbudget. FY 92 shows a small overbudget. On average, for the entire period, the Corps

---

[1]U.S. Army Corps of Engineers, Institute for Water Resources, Ft. Belvoir, VA 22060

The views expressed in this paper are solely those of the authors and do not necessarily reflect the policy of the U.S. Army Corps of Engineers.

overbudgeted by about 9.5%. This amount of overbudgeting, given the downward trend in overbudgeting over the five years of data, is not a cause for alarm, but it raises the issue of uncertainty and risk inherent in forecasting maintenance dredging requirements and budgets.

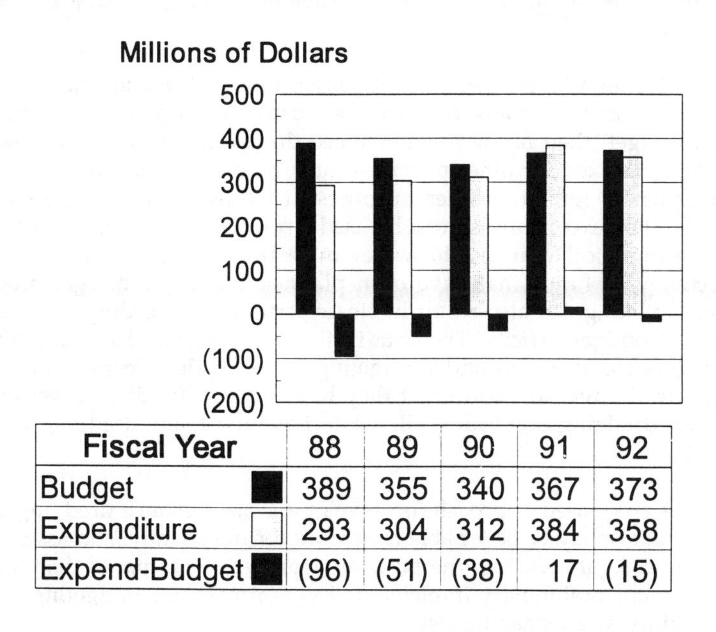

Fiscal Year	88	89	90	91	92
Budget	389	355	340	367	373
Expenditure	293	304	312	384	358
Expend-Budget	(96)	(51)	(38)	17	(15)

Figure 1. Corps Maintenance Dredging and Survey Budget, Expenditures, and Difference from COEMIS and ABS

In developing a budget for maintenance dredging, Corps dredging managers must deal with considerable uncertainty and associated risk. First of all, the budget forecast must be made nearly two years in advance of the maintenance dredging. This long lead time makes it difficult to predict dredging requirements when sedimentation patterns and shoaling rates are understood and stable, and nearly impossible when they are not. The amount of material to be dredged can vary considerably from year to year at a dredging site. Another factor contributing to budget uncertainty is the dredging contractor marketplace. Most Corps dredging is done under contract. Market conditions, amount of work to be done, competition, and general economic conditions affect the bids that the Corps may receive on a given dredging project. Predicting the bid cost is fraught with uncertainty. Finally, the Corps budgets for about five hundred dredging sites each year, but these sites are chosen from a much larger pool of possible dredging sites. Some sites are dredged each year, but many sites are dredged on an irregular basis. Corps managers must choose which sites must be dredged as well as the amount of dredging required. It is no wonder that at the dredging site level, the actual dredging required can differ markedly from the forecast. Often, budgeted dredging sites are not dredged at all, while unbudgeted dredging sites are dredged.

The high level of uncertainty associated with maintenance dredging requirements entails risk for the Corps. If Corps O&M managers underbudget, then needed maintenance dredging may not get done, or funds earmarked for other purposes must be diverted to dredging, with consequences for those other purposes. If Corps O&M managers overbudget, then excess funds must be used elsewhere. Funds made available during the fiscal year in which they must be spent may not be used as effectively as funds that have been planned for in the budget process. Overbudgeting can also result in closer scrutiny of the Corps budget in times of budget deficits. The data in figure 1 indicate that Corps O&M managers are averse to underbudgeting. It seems that Corps O&M managers overbudget to ensure that they have enough funds to cover maintenance dredging uncertainty. Better to have too many dredging dollars than too few.

A risk-based approach for developing maintenance dredging budget estimates may be better able to account for uncertainty, and assess and manage risks inherent in the budgeting process. This paper discusses a concept and preliminary implementation approach for budgeting under uncertainty and managing risk.

Risk-Based Concept for Maintenance Dredging Budgets

The Corps O&M budget is developed via a bottom-up process constrained by budget targets set by headquarters in consultation with the Secretary of the Army and the Office of Management and Budget. Each discrete O&M work task at each project is budgeted. In the case of maintenance dredging, the budget for each dredging site within a designated civil works project is estimated. These work tasks are reviewed at each organization level—district, division, and headquarters—and finally aggregated to form the Corps budget request.

Each maintenance dredging work task is budgeted by giving a single number estimating the total cost. There is no measure of the uncertainty of that number. Figure 2 depicts a budget estimate of $150,000 for a hypothetical dredging work task. One can only assume from the information given that the only possible budget estimate is $150,000.

However, might not the actual budget required be as much as $200,000 or as little as $100,000? How likely is it that the budget amount will be $200,000 or $100,000? Again, there is no indication from the point estimate.

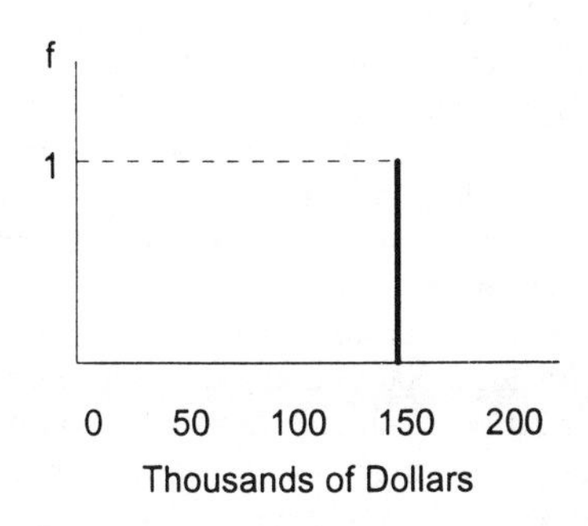

Figure 2. Point Estimate for Maintenance Dredging Budget at a Dredging Site

Figure 3 depicts a budget estimate for the same hypothetical dredging project that includes a measure of uncertainty. In this case, for the same dredging site, three budget numbers are estimated. These are the minimum likely budget, the maximum likely budget, and the most likely budget that is forecast to be required to meet the dredging requirements.

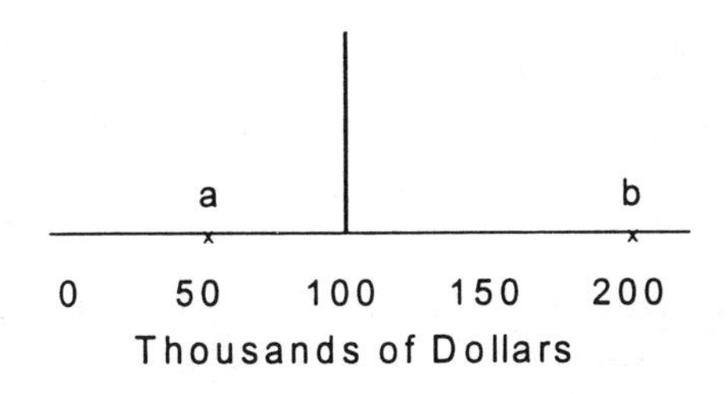

Figure 3. Maximum, Minimum, and Most Likely Estimates for Maintenance Dredging Budget at a Dredging Site

This budget estimate conveys much more information than the previous single-point estimate. The minimum budget forecast designated by "a" is $50,000. That means that physical, economic, and institutional factors could occur that would require maintenance dredging costing $50,000. Likewise, the maximum amount of dredging funds that could be required is $200,000, as designated by "b." Perhaps, increased shoaling rates are possible that would increase the maintenance dredging requirement. Note that the most likely dredging budget is $100,000. Figure 1 showed a point estimate of $150,000. This estimate indicates that the dredging manager for this site was averse to the risk of underbudgeting, so he added $50,000 to the most likely dredging budget estimate to help ensure that he would not be underbudgeted. The key concept is to add information that reflects the uncertainty in the budget estimate. If the maximum expected budget required in figure 2 was only $150,000 instead of the $200,000 shown, that would mean that the point estimate in figure 1 eliminated all risk associated with underbudgeting. If the maximum budget that might be needed is indeed $150,000, then there is no possibility of underbudgeting. However, it is certain that the dredging site will be overbudgeted!

If one assumes a distribution for possible budget estimates, then it is possible to determine the relative frequency or probability for these estimates. Figure 4 depicts a triangular distribution of possible budget requirements that is estimated from the three values given in figure 3.

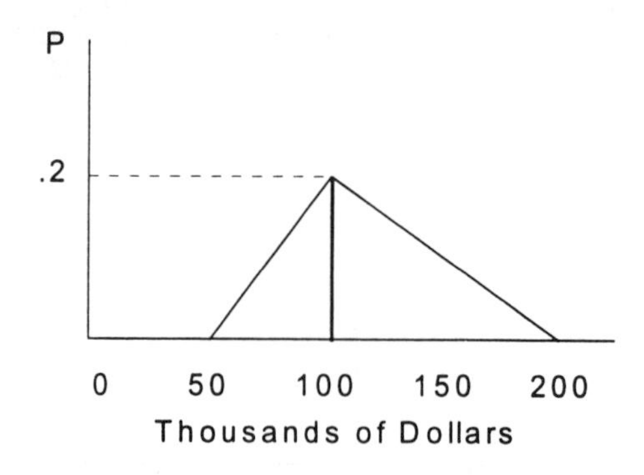

Figure 4. Triangular Distribution of Possible Dredging Budget Requirements at a Dredging Site

Using this distribution, one can determine how likely it is that the actual budget needed will be less than a threshold value. There is more information with which to make a decision about budgets for this particular project. If distributions are developed for all Corps dredging sites, there will be better information on which to base the overall Corps budget for dredging. There are many ways to estimate the uncertainty associated with budget estimates for a dredging site. Figures 3 and 4 depict methods that rely on the judgment of the dredging site manager. Another method is to examine the historical expenditures at a site, select an appropriate distribution, and calculate the parameters of the distribution, such as the mean and standard deviation. The presumption here is that the historical data is representative of the population of possible budget requirements at the dredging site. Figure 5 depicts a possible normal distribution for the budget requirements for the hypothetical project.

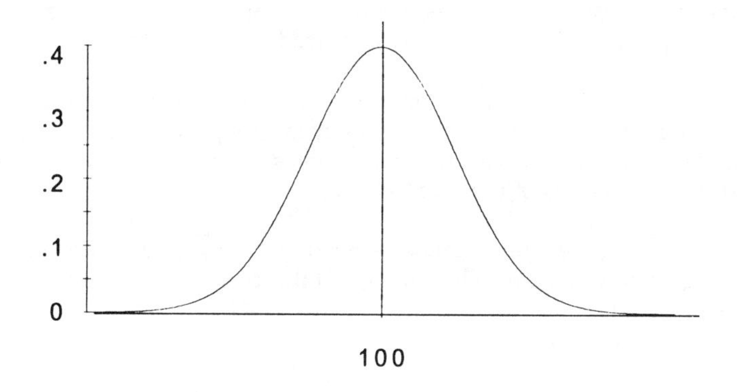

Figure 5. Normal Distribution of Possible Budget Requirements at a Dredging Site

Proposed Implementation of Risk-Based Approach

The implementation of a new methodology for estimating budget requirements for maintenance dredging that explicitly accounts for uncertainty, and attempts to manage the risk associated with the uncertainty, must be done with care and must proceed step by step. The Corps is currently considering a three-phase implementation that slowly introduces uncertainty analyses and risk management concepts into the budget process. Each phase decomposes uncertainty into more elementary units.

In Phase 1, there is no decomposition. Sources of uncertainty are simply identified but not quantified. Historical dredging expenditures are used to develop distributions of possible budget requirements. The parameters of these distributions are used to set a division target for maintenance dredging budgets while allowing flexibility at the project level in setting budgets. Initially, the division target is to be set at a five-year adjusted average of expenditures for maintenance dredging within the division. Budgets for individual projects can vary from projected five-year averages, but the division total must be within the division target.

In Phase 2, uncertainty is decomposed into variable- and fixed-cost factors. The amount of material to be dredged, the hours required to dredge, the dredging rate as $/cubic yard or $/hour, and other cost factors, such as mobilization and demobilization, are the important cost factors determining the dredging budget requirements. Each of these factors must be examined to develop distributions that capture the uncertainty around them. Monte Carlo techniques can be used to combine the cost factors to develop overall measures of uncertainty for the dredging budget at the project, district, division, and headquarters levels.

Phase 2 initiates the development of a budget estimating model. The model should identify all the relevant cost factors that affect the dredging required for each project. Estimates of uncertainty will be made about the parameters of the cost model.

The budget estimating model for developing budget estimates for dredging projects would follow the general form

$$TC = VC + FC,$$

where TC = Total Dredging Cost
VC = Variable Cost
FC = Fixed Cost.

Both VC and FC can be treated as random variables with a distribution that is estimated from cost data from individual projects. Monte Carlo modeling can be used to estimate a distribution for TC for each project.

One difficulty with using this approach is the lack of data for VC and FC. Existing databases must be examined to see if it is possible to come up with a good set of data to estimate the distributions.

Another difficulty is that the VC of dredging depends on the rate ($/cu. yd.) and the amount of sediment to be moved (cu. yd.), the type of sediment (rock, clay, silt, sand), and the disposal method (open water, upland). There are many cost factors that must be examined for the impact on *TC*. For example, one possible budget estimating model could take the form

$$TC = \text{Rate } (\$/\text{cu. yd.}, \$/\text{hr.}) * \text{Quantity } (\text{cu. yd., hrs.}) + S\&A + E\&D + DC + SC \,,$$

where *S&A* = supervisory and administrative costs
E&D = engineering and design costs
DC = all costs associated with disposal
SC = all study costs for planning and monitoring dredging projects.

Some portion of these costs must be allocated to each dredging project or accounted for separately to come up with a good budget estimate.

Finally, Phase 3 attempts to incorporate the underlying physical, economic, and institutional processes that contribute to the uncertainty associated with dredging. If successfully developed, these models will enable the Corps to examine uncertainty with all the factors that influence the need for dredging and its associated costs.

This more complex model would relate other events, such as floods, to the quantity of sediment to be moved or the rate cost of dredging. Thus, the variable cost of dredging would depend on sedimentation rate, flooding events, and other factors that are judged to be important. Correlating these other events to impacts on dredging will be difficult. Using this approach will allow the Corps dredging program to allocate funding based on both the uncertainty in costs and economic performance. This will assist in minimizing the navigation impacts of constrained budgets.

We expect to begin by looking at past cost data and using simple Monte Carlo methods to develop distributions for total costs of dredging at each project. Experience with simple models will dictate more complex modeling approaches.

Issues

The risk-based approach and implementation plan are preliminary and much work remains to implement a practical set of procedures within the budget process. Several issues must be addressed.

Data comparability

All divisions must use the same base data for each of the phases. Expenditure data should be from COEMIS, not local records, for comparability across the Corps. *E&D* and *S&A* must be added back in to expenditure data.

Policy on risk

The decision to employ risk-based models to forecast maintenance budgets will require an examination of policies to ensure that they are relevant to risk-based methodologies. The Corps may have to develop policies describing the acceptable risk factors. How much risk of overbudgeting or underbudgeting will be tolerated?

Considering benefits

All the phases concentrate on the cost side of the equation. Decisions about which projects to dredge should consider the benefit side of the equation as well. Projects with the largest net benefits (benefits-costs) should be given higher consideration. However, benefits are uncertain, so risk analysis is needed to consider the distribution of benefits and costs to aid decisions about which projects to dredge.

Conclusions

The Corps of Engineers is committed to using risk-based approaches to help quantify uncertainty and assess and manage risk. A risk-based budgeting approach for maintenance dredging may offer a better way for the Corps to develop dredging budgets and meet the need for dredging with minimal risk and least cost.

Developments in Risk Analysis Commercial Software

Charles Yoe[1]

Abstract

There has been a marked increase in the number and type of commercial software packages available to support risk and uncertainty analysis. Monte Carlo simulations, the first such tool, are more numerous and powerful. Additional capabilities including distribution fitting, what-if analysis, analytical hierarchy processes, and decision analysis software are some areas where impressive new capabilities have been developed. Selected new software packages are reviewed in this article.

Introduction

Not too many years ago, but quite a few more than many of us want to admit, fledgling scientists and analysts waited in rooms with green cinder block walls for the next available card punch machine. With any luck at all you'd get a machine with a ribbon that had not completely faded. This made searching for the inevitable errors possible.

Submitting a batch job was no less an ordeal. Boxes of cards with colored diagonal stripes across the tops of the cards (the easier to reassemble them in proper order if dropped) were handed to the first generation of disinterested, overworked computer nerds. Some time later, the length of time depending on the times of day and semester, you'd go, for the first time, to a numbered cubbyhole looking for the output. The hope in your heart was that something would be there and it would be thicker than that single page of output that always accompanied a program that did not execute properly.

---

[1]Associate Professor of Economics, College of Notre Dame of Maryland, 4701 North Charles Street, Baltimore, MD 21210-2476

What a miracle it was, our professors said, as they frightened us with stories of multiple regressions done by hand. We are so fortunate, they said, to live at a time when we could do in a day what once took them a week. We enjoyed unimaginable good fortune.

At the Engineering Foundation's conference, "Risk-Based Decision Making for Water Resources Management VI," the memories of those miracle days had faded. The conferees bemoaned the fact that little of today's software can be linked to our other models, rendering it less useful than the software could be. Object-oriented programming, the new rage, frustrates us. Even as it opens new doors to us it does not open them wide enough. What we want are truly smart objects.

Never mind the advances in decision-science software that permit huge leaps forward with standard decision, value, and chance nodes. We want to be able to define an object that acts like a hydropower generator. When they fail in our simulations we want them to crack, smoke, and burn on screen to make the point of failure most vivid. We want the technical sophistication in our models, along with the ease of programming, animation, easily generated reports, and the flexibility to link with other applications and models. We want it all and we want it now.

Success and rapid development in information management systems have spoiled us. Where once lack of imagination constrained us, today active imagination frustrates us. We've seen the advances of recent years and we are already imagining the advances that await us.

We're far closer to a future of linkages, smart objects, and ever-increasing speed than we are to a past of card punches and interminable waits. This is certainly evident in the field of risk and uncertainty analysis. Though the acolytes of risk analysis deserve the most credit for extending concepts, principles, and methods of risk analysis within their own disciplines, developments in computers and computer software are a close second in causing the spread of risk analysis throughout the academic and practical landscape.

This paper takes a look at mainstream software developments that aid risk analysis. The software packages discussed here represent a sampler of the commercial software packages available that can be used to make risk analysis and risk communication easier.

The software packages discussed in this paper are mainstream commercial software packages. They are produced, marketed, and supported by reputable companies. Nevertheless, none of these software packages will be found in your local computer warehouse outlets. They are specialty products. Following a brief review of the beginnings of quantitative risk analysis, this paper turns to a discussion of various software types.

The Beginnings

During the early 1970s, when the Water Resources Council established the "Principles and Standards for Planning Water and Related Land Resources" (P&S) in September 1973, water resources planning was changed forever. The P&S strengthened and formalized the water resources planning process. Risk and uncertainty analysis was explicitly required in the planning process for the first time.

It would be another decade and a half, however, before attention was seriously focused on the conduct of risk and uncertainty analysis. Why did it take so long for risk and uncertainty analysis to catch on and be taken seriously? The possible explanations are many.

Changes to the water resources planning process introduced by the P&S were major. The P&S advocated a formal four-account planning process, iterative planning, a prescribed planning process, reporting requirements, alternative plans, trade-offs between national economic development and environmental quality, and other significant changes. Requirements for risk and uncertainty analysis were, in the scheme of things, minor concerns to planners trying to cope with the massive changes wrought by the P&S. There were so many first-order changes that attention to risk and uncertainty languished for years. Through the 1970s and '80s practical attention to risk and uncertainty analysis was relegated to a few words about different population projections and the project benefit-cost ratio's sensitivity to the interest rate.

The initiation of this Engineering Foundation conference series in the early 1980s provided a benchmark for the growing academic and professional interest in issues of risk and uncertainty analysis for water resource planning and management. The proceedings from this series provide an excellent overview of the evolution of thought on this subject. As interest in the subject area grew, the theory developed and applications to water resource problems became more numerous. The evolution of thought and the development of methodologies were necessary and significant steps in the evolving acceptance of risk and uncertainty analysis. It was the research community that kept the interest in risk and uncertainty analysis alive while planners' energies were directed to more immediate issues.

The advent of cost-sharing changes in the Water Resources Development Act (WRDA) of 1986 (P.L. 99-662) provided a unique window of opportunity for incorporating risk and uncertainty analysis in federal project planning. The WRDA of 1986 increased the cost of federal water projects to local sponsors of these projects. Once they became responsible for a larger share of the costs, they became more interested in the planning process. The downsizing of projects became a new phenomenon.

Dollars and cents motivated federal and nonfederal partners to consider the explicit and implicit risks of different project designs. It took researchers and policymakers to put the tools of risk and uncertainty analysis on the table. It took dollars and cents to motivate planners to pick them up.

At about the same time, a risk analysis research program directed by the U.S. Army Corps of Engineers Institute for Water Resources (IWR) was gaining momentum within the agency. A series of risk analysis workshops was taken to several Corps divisions to introduce Corps personnel to the basic principles of risk analysis. This "traveling road show" became the basis for what has become one of the Corps' most popular training courses. Thus, the education efforts of IWR benefited from persistence and fortuitous timing. It spoke the gospel of risk analysis at the same time economic circumstances were winning over converts.

The confluence of this growing agency interest in risk analysis with advances in personal computers and software created a coincident flood of interest and ability that has seen practical application of risk and uncertainty analysis grow at an unprecedented rate in recent years.

It is the last piece of this puzzle, computer software, that is the focus of the remainder of this paper. The concepts of risk and uncertainty analysis do not come easily to the practitioner who lacks formal training in this area. Probability distributions, Monte Carlo simulations, and the like are second nature to the research community, but they are often distant memory traces for many practicing professionals. Until recently, a Ph.D. and a mainframe were considered necessary for risk and uncertainty analysis.

Since the late 1980s, commercial software companies have begun to put what was once mainframe analytical power into the hands of any practitioner with a PC and a couple hundred dollars. Though the advances may have not gone far enough fast enough to satisfy the research and academic communities, they have advanced practitioners farther down the road of practical risk and uncertainty analysis than another other single innovation or event.

The discussion that follows considers a sampling of commercially produced software packages that can be useful in conducting quantitative risk analysis. There is a great deal of software that is commercially produced. This paper is less interested in the software that is available to the research community than it is in the software that is more generally accessible. Thus, commercial software is defined as those packages marketed to a broad range of users but that require a more experienced user. The software packages discussed below vary in nature and applicability, from packages developed explicitly for risk analysis to packages that are only tangentially useful for risk analysis.

Monte Carlo Simulation Software

During the 1970s, the Corps of Engineers provided guidance to agency personnel, suggesting that a project's sensitivity to a particular value might be tested by repeating the calculation 10,000 times. This informal guidance alluded to an exotic-sounding technique, Monte Carlo simulation. The guidance pointed out that such calculations were being done easily at many universities.

Today, products like @RISK[2] and Crystal Ball[3] put Monte Carlo simulations at the fingertips of analysts around the world. These products are spreadsheet-based packages that allow the user to vary any cell values. Built-in probability distributions allow the user to specify distribution parameters to describe the uncertainty or risk in a value. Simulations of a varying number of iterations can then be run based on Monte Carlo or Latin hypercube sampling methods. Simulation results can be displayed in graphic or numeric forms.

Crystal Ball runs with the Excel and 1-2-3 for Windows software. @RISK runs with Excel and Lotus for Windows and DOS versions of Lotus. There is also a Mac version of @RISK.

---

[2]@RISK is produced by Palisade Corporation, 31 Decker Road, Newfield, NY 14867, (800) 432-7475.

[3]Crystal Ball is a product of Decisioneering, Inc., 1727 Conestoga Street, Boulder, CO 80301, (303) 292-2291.

Users tend to like the graphics capabilities of Crystal Ball, particularly the graphic summaries of the simulation that appear in real time. Visual "mouse control" of distribution extremes makes it easy to create truncated versions of all the built-in distributions. The "custom" distribution allows the user to create a probability distribution visually that does not correspond to any of the built-in distributions. Crystal Ball's Advanced Macro Interface (AMI) provides the user the opportunity to build customized applications and integrate Crystal Ball with other tools.

@RISK was originally developed as a true Lotus add-in. This makes it a powerful programming tool. The @RISK distribution functions can be used as arguments in the regular Lotus functions. This enables the user to build models that are more complex than can be built with Crystal Ball.

As the Monte Carlo simulation capabilities have been developed for the spreadsheet environment, it has become more commonplace for the spreadsheet packages to include probability distribution functions with the basic package.

Distribution Fitting Software

Expressing uncertainty and risk in terms of probability distributions is an essential step for Monte Carlo simulation. Perhaps a natural step in the development of risk analysis tools is the provision of distribution fitting capabilities that will help practitioners describe their data as probability distributions.

Distribution fitting capability has long been available to users of mainframe statistics packages. DOS-based STATGRAPHICS[4] has included distribution fitting capability since its earliest software packages for the personal computer. More recently, Best Fit[5] has been introduced as a distribution fitting software for the Windows environment. Best Fit is a stand-alone package that offers the user chi-square, Kolmogorov-Smirnov, and Anderson-Darling tests as goodness-of-fit measures for eighteen families of distributions.

[4]STATGRAPHICS is a product of Manugistics, 2115 East Jefferson Street, Rockville, MD 20852, (800) 592-0050.
[5]Best Fit is produced by Palisade Corporation, 31 Decker Road, Newfield, NY 14867, (800) 432-7475.

Distribution fitting capability has two major uses in risk analysis. First, it allows the user to gain insight into how existing data are distributed. This information can be used to specify probability distributions for use in the construction of simulation models. A second use of this capability is to fit distributions to simulation model results. The results of a complex spreadsheet simulation model may frequently be described as a single probability distribution. Software like Best Fit helps the analyst determine how the simulation output data are distributed. This is particularly useful when the output of one model is needed as an input to a subsequent model.

What-If Software

The widespread use of spreadsheet software led, quite naturally, to the advent of what we call here "what-if" software. What if we don't sell as many units of the new product as we expect? What if development costs are higher than expected? What would happen to profits in these and other scenarios in an uncertain environment?

What-if analysis could be done in a spreadsheet by simply changing the values of cells that could take on a range of values. The Microsoft Excel[6] "What-if add-in macro" is a typical spreadsheet enhancement developed to facilitate such rudimentary uncertainty analysis. To use this feature a single spreadsheet model is created. A range of input values is specified for each cell value that is to be varied. These values are listed in a data sheet or data table. The what-if feature substitutes every combination of the input variables possible either one step at a time or by cycling through all the possibilities. The number of combinations varies with the software package, but it is relatively limited.

What-If Solver[7] provides optimization capability for DOS-based versions of Lotus 1-2-3 and Symphony. What-If Solver goes beyond the simple what-if questions to provide a powerful optimization capability for the spreadsheet environment. Problem solution requires four basic steps. First, the what-if cells are selected. These cells contain the values that are to be adjusted, i.e., the endogenous variables. Second, the constraint cells are identified and the constraints are specified. Third, an

---

[6]Microsoft Excel is a product of Microsoft Corporation, One Microsoft Way, Redmond, WA 98052-6399, (800) 426-9400.

[7]What-If Solver is a product of Frontline Systems, Inc., 140 University Avenue, Suite 100, Palo Alto, CA 94301, (800) 451-0303 ext. 55.

optimum cell, such as net benefits or project costs, may be identified. The value of this cell may be maximized or minimized. Finally, the What-If Solver is executed.

Linear and nonlinear models can be optimized. External knowledge of the problem is required to find a global optimum. If there are many possible local optima, the model must be seeded with cell values that "place" the problem in the vicinity of the global solution. Lagrange multipliers can be generated for constrained optimization problems. These are useful for estimating shadow prices. Reduced gradients can also be estimated to examine the effect of changes in nonbasic variables in linear programming problems.

TopRank[8] is another software package that has taken this existing spreadsheet capability and extended it. It offers automatic identification of "input variables," automated what-if analysis, and advanced data tables. One of the critical needs in uncertainty analysis is to identify those variables that are most important to the decision process. With TopRank the user identifies "output variables" and TopRank automatically identifies all the cells or "input variables" in a spreadsheet that affect the output variable/cell.

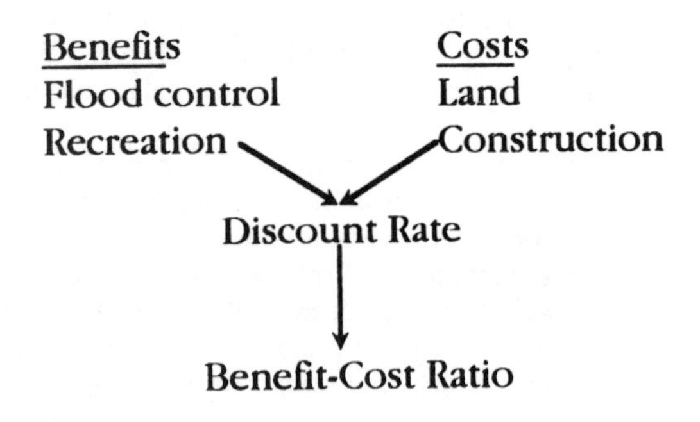

Figure 1. BCR Model with Five Variables

[8]TopRank is produced by Palisade Corporation, 31 Decker Road, Newfield, NY 14867, (800) 432-7475.

The what-if analysis allows the user to specify a minimum and maximum range for input variables. TopRank changes the variable value across this range for one variable, all other variables equal, and keeps track of the output value calculated. The input variables are then ranked according to the magnitude of the change they've caused for each output variable. For example, consider a simple benefit-cost analysis computation as outlined in figure 1. There are two types of benefits, flood control and recreation, and two types of costs, land and construction; for simplicity, assume these values are all expressed as accumulated present values. The fifth variable in this model is the discount rate. The discount rate is needed to convert all monetary values to their equivalent annual values over a fifty-year period.

The benefit-cost ratio (BCR) is the only output variable. Each of the five input variables have been identified as affecting the BCR. For the what-if analysis, benefits were allowed to vary by ±30%. Costs were varied by ±15% and the discount rate was allowed to vary asymmetrically from a minimum of 6% to a maximum of 14%. The values are summarized in table 1.

Item	Minimum	Most Likely	Maximum
Flood control	70000	100000	130000
Recreation	35000	50000	85000
Land	16500	20000	23500
Construction	59500	70000	80500
Discount Rate	6	8	14

Table 1. BCR Variable Ranges

The tornado graph of figure 2 shows the results of the what-if analysis. Given the parameters of this simple model, the value of flood control benefits is clearly the most influential variable for the BCR. The BCR swings from a low of about 1.3 to a high of 2.0 with the values of flood control benefits. TopRank also provides a numerical summary of the analysis as well as other graphical formats for displaying the results.

The advanced data table feature allows the user to specify variable values for up to a nine-way table. In the sample model, nine values were used for each of the five variables, though there is no requirement that equal numbers of values be used for each variable in the table. The data table analysis combines each value for flood control benefits with every possible combination of values for the other variables and then repeats this process for each value in the data table. TopRank can identify the input variables that have the most effect on the benefit-cost ratio, in this instance, flood control benefits.

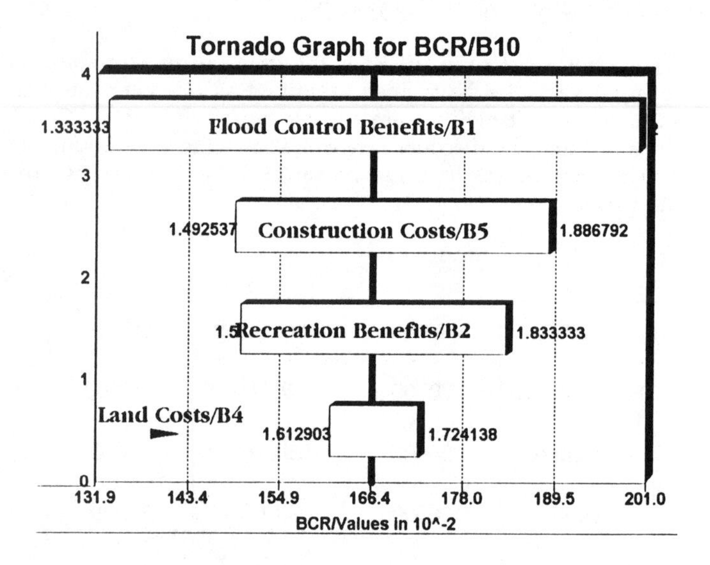

Figure 2. Results of the What-If Analysis

As water resource planners deal with risk and uncertainty analyses, one of the most frequently asked questions is which variables are most important to the decision process. TopRank appears to be a useful spreadsheet environment tool for gaining insight into the answer to that question. The systematic sensitivity analysis that can be done with TopRank can be very useful when there are critical values of variables that need to be evaluated. Critical combinations of values can likewise be evaluated and response surfaces, for use in other models and applications, can be generated from the outputs of such analyses. Data tables also appeal to decision makers who are less comfortable with the probability distributions of Monte Carlo simulations.

Decision Analysis

The application of decision science to real-world problems through the use of systems analysis and operations research has grown into the discipline of decision analysis. Decision analysis provides structure and guidance for thinking systematically about decisions made under conditions of uncertainty. Clemen, in his text *Making Hard Decisions: An Introduction to Decision Analysis* (1991), offers the decision analysis process flowchart shown in figure 3. This process is generic enough to lend itself to any number of decision analysis tools. There are several new products on the market that are quite useful applications of this decision process, including multicriteria tools and decision trees.

Multicriteria Tools

Expert Choice[9] is a stand-alone tool for the Windows environment that helps the user to organize a decision problem in a hierarchical structure, allowing him or her to see a problem as a whole system and not a set of isolated parts. The elements of the problem are organized into a hierarchy similar to a family tree. The top level consists of the goal node. Intermediate levels represent factors such as scenarios, objectives, and subobjectives. The bottom level of the hierarchy usually contains the alternative choices that are rated against each other.[10]

Good decisions are based on the attainment of goals. Criteria and objectives can be specified to measure how well the goals are achieved. Decisions are complicated by the trade-offs we must make among competing criteria and objectives. Uncertainty and competing interests add to the complexity of decision making. With Expert Choice, once the decision structure is specified, data, knowledge, intuition, and experience are applied to derive priorities from among the many objectives, criteria, etc. This is done through pairwise comparisons. These pairwise judgments can be expressed verbally, numerically, or graphically.

Consider a simple model where decision makers must determine how a limited maintenance budget is to be distributed among competing projects. For simplicity, suppose four maintenance projects differ in their

---

[9]Expert Choice is a product of The Decision Support Software Company, 4922 Ellsworth Avenue, Pittsburgh, PA 15213, (412) 682-3844.
[10]The bottom level of the hierarchy can also be expressed as intensities of various standards. Alternatives can then be rated against standards rather than against each other.

costs, the political interest in their accomplishment, and their environmental impacts. To complicate matters, each project produces benefits that have never been estimated.

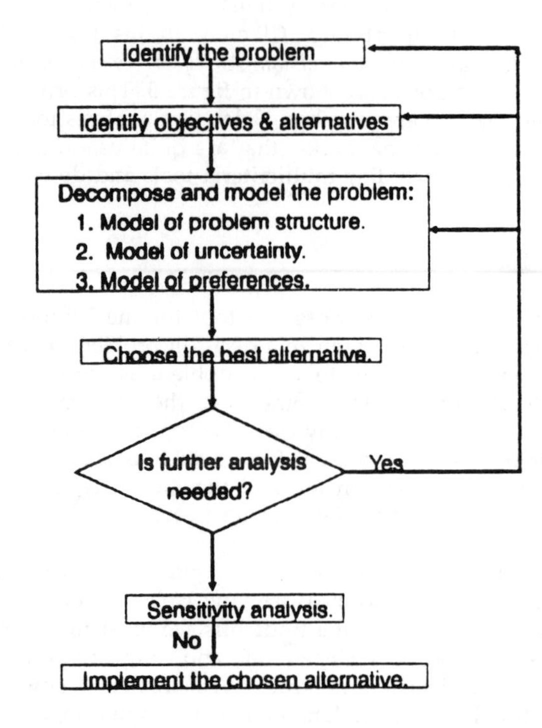

Figure 3. Decision Analysis Process Flow Chart

Figure 4 shows a hierarchical model for such a decision problem. The goal is to identify the best projects based on consideration of the four decision criteria shown beneath the goal. The four alternative projects are listed beneath each of the criteria. The analytical hierarchy process (AHP) proceeds by having the decision makers compare the criteria two at a time for a total of six different comparisons (politics versus cost, politics versus benefits, etc.). Comparisons can be made verbally (i.e., political considerations are moderately more important than environmental impacts), numerically, or graphically. Figure 5 presents a questionnaire that can be generated by Expert Choice showing the comparisons to be made at the first criteria level.

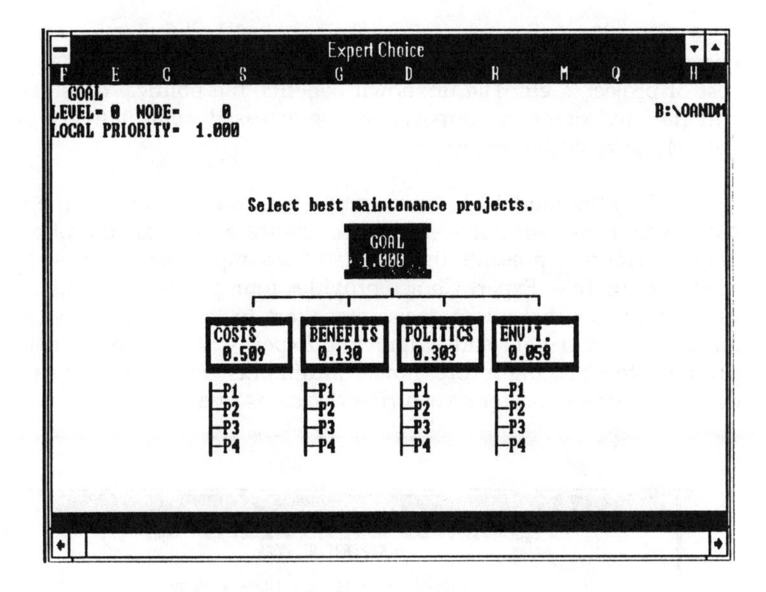

Figure 4. Hierarchical Model Used for a Decision Problem

Compare the relative importance with respect to: GOAL

Circle one number per comparison below using the scale:
1=EQUAL 3=MODERATE 5=STRONG 7=VERY STRONG 9=EXTREME

1	COSTS	9	8	7	6	5	4	3	2	1	2	3	4	5	6	7	8	9	BENEFITS
2	COSTS	9	8	7	6	5	4	3	2	1	2	3	4	5	6	7	8	9	POLITICS
3	COSTS	9	8	7	6	5	4	3	2	1	2	3	4	5	6	7	8	9	ENVIRONMENT
4	BENEFITS	9	8	7	6	5	4	3	2	1	2	3	4	5	6	7	8	9	POLITICS
5	BENEFITS	9	8	7	6	5	4	3	2	1	2	3	4	5	6	7	8	9	ENVIRONMENT
6	POLITICS	9	8	7	6	5	4	3	2	1	2	3	4	5	6	7	8	9	ENVIRONMENT

GOAL: Select best maintenance projects.

BENEFITS --- Benefits of each maintenance project.
COSTS — Expenditures required for each maintenance project.
ENVIRONMENT --- Environmental impacts of each maintenance project.
POLITICS --- Political interest in each maintenance project.

Figure 5. Sample Questionnaire for the First Criteria Level

The alternatives are then ranked against each other for each of the criteria. Thus, the cost of project 1 can be compared numerically to the cost of project 2, etc. The unknown benefits, the political considerations, and the environmental impacts can be assessed quantitatively or qualitatively to rank the project.

Once the model has been completed, the synthesis of the various judgments comprising it yields a best choice and a ranking of all alternatives. Figure 6 presents the results of a sample analysis based on the model in figure 4. Expert Choice provides four graphical sensitivity analysis techniques that allow decision makers to test the sensitivity of the outcome to their data and judgments. Expert Choice offers considerable promise for making professional judgments an explicit part of the decision process and for structuring decisions in a rational fashion.

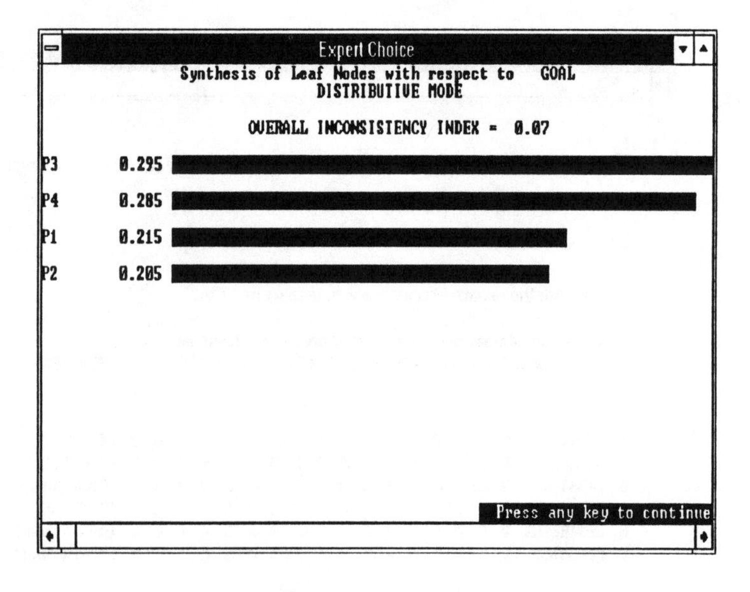

Figure 6. Synthesis of Completed Model

Influence Diagrams and Decision Trees

Among the tools developed by decision analysts are two particularly useful aids for structuring problems: influence diagrams and decision trees. Influence diagrams, such as the one shown in figure 7, can be used to specify the basic relationships among decisions, uncertainties, and values. Decision trees, like the one in figure 8, show the precise sequence of events in a decision problem.

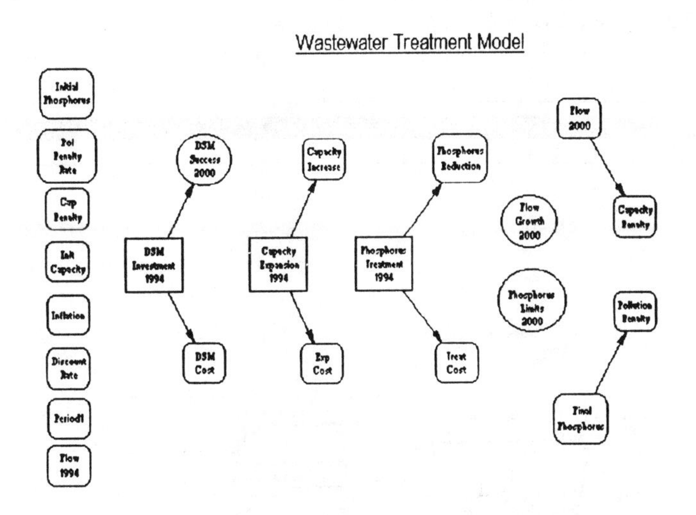

Figure 7. Influence Diagram of a Wastewater Treatment Model

A number of decision-tree software packages are now available. Supertree[11] provides a textbook with a DOS-based decision-tree software package that is quite well suited for student use. The text, *Decision Analysis with Supertree* (McNamee and Celona 1990), provides a good basic explanation of probability, uncertainty, and risk in an influence-diagram/decision-tree context.

DATA[12] is a Mac-based decision-tree software package. Payoffs defined in DATA models can specify probabilistic events with distributions and outcomes as ranges of values. The expected value of any path through the decision tree can be estimated.

---

[11]SuperTree is a product of The Scientific Press, 651 Gateway Boulevard, Suite 1100, South San Francisco, CA 94080-7014, (415) 583-8840.
[12]DATA is a product of TreeAge Software, Inc., P.O. Box 32[illegible], Boston, MA, (617) 426-5819.

DPL[13] is a stand-alone Windows-based software package that combines the strengths of influence diagrams and decision trees. With DPL, structured, deterministic, and/or probabilistic models can be built; the models can be evaluated; and the results communicated. DPL appears to offer the widest array of options for such decision analysis because it offers the simplicity of the influence diagram and the detail of the decision tree in a synthesis of the two tools. The graphics in figures 7 and 8 were both created with DPL.

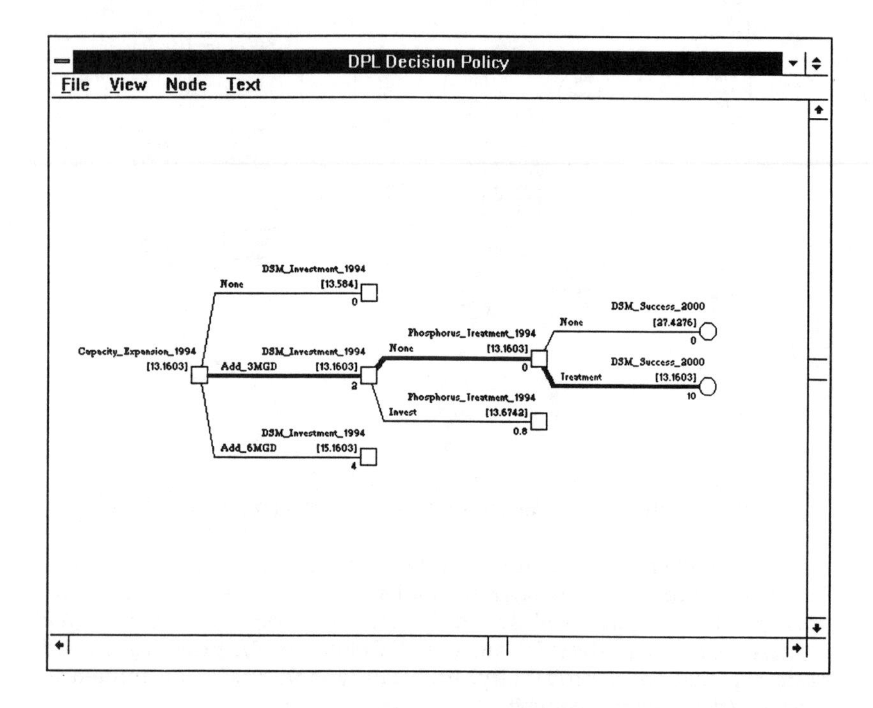

Figure 8. Decision Tree Showing Sequence of Events in a Decision Problem

DPL expressions used in model building make use of a full range of arithmetic and spreadsheet functions. The spreadsheet functions include mathematical, financial, and statistical tasks. In addition, logical and selection functions (i.e., lookup functions) are also available. These

---

[13]DPL is a product of ADA Decision Systems, 2710 Sand Hill Road, Menlo Park, CA 94025, (415) 926-9251.

built-in functions make it possible to create rather complex arguments and expressions for the decision trees. Twenty-one named distributions are also available for use in expressing basic model uncertainty.

The DPL decision model can be solved in a number of ways. In the first phase, the DPL algorithm provides the path outcome. Multiple outcomes are allowed. In the second phase, two optional transformations are possible. Multiple outcomes may be combined using a default or a user-defined objective function. Also, the output of the objective function can be transformed using a built-in utility function. In the third phase, the output of the objective function is used to produce an optimal expected value decision policy from the tree. The output of the utility function, if used, produces an optimal, certainty-equivalent decision policy. Expected values and certainty equivalents are computed for each chance node.

The sensitivity analysis available within DPL is one of the software package's strengths. Sensitivity analysis can be conducted on values, probabilities, and events. Several graphics options are available for displaying the results of the sensitivity analysis.

Neural Networks

Though not explicitly a risk analysis tool, neural networks can be useful in recognizing nonlinear statistical relationships among data. To the extent that uncertainty results from our inability to recognize patterns in complex situations, neural nets, as they are sometimes called, can be valuable aids to an uncertainty analysis.

Most neural nets use some version of the following steps. First, the user defines what is to be estimated, i.e., the output or dependent variable. This is often a binary success/failure variable, though it need not be. Second, all possible influences on the output are identified. These are the independent variables. Third, data are collected. Fourth, the structure of the neural network is specified. This entails identifying the inputs and outputs, arraying neurons to form interconnected layers,[14]

[14]Layers are either connected to inputs (input layer), other layers (intermediate or hidden layers), or to outputs (output layer). No layer can be connected to another hidden layer and either the input or output layer at the same time. The input layer is a distributor of signals from the outside world. Hidden layers detect features of such signals. The output layer collects the features and produces the response to them.

and constructing processing elements where the inputs from previous layers are weighted. The structure of the network may be determined by the user or by the software. Fifth, the neural network is trained. The error, or difference between the desired response and the actual response, is calculated and used to adjust the input weights and threshold values of the neuron so the error in the network response will be smaller the next time. Finally, the trained network is run, to make the forecasts or estimates for which the system was designed. Most neural network software has been developed for the spreadsheet environment.

@BRAIN Neural Network Development System (NNDS)[15] is a DOS-based Lotus 1-2-3 add-in that creates, manages, and runs neural networks from spreadsheet cells. With 1-2-3 commands, the user can preprocess neural network inputs, postprocess neural network outputs, manage the training and testing of data, and graphically evaluate the performance of networks. @BRAIN uses the classic feedforward neural network.

Neuralyst[16] is a Windows-based software package developed for use with the Excel spreadsheet. Like @BRAIN, Neuralyst applies the technology of neural networks to facilitate emulation, analysis, prediction, and association in a spreadsheet environment. Suppose we had some data about people who did and did not evacuate their homes during a recent flood and we want to use this information to forecast the behavior of other people. Table 2 presents input and output from such a hypothetical neural network run on Neuralyst.

Column 1 identifies people in a flood plain. Each person's age is given in column 2. Males are coded 1, females 0. Whites are coded 1, nonwhites 0. Column 5 indicates whether the person heard the flood forecast and warning given for a flood, with a 1 for those who did and 0 for those who didn't. In column 6, those who evacuated are coded 1, those who didn't are coded 0. The network was trained using the data for persons A through P and was used to forecast the behavior of individuals X1, X2, and X3.

The model output is contained in the last column. Using the network weights and the input data, the model has estimated values for the data set. Values with outputs greater than 0.5 would be forecast as

---

[15]@BRAIN is a product of Talon Development Corporation, P.O. Box 11069, Milwaukee, WI 53211-9931.

[16]Neuralyst is a product of Epic Systems, P.O. Box 277, Sierra Madre, CA 91025-0277, (818) 355-2988.

Person	Age	Sex	Race	Heard	Left	Output
A	34	1	0	1	1	0.849
B	64	0	1	1	1	0.9952
C	41	1	1	0	1	0.9988
D	50	1	1	1	1	0.9998
E	31	0	0	1	0	0.0079
F	41	0	1	0	1	0.9979
G	29	1	1	0	0	0.0056
H	57	0	1	0	0	0.0251
I	29	1	1	1	0	0.064
J	31	1	0	1	0	0.1086
K	71	0	0	1	0	0.0005
L	46	0	1	0	1	0.98
M	33	1	1	1	1	0.9063
N	39	0	1	1	1	0.9997
O	21	1	0	0	0	0.0002
P	48	1	1	1	1	0.9998
X1	29	1	1	1	0	0.064
X2	61	0	1	1	1	0.9986
X3	41	1	1	0	1	0.9988

Table 2. Hypothetical Neural Network

"evacuated"; those below 0.5 would be forecast as "did not evacuate." The model forecast each incident accurately. Neural networks add a convenient, alternative nonlinear estimation technique to nonlinear regression and other estimation techniques.

Presentation graphics software has little to do with uncertainty analysis. However, it is worth noting that the ability to do what-if analysis has begun to find its way into presentation graphics packages. Stanford Graphics[17] has added an Intelligent Data Cursor (IDC) that provides a visual link between the curve on screen and the spreadsheet. The cursor allows the user to change data without reentering the spreadsheet. By clicking on any data point in the curve, it can be dragged to a new position. The data for that new point will be entered into the spreadsheet. The IDC provides analysts with a method for visually determining the data values needed to produce a curve with the desired properties.

[17]Stanford Graphics is a product of 3-D Visions, 2780 Skypark Drive, Torrance, CA 90505, (800) 729-4723.

Summary

Tremendous strides have been made in the development of a diversity of commercial software useful for risk and uncertainty analyses. In the two years since the last Engineering Foundation conference in this series, numerous new products have appeared on the market. The advances have come in the form of the introduction of new capabilities and increased competition among existing capabilities. Both bode well for risk analysts. Given the rapid advances in the recent past, it is not difficult to imagine that in the not-too-distant future, smart objects and linkages, the frustrations of analysts in 1993, will become available, clearing the way for our next generation of imaginative frustrations.

References

Clemen, R. T. 1991. *Making hard decisions: An introduction to decision analysis*. Boston: PWS-Kent Publishing Company.

McNamee, P., and J. Celona. 1990. *Decision analysis with Supertree*. 2d ed. San Francisco: The Scientific Press.

Uncertainty versus Computer Response Time

William D. Rowe[1]

Abstract

Interactive on-line presentation of risk analysis results with immediate "what if" capability is now possible with available micro-computer technology. This can provide an effective means of presenting the risk results, the decision possibilities, and the underlying assumptions to decision makers, stakeholders, and the public. However, the limitation of computer calculational power on microcomputers requires a trade-off between the precision of the analysis and the computing and display response time. Fortunately, the uncertainties in the risk analysis are usually so large that extreme precision is often unwarranted. Therefore, risk analyses used for this purpose must include trade-offs between precision and processing time, and uncertainties introduced must be put into perspective.

Background

There are a variety of uncertainties in carrying out risk analyses, including measurement, model, and future event uncertainties. In the cases where these uncertainties have been addressed adequately, and means found to arrive at reasonable decisions among alternatives, the problem of presenting the decision and the underlying assumptions to decision makers, stakeholders, and the public remains. Miscommunication, overly complex presentations, and condescending presenters are just some factors that contribute to the problem. When the analysis depends on a great deal of subjective judgments, particularly the judgment of experts, even the basic assumptions can be challenged by decision makers and the public.

---

[1]President, Rowe Research & Engineering Associates, and Chief Scientist, Automation Research Systems, Ltd., 309 North Alfred Street, Alexandria, VA 22314

The successful presentation of results of competent risk analyses requires a display of both the uncertainties involved and the underlying assumptions used. This becomes even more critical when subjective elements and the judgment of experts are used, when the analysis will be used in an adversarial process, or when the credibility of the sponsors or analysts is questionable. Both the results of the analysis and the decision made with the results must be presented effectively.

An approach has been developed, using range/confidence estimates, that makes the uncertainties, margins of safety, and underlying assumptions visible in a manner that can be easily presented. Computer technology has made it possible both to exercise risk models rapidly and to present the results using multimedia techniques. The approach developed here seeks to extend this process by simultaneously calculating risks using the model and presenting the results to stakeholders, decision makers, and the public. An important part of the approach is to allow the assumptions used in the analysis to be challenged, alternative assumptions to be entered, and results to be made available almost immediately and compared with previous cases. An on-line "what if" analysis is the objective.

In attempting to carry out this objective, several trade-offs are necessary. Rapid processing and high data precision are incompatible for extensive analyses, even using the latest technology in hardware and software. This is particularly true when portability and ease of demonstration are important. One way out of this problem is to reduce selectively the precision of results such that the loss of information does not significantly affect results and any error introduced will be very small compared to the other uncertainties and error.

An example of a range/confidence risk analysis follows. After the example is shown, the critical delay process is examined, and a work-around is addressed along with a discussion of the impact on results. First, the basis for range/confidence risk analysis is described; then an example, using arsenic in well water, is presented.

Range/Confidence Risk Estimates

Standards for acceptable risk are usually set in terms of a probability of a consequence of given magnitude in a given time period. For example, the U.S. Environmental Protection Agency established a probability of a one-in-a-hundred-thousand chance of additional cancers over a seventy-year lifetime for acceptable operation of hazardous waste incinerators. While there are major problems in establishing such risk-acceptance standards, the focus here is on determining if a comparable "risk performance level" has been achieved in the design of a system or subsystem. Since the "risk performance level" cannot be measured directly, how does one know if the standard has been achieved?

Typically, a level of probability, expressing the performance level and calculated from models, is compared to the acceptance standard to determine if the performance level for the risk-acceptance standard has been met. This calculated level of probability, termed an "achievement standard," often has very large uncertainties in its calculation, arising from a variety of sources. These uncertainties range from choices among competing models to lack of data from measurements. Unfortunately, the results of calculations of the models used to estimate the achievement level are reported as point estimates. Point estimates hide the margins of safety and contingencies used in making estimates about uncertain conditions.

When dealing with uncertainty in measurements, in understanding complex system relationships, and in predicting future events, contingencies are often added to account for our uncertainty. Seldom are these contingencies made explicit, and often they are hidden. As estimates are combined in models used for achievement standards, the contingencies multiply, leading to large, undue, hidden conservatism in our estimates. When these estimates are used in estimating risks in regulatory decisions or cost to achieve the risk levels, management is preempted from understanding the consequences of overconservatism, its quantification, and its costs. Essentially the decision is made at the technical level, not at the management level.

The use of range estimates in place of point estimates represents an important conceptual breakthrough for making visible the extent of uncertainty in risk estimates. Range estimates are not new, but the particular type of range/confidence estimates to be described in this paper is different. It may appear to be a very simple extension of what has been used, but it affords significant progress in addressing and working around uncertainty.

There are many ways to make range estimates. However, the method of making range estimates used here has been shown to be highly useful. Three estimates of a parameter (e.g., probability of an event with a given consequence magnitude) are made using alternative models or alternative levels of parameters within a model.

A "bare estimate" is made with *no margins of safety*. This is, by definition, a best estimate of what is likely to occur, but the actual risk that is being estimated may be either higher or lower. This type of estimate is very difficult to make with engineers who are trained to build in margins of safety. However, it is necessary to force this type of analysis if the margins of safety are to be made subsequently visible and used as a tool for addressing uncertainty.

A "conservative estimate" is made using margins of safety high enough to provide confidence (a reasoned subjective judgment) that actual risk lies below the estimate. This type of estimate is consistent with conservative engineering design. If the underlying process were a normal distribution,[2] then an illustrative example would be the ninetieth to ninety-fifth percentile.

A "worst-case estimate" attempts to make sure that the actual risk is always below the model estimate. The estimate takes into account all reasonable, imaginable excursions. In a normal distribution it would exceed the ninety-ninth percentile. It is also possible to use more than three estimates when warranted.

The three (or more) estimates make up a range estimate. Any one of the three estimates cannot be taken out of context with the other two. To do so invalidates the framework. The distance between the bare- and the worst-case estimate of a parameter is a measure of its uncertainty, and an uncertainty index may be derived by dividing the result of the worst-case estimate by that of the bare case:

Uncertainty Index (UIX) = Worst-Case Result/Bare-Case Result

The magnitude of the index is used to make comparisons about the relative uncertainty of range estimates. A higher index implies higher uncertainty in the range estimate.

An important aspect of these estimates is the recognition that the explicitly expressed confidence levels are subjective. They represent the subjective evaluation of the estimator and are subject to challenge, requiring explicit documentation of the assumptions used to make the estimates. The subjectivity employed has always been there, but was invisible to the decision maker. This process makes it visible and manageable.

Challenges to particular subjective estimates are best addressed by sensitivity analysis when both the risk estimates and confidence-level

---

[2]This seldom occurs, and we may not be able to discern such a distribution if it did occur.

estimates are combined into the achievement standard range estimate. Experience has shown that increasing range values at high levels of confidence is relatively insensitive to the decision.

Combining Range/Confidence Estimates in Models

All parameters in models are estimated using ranges and are combined according to model requirements. The new combined range estimate for the achievement standard is presented as a plot of increasing risk on the abscissa versus the expressed combined confidence level on the ordinate for all combinations of confidence levels. The resulting curve usually has a very sharp knee, indicating very high levels of overconservatism, and provides a totally new perspective on deciding on achievement levels and acceptable confidence levels. With multiple-parameter achievement models, the achievement-level calculations are multimodal, leading to cases where reducing the risk level can lead to higher rather than lower confidence.

The advantageous use of range/confidence estimates becomes evident when multiple parameters are combined in a risk assessment model. Since there are three estimates for each parameter and all combinations of estimates must be made, there are many more computations to be made. For example, if there are two parameters, nine estimates of risk are required. In general, the number of risk calculations to be made, N, is determined by

$$N = 3^n ,$$

where n is the number of variables.

Since all the calculations are arithmetic, they are easily automated. Spreadsheets work very well, and a number of templates have been designed for this purpose.

Example—Risk Analysis of Arsenic in Well Water

An example of a multiplicative model is the determination of the risk of ingestion of a chemical considered as a carcinogen in well water.

Arsenic is found in well water in varying concentrations with a database of twenty samples from each of forty wells. A maximally exposed individual uses one well for drinking, and the pathway is direct ingestion of well water. For the example, the problem is constrained to three factors:

1. <u>Arsenic concentration</u>—Uncertainty in measurement
2. <u>Arsenic potency-slope model</u>—Uncertainty in both the model and measurement
3. <u>Exposure through ingestion</u>—Uncertainty in future events

Table 1 details the results of range estimates for each parameter. EPA Regulatory Assessment Guide values are indicated in the table by "(RAG)."

CASE	ASSUMPTION	LEVEL	CONF.
FACTOR #1	**Arsenic Concentration [micrograms/liter]**		
Bare Case	Mean of all samples (RAG)	0.3 ug/l	50%
Conservative Case	90th percentile of all samples (RAG)	1.2 ug/l	90%
Worst Case	Highest sample value	3.2 ug/l	99%
FACTOR #2	**Arsenic Potency-Slope [1/(mg/kg/day)]**		
Bare Case	50% confidence slope	0.015	50%
Conservative Case	95% confidence slope	15	95%
Worst Case	98% confidence slope	30	98%
FACTOR #3	**Exposure Scenario**		
Bare Case	The wells are sampled monthly	30 days	50%
	Sampling cost = $100/sampling	*$1,200/yr.*	
	Ingestion (RAG)	2 liters/day	
	Weight (RAG)	70 kg	
Conservative Case	The wells are sampled yearly	365 days	95%
	Sampling cost = $100/sampling	*$100/yr.*	
	Ingestion (RAG)	2 liters/day	
	Weight (RAG)	70 kg	
Worst Case	The wells are not sampled	25,550 days	99.99%
	Sampling cost = none	*$0*	
	Ingestion	3 liters/day	
	Weight	50 kg	

Table 1. Results of Range Estimates for Each Parameter

The lifetime cancer risk model is shown in equation (1). In addition, if this risk exceeds one in ten thousand over a seventy-year lifetime, then the wells must be pumped at a cost of $4,800 per year.

Lifetime Cancer Risk Increase = LC

$$LC = I \times C \times D \times P/(W \times 1000 \times 25550)\,, \tag{1}$$

where

I = Ingestion in l/day	Measurement Uncertainty
C = Concentration in ug/l	Measurement Uncertainty
D = Days of exposure	Measurement Uncertainty
P = Potency-slope value in 1/(mg/kg/day)	Model Uncertainty Measurement Uncertainty
W = Weight in kg	Measurement Uncertainty
1,000 = 1,000 ug/kg	Conversion Factor
25,550 = No. of days in 70 yrs.	Conversion Factor

There are five parameters in the multiplicative model, of which two, I and W, represent distributions of the general population. Range/confidence estimates are made for each parameter leading to $(3)^5 = 243$ calculations or vertices. One means of simplification here is to use only three parameters, as shown in table 1, where only selected values of I and W are used. This is one aspect of the simplification, requiring only 27 calculations rather than 243. The combined confidence levels for the three-variable case is shown in equation (2).

$$C = 100 / pqr \int_X^p \int_Y^q \int_Z^r ((X + a)(Y + b)(Z + c) - XYZ)\,da\,db\,dc \tag{2}$$

where, for a uniform confidence distribution as a worst case,

$$p = q = r = 1\,,$$

$$X = pC_1,\ Y = qC_2,\ Z = rC_3\,,$$

and C_1, C_2, and C_3 are the confidence levels for the three parameters. The error functions for each parameter are a, b, and c, respectively, and may be positive or negative in value.

This requires a fivefold integration in the case of all five parameters. The integration by Simpson's rule takes anywhere from several minutes to several hours for each calculation, depending on the number of variables, on a 66 MHz 486-chip personal computer. This then becomes a crippling limit on the speed of update for presentation.

To find a more rapid means of calculation, it has been shown by repeated calculation that the confidence levels calculated for uniform and unimodel distributions will never exceed

$$C = 100[1 - 2^n \prod_{1}^{n} (1 - C_n/100)], \tag{3}$$

where n is the number of parameters.

This equation is always conservative in understating the confidence level for every case except when the confidence levels for each parameter are at the 50% level. The confidence level in this case becomes degenerate, and the actual result of the integration is entered for this one case as a constant.

With this approximation, the results are easily calculated on the fly in a simple spreadsheet calculation or similar programs. Table 2 shows the first nine entries of a Microsoft EXCEL 4.0 program model for

20 wells with 40 samples from each		**Bare Case** Average value of all wells	**Cons. Case** 90th percentile of all wells (RAGs)	**Worst Case** Maximum sample value
Potency-Slope /(mg/kg/day)				
30	Level ug/l	0.3	1.2	3.2
98%	Conf Lev.	*50%*	*90%*	*99%*
Bare Case		**3.02E-07**	**1.21E-06**	**3.22E-06**
Weight (kg)	70			
Ingestion (liters/day)	2			
Exposure (days)	30			
Confidence Level	50%	**96**	**99.2**	**99.92**
Conservative Case		**3.67E-06**	**1.47E-05**	**3.92E-05**
Weight (kg)	70			
Ingestion (liters/day)	2			
Exposure (days)	365			
Confidence Level	95%	**99.6**	**99.92**	**99.992**
Worst Case		**5.40E-04**	**2.16E-03**	**5.76E-03**
Weight (kg)	50			
Ingestion (liters/day)	3			
Exposure (days)	25,550			
Confidence Level	99.99%	**99.9992**	**99.99984**	**99.999984**

Table 2. Risk and Confidence Levels for Arsenic in Well Water—Worst Case for Potency

making both the risk and confidence level calculations. There are a total of three such tables for all twenty-seven variables in the example above, but only one is shown. Table 3 shows the results of the calculations in a single table, ordered by the level of risk found. The smallest risk level is at the top and the highest risk level at the bottom of the table. Examination of the results shows a risk range of over seven orders of magnitude of uncertainty, from 1.51E-10 to 5.76E-3. Confidence levels range from 50% to 99.999992%. There are only two values below 95%, showing that the simplification has not affected the confidence calculation in any significant way.

Lifetime Risk	Confidence	Cost ($)	Case
1.51E-10	62.5	1,200	BBB
6.04E-10	80	1,200	BCB
1.61E-09	98	1,200	BWB
1.84E-09	90	100	BBC
7.35E-09	98	100	BCC
1.96E-08	99.8	100	BWC
1.51E-07	90	1,200	CBB
2.70E -07	99.98	0	BBW
3.02E-07	96	1,200	WBB
6.04E-07	98	1,200	CCB
1.08E-06	99.996	0	BCW
1.21E-06	99.2	1,200	WCB
1.61E-06	99.8	1,200	CWB
1.84E-06	99	100	CBC
2.88E-06	99.9996	0	BWW
3.22E-06	99.92	1,200	WWB
3.67E-06	99.6	100	WBC
7.35E-06	99.8	100	CCC
1.47E-05	99.92	100	WCC
1.96E-05	99.98	100	CWC
3.92E-05	99.992	100	WWC
2.70E-04	99.998	4,800	CBW
5.40E-04	99.9992	4,800	WBW
1.08E-03	99.9996	4,800	CCW
2.16E-03	99.99984	4,800	WCW
2.88E-03	99.99996	4,800	CWW
5.76E-03	99.999984	4,800	WWW

Table 3. Results of the Calculations

Figure 1 is a direct EXCEL plot of the confidence levels on the ordinate versus the risk levels on the abscissa. It is evident that one reaches very high confidence at very low levels of risk, directly showing the large margins of safety built into the single value estimates. The combined confidence estimates take into account the simultaneous conditions of the parameters where coincident conditions at the high confidence levels are extremely rare. The wiggles in the curve, where one can

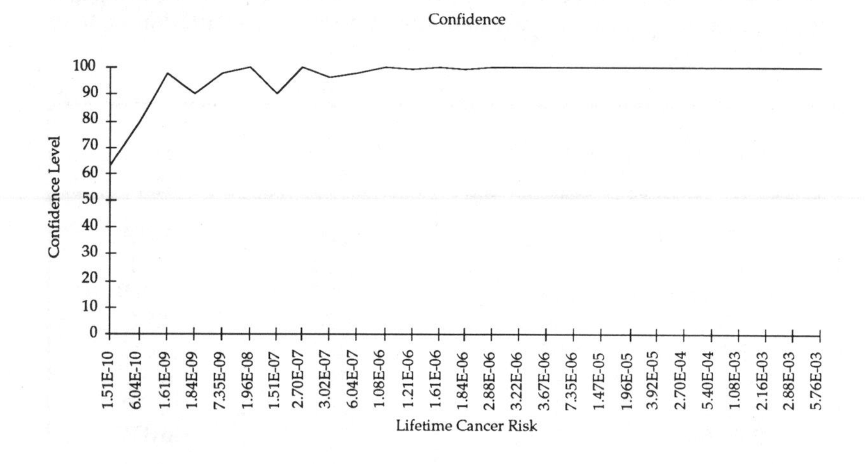

Figure 1. Plot of the Joint Confidence and Risk Levels

go to a lower risk estimate with higher confidence, are an artifact of the ordering of the risk levels and the different sensitivities of the three parameters. On a risk basis alone, one can assure very low risk levels with rather high confidence. However, figure 2 additionally shows a plot of cost on the left-hand ordinate versus the same scale of risk as in figure 1 on the abscissa, overlaid on the confidence scale now shown on the right-hand side. The cost data is quite lumpy.

However, the superposition of the two graphs, as shown in the figure, provides an evident means to select a low risk level with very high confidence while assuring a very low cost. For example, a risk level of 1.08E-06 lifetime at a confidence level of 99.998% leads to a zero-cost solution.

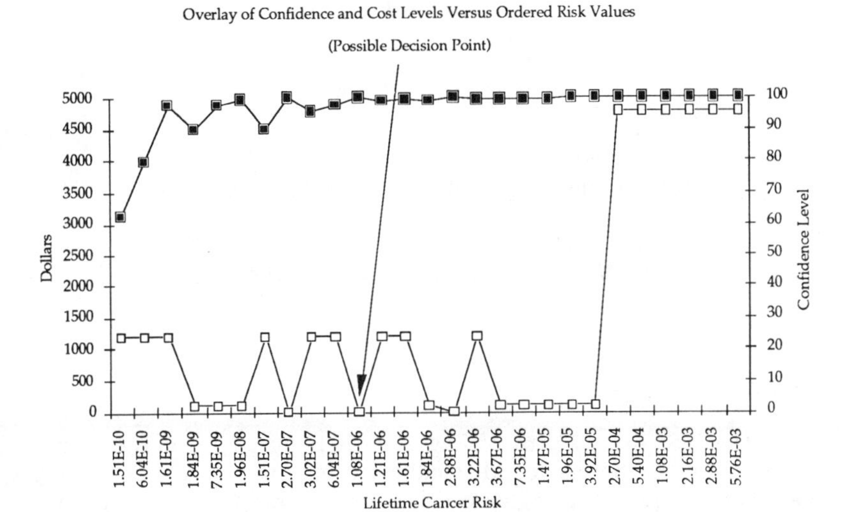

Figure 2. Cost and Confidence Levels versus Risk Level

Since all the assumptions and levels are stated, it is easy to change these and immediately see the results. It can then be determined if such changes significantly affect the results or have any impact on the decisions that might be made. This provides a very powerful means of presenting risk data as it affects the results. The ability to make these changes immediately visible on a "what if" basis now becomes a meaningful and decision-rich environment.

Summary

Approximations, made to display the results of risk analyses using range/confidence estimates rapidly, can be established without affecting the results significantly. The wide margins of safety and subjective evaluations have much more impact than the small adjustments made for rapid processing. The ability to communicate with decision makers, stakeholders, and the public is significantly enhanced.

Risk Simulation Model for Rehabilitation Studies

Richard Males,[1] David Moser,[2] Michael Walsh,[3]
Walter Grayman,[4] and Craig Strus[5]

Abstract

The U.S. Army Corps of Engineers is required to analyze major rehabilitation proposals for all facilities using risk-based benefit-cost analysis techniques. Existing spreadsheet-based tools are slow, inflexible, and highly problem-specific. A generalized C++ program using object-oriented techniques has been developed and applied to the economic risk analysis problem. The program provides significant advantages in the analysis effort, including greatly increased speed, greater ease of use, and more powerful modeling capabilities. Further enhancements to the model are being developed.

Introduction

The Corps of Engineers operates and maintains some 300 locks and dams, 350 flood control structures, and 75 hydropower projects across the nation, representing billions of dollars of capital investment. Most of these projects are greater than thirty years old. As these projects age, repair and rehabilitation are required. Major rehabilitation proposals must use risk-based economic analysis approaches to identify the best al-

[1]President; RMM Technical Services, Inc.; 3319 Eastside Avenue; Cincinnati, OH 45208
[2]Economist, U.S. Army Corps of Engineers, Institute for Water Resources, Ft. Belvoir, VA 22060
[3]Civil Engineer, U.S. Army Corps of Engineers, Institute for Water Resources, Ft. Belvoir, VA 22060
[4]Principal, Walter M. Grayman Consulting Engineer, 730 Avon Fields Lane, Cincinnati, OH 45229
[5]Senior Programmer/Analyst; Planning and Management Consultants, Ltd.; Route 9, Box 15; Highway 51S; Carbondale, IL 62901

ternative for the rehabilitation of a given project. Rehabilitation efforts are designed to improve the reliability and/or efficiency of existing Corps water resource projects. Reliability improvements reduce the risk that a project will perform unsatisfactorily. Decisions about rehabilitation proposals must rely on risk-based benefit-cost analysis, because it is impossible to predict with certainty when a structure or piece of equipment will fail to perform satisfactorily.

Economic Analysis Model

The economic analysis approach uses a life cycle cost analysis. A base case (repair when necessary) and alternatives (advance rehabilitation, then repair when necessary) are specified. The life cycle costs for the base and alternatives are calculated. Four types of costs are identified: investment costs (cost of the rehabilitation that is done at the beginning of the period to improve reliability and reduce the likelihood of unacceptable performance in the future); maintenance costs (considered fixed for the facility); repair costs (dependent on the failure modes and frequencies); and opportunity costs (value of the lost output).

Each project or "feature" (e.g., a facility for which the analysis is to be performed, such as a hydropower generating plant) is viewed as being composed of subfeatures (e.g., generating units), which are themselves composed of components (e.g., turbines and generators) that can degrade and perform unsatisfactorily in a variety of modes. Estimates of the probability of unsatisfactory performance (PUP) for a given component are made, as well as an estimate of the change in the PUP as the component degrades over time. At the same time, a cost of repair and a revised value of the PUP after repair are developed for each component.

Monte Carlo simulation techniques are used to calculate life cycle costs for the facility. At each time period, a random number between 0 and 1 is generated for each component. The random number is checked against the PUP that has been determined for each component. If the random number is greater than the PUP, then the component is assumed to perform unsatisfactorily for that time period. Repair costs and other opportunity costs during the unplanned outage are calculated, the component is taken out of service for a set period of time for repair, and the PUP is reset to a new value based on the repair. If the random number is less than or equal to the PUP, then the component is assumed to perform satisfactorily and the PUP is degraded some amount to reflect some deterioration over the time period. All costs are discounted and a present value of costs is calculated for each iteration of the simulation. An iteration is typically one fifty-year cycle. Usually, several thousand iterations are required to develop a distribution and expected value of risk costs for the alternative. Similar analyses are developed for the base case and for advance rehabilitation options. Alternatives are judged based on their benefits and costs. The benefits are calculated as the reduction of costs below the baseline. The costs are the capital investment, the opportunity costs due to lost service during the rehabilitation, and the other expenditures required to implement the alternative.

Model Implementation

The economic analysis model described above was initially implemented through spreadsheet modeling. The models that have been developed to perform this analysis usually require at least eight hours to complete a 5,000-iteration simulation for a moderately sized hydropower project, using an IBM-compatible 486/33 computer. In addition to being slow, they are inflexible and not easily understood. The spreadsheet models are specific to a facility and set of components, rather than being generalized. Of necessity, the spreadsheet models are simplified representations.

The spreadsheet models are too slow to analyze many alternatives for rehabilitation at a project. Significant speed increases are needed to allow for interactive analysis by Corps personnel. While increases in speed were the primary motivation for developing an alternative modeling approach, a general model, with greater flexibility, ease of use, more sophisticated user-interface features, and more complex representations, was also desired.

A technology search was carried out to determine previous work in the area. While some commercial products have been used for risk-based simulation (primarily as spreadsheet add-ins), none were deemed suitable due to the complexities of the current problem, in particular the changes in PUP over time for a component. Accordingly, the decision was made to develop a computer program using a programming language. The strongly physical nature of the problem suggested that object-oriented modeling approaches might be valuable, and the C++ language was selected for implementation. An iterative, rapid-prototyping methodology, in which a series of working programs is generated, each with increasing complexity, was adopted.

The developed program, referred to as REPAIR (Risk-based Economic Program for the Analysis of Investments for Rehabilitation), is implemented as a set of C++ programs for IBM-compatible computers, including a user interface, the simulation program itself, and a graphics and statistics postprocessor. These programs communicate through standard-format data files (ASCII files and dBase-format database files). This modular approach allowed for team programming and enhancement and replacement of individual modules as needed. Thus, although the user interface is currently implemented in a DOS character-based environment, it could readily be replaced by a Windows-oriented front-end, without requiring any changes to the simulation model itself. Similarly, enhancements to the postprocessor modules of graphics and statistics can be carried out independently.

REPAIR has been tested on problems that were formerly run through the spreadsheet analysis and provides comparable results, with a hundredfold increase in solution speed—problems that required eight hours under the spreadsheet technique are simulated in five minutes. As contrasted with the spreadsheet model, which requires recoding to change the number of components or subfeatures, the existing program is entirely data driven.

Object-Oriented Formulation

The prototype makes use of object-oriented programming (OOP) approaches. Under this approach, the program is viewed as a set of interacting objects. Each object has a certain set of capabilities and responds to requests with certain behaviors. An object can be thought of as something that "knows some things and knows how to do some things." By defining what the object knows, and what it can do, and then by combining these objects in desired ways, the overall endpoint is achieved. For the case of a Monte Carlo simulation, where real-world objects are being modeled, the approach is particularly valuable.

In OOP, objects are developed as instances of a class; that is, the knowledge and behaviors are defined for a generic group (the class). Different objects belonging to the same group are limited to the same set of knowledge and behavior (as defined by the class), but each object may be in a different state from other objects of the class at any given time. A class defines a general category of behavior. An "object" is a specific instance of the class—it behaves as the class behaves, but with its own data.

In the current example, the hierarchical framework of the spreadsheet model for representing a facility was retained and generalized. The parts of the facility are grouped in an inverted tree structure with a single feature at the highest level. The feature comprises subfeatures (currently up to ten), which themselves comprise components (up to ten per subfeature). A hydropower facility with five operating units would have five subfeature objects, each a member of the subfeature class. Each subfeature object might be represented as containing two components, a turbine and a generator. Thus, there would be ten component objects, each a member of the component class, but with different data.

A class is defined for each entity. A simplified description of the behavior of each class is as follows:

Feature Class—contains subfeatures, incurs opportunity costs based on the state of the subfeatures, reports on its own state

Subfeature Class—contains components, determines its own state based on the state of components, reports on state

Component Class—degrades, fails, repairs, and incurs repair costs based on a repair policy, reports on state

As an example of class data, among the data items for the component class are initial PUP; PUP degradation rate; time to repair (number of cycles); repair cost; postrepair PUP; current value of PUP; status of component; number of cycles since component was last out of service; total number of outages; and total time out of service.

Additional classes are used in the model. A simulation class takes user input, runs iterations, summarizes statistics, and writes an output file. An iteration class takes a feature through an entire life cycle and accumulates costs. A random-number generator class generates random numbers as needed, and a database class is used to manage and supply parameter data for the subfeatures and components.

Alternative repair policies can be specified for a given simulation. Each component is "aware" of the policy in effect, and responds appropriately when repair is needed. Policies currently implemented are repair when broken; do not repair if broken; and repair all components in a subfeature when any component is broken. Other, more complex, policies can be envisioned and are readily incorporated into the approach.

Issues

The model provides a definite improvement over the existing spreadsheet approaches. The object-oriented framework is valuable in conceptualizing and understanding the problem and in directing further enhancements. In particular, more complex behaviors and rehabilitation policies can be defined and easily incorporated into the model by changing class behavior. However, each such change may require additional data, which may be difficult to obtain in the real world. While highly sophisticated models can be developed, the parameters for these models may be difficult to define using available data.

Also of concern for the continued development of the model is the issue of whether to continue development as a generalized model or to recognize that problems are sufficiently site-specific, thereby making development of site-specific programs cost-effective. At present, the hydropower problem seems to be modeled reasonably well by the existing program. Other types of problems (locks and dams, breakwaters) that appear to be more site-specific are being examined for commonalities in hopes of either further generalizing the model or creating other problem- or site-specific models, using the existing model as a starting point and framework.

The speed of the model and the great amount of detailed output that can be obtained dictate the need for additional postprocessing capability, particularly in the arenas of statistical analysis, graphics, and comparison of alternatives.

The existing object-oriented framework can likely be improved. At present, it is the responsibility of the component to determine the rehabilitation policy that is in effect, and "repair itself" accordingly. A preferred approach would be to have the component request action of a "fixer" object that would have global knowledge of all component states and would be able to respond to more complex policies. The fixer object would inform the component object of the action to be taken (repair, do not repair, etc.), based on more knowledge than is available to the component itself.

Current Status

As of December 1993, the Phase I implementation of REPAIR is complete, and a report and documentation of the phase are expected to be available shortly from the U.S. Army Corps of Engineers, Institute for Water Resources, Ft. Belvoir, Virginia. Enhancements to the model have been designed, and a Phase II implementation is due for completion in July 1994. Among the enhancements in the Phase II model are development of a Windows interface; incorporation of approaches to handle more complex simulation issues and policies; and development of additional postprocessors to handle comparison of alternatives.

Environment-Irrigation Trade-Offs and Risks

Norman J. Dudley[1]

Abstract

Allocation of scarce water resources between environmental and commercial uses is an extremely important problem in many parts of the world. A study is under way in Australia to examine a water allocation and property rights system in which both wetlands managers and irrigators participate in a market for reservoir water in a highly variable climatic environment. This system allows both commercial and environmental water users to make their own trade-offs between maximizing the expected outcomes and minimizing the risks they face, while exposing their decision processes to the changing value water has to other users over time. In turn, the regional trade-offs between long-term allocations of water resources to environmental versus commercial uses can be quantified and minimized. This information is vital if policy-makers are to make informed decisions about the appropriate levels of environmental amenities to maintain.

Introduction

Figure 1 is a schematic approximation of the Gwydir River system. It shows the wetlands on the main river, and the recently developed irrigation areas to the north and south of the wetlands. The irrigation areas are located on delta streams of the main river. Before regulation, those rivers only flowed in flood times when the main river was sufficiently high to overflow into them. But since regulation, weirs have been used to divert most of the flow from the main river into the northern and southern rivers to supply the irrigation areas. Consequently, the size of the wetlands area has decreased, resulting in moves to reallocate

[1]University Fellow, Centre for Water Policy Research, University of New England, Armidale, NSW, Australia 2351

water to the wetlands. How best to share the available water supplies between the main uses of irrigation and wetlands maintenance is a major concern.

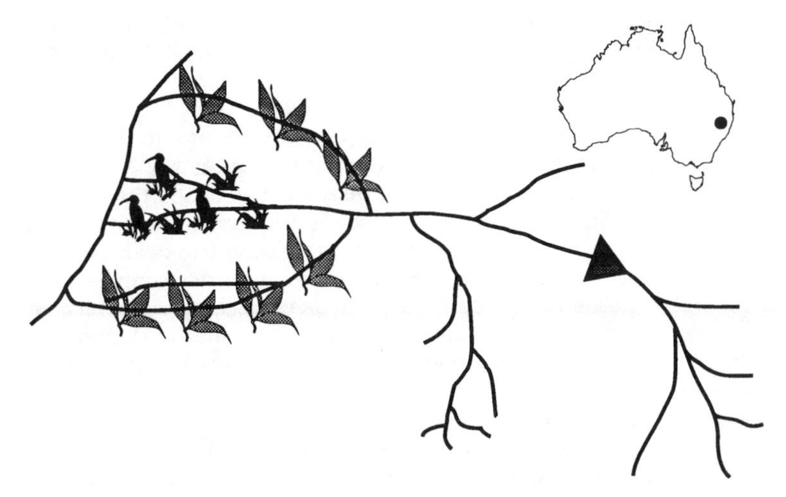

Figure 1. Schematic Approximation of the Gwydir River System

The author is involved in a project to quantify and minimize the trade-offs involved in sharing the scarce water between irrigation and wetlands. The water resources of the river system are shared by capacity sharing in the project.

Capacity Sharing

The concept of capacity sharing was described in detail at the last of these conferences (Dudley 1992a). It was noted there that "priority" capacity sharing may be preferred in some cases to the more commonly discussed "percentage" capacity sharing. Only the latter is considered in this paper, with the concept being reviewed briefly in the following.

Capacity sharing is an institutional arrangement or property rights structure which provides each user, or small group of users, of reservoir water with long-term or perpetual rights to a percentage of reservoir inflows and a percentage of empty reservoir capacity or space in which to store those inflows, and from which to control releases. It is as if each user (individual or group) owns their own small reservoir on their own small stream. The shareholders' reservoir releases through time can be managed according to their particular supply reliability preferences, with minimum interference from other reservoir users. User rights to percentage shares of tributary flows downstream of the reservoir can also be included. Under capacity sharing, each individual decision-making entity makes their own supply-side and demand-side decisions. The probabilities of (nontransferred) water supply can be calculated directly from historical or synthesized streamflow data alone (Dudley 1990; Dudley 1992a).

Whereas Dudley (1992a) described capacity sharing in detail with the aid of diagrams, the basic concept is introduced here by analogy. Consider the analogy of the reservoir to a money savings bank in which each account has a maximum to which the account can accrue (which is a percentage of the maximum sum of deposits which the bank can contain); negative balances are ruled out; and the account holder controls withdrawals but deposits are made by a stochastic process over which the account holder has no control, although the probabilities of deposits are known. Each depositor seeks a preferred balance between running out of funds periodically, due to keeping average account funds low; and forgoing potential deposits periodically, due to high average account contents. Each makes the choice and each reaps the consequences. Funds can be transferred from one account to another by a joint request by the affected account holders. Also, two account holders may transfer, from one to the other, all or a portion of one's rights to deposits and/or account capacity. Monitoring withdrawals, transfers, interest, and deposits with the aid of computers keeps track of the contents of each account through time. Should one user's account have insufficient space to contain that user's share of inflows, the surplus is "spilled" equally into the remaining unfilled accounts until all are filled. These have become known as "internal spills." Once all accounts are full, the bank is unable to accept any further money.

Similarly, with a water reservoir on a stream, users' entitlements to reservoir inflows and internal spills would be added to users' reservoir "accounts" (unless full), withdrawals and share of evaporation and seepage losses subtracted, and adjustments made periodically to ensure that the sum of "deposits" so computed equals the actual reservoir contents (Dudley and Musgrave 1988).

The concept of capacity sharing overcomes limitations of the current water allocation methods in Australia, especially in the predominantly summer and uniform precipitation areas featuring very unreliable water supplies. In these areas an almost continuous task of water management is the reassessment of the desirable balance between allocating currently stored water to the immediate season versus saving some for future use in the form of reservoir "carryover." The more water saved for future use, the greater the reliability of water supply from the system, but at the cost of reduced average supplies because of increased reservoir spills and evaporation losses. This trade-off between mean quantity and reliability of supply is very important in most of Australia. The main problem with the traditional methods of sharing water supplies among users is that all but a few special users are forced to share equally in whatever trade-off point is chosen, regardless of their preferences (Dudley 1992a). It is extremely difficult to tailor supply reliability to individual user needs on a large scale, under traditional allocation methods, while preserving flexibility to take emerging demands into account. However, capacity sharing easily allows the individual users or groups to choose their own mean/reliability trade-off at a point in time, and to rechoose through time in light of changing conditions and preferences (Dudley 1990; Dudley 1992b). All types of downstream-of-reservoir users and user groups can be included: irrigators, domestic and industrial users, instream environmental and recreational users, as well as users of offstream and instream wetlands.

Alaouze (1991) and Musgrave, Alaouze, and Dudley (1989) have shown algebraically that capacity sharing is at least as good as the commonly used method of release sharing for achieving an efficiency objective in the predominantly winter rainfall area of southern Australia. Recent legislation in a southern state, Victoria, incorporates provision for its use (Dudley 1992b).

The capacity sharing concept includes water markets operating at two levels, (i) to transfer water already in storage, streams, or channels, and (ii) to transfer the long-term rights to parcels of shares in reservoir capacity and streamflows. Although these users can choose to manage their own shares of reservoir capacity and inflows independently from each other, except for internal flows, they can also interact in the short run via a market for reservoir water. Hence, users cannot lose water rights to governments or other users without market compensation which they are willing to accept. This provides two very important requirements for efficient and sustainable use of water and associated resources in the long term—*security* of tenure of supply rights of known reliabilities to users, and *flexibility* for water resources to move into alternate uses as conditions change.

In this way, the water management of all the users is integrated by their interaction in the water market. In the short run, each user's water-use choices are confronted with the current and expected future value of that water to other users and uses, as expressed in the current water price. Similar integration takes place in the long run by the users' joint participation in markets for long-term rights to inflows and reservoir capacity.

However, in the following discussion, the trade-offs between irrigation and wetlands objectives are first considered without water marketing, which is introduced later in this paper.

Aims of the Study

The aims of the study are to quantify and minimize the trade-offs between effectiveness of wetlands water supplies and the opportunity costs of those supplies, where

a) *effectiveness* is defined as the degree of approximation to natural (i.e., unregulated, historical) flows to the wetlands,
b) *opportunity costs* are irrigation industry benefits forgone, which are determined by the above models which optimize integrated farm and reservoir management decisions, and
c) *trade-offs* are minimized by optimizing the allocation of water between irrigation and wetlands through time under the great uncertainty resulting from widely fluctuating and highly unpredictable streamflows and precipitation/evaporation ratios throughout the whole year in the study area characterized by predominantly summer precipitation. The optimal allocation covers both the long-run sharing of rights to reservoir capacity, reservoir inflows, and downstream tributary flows, and the short-run trading of water in a market for reservoir water.

Minimization of trade-offs results from the integration of wetlands water supply and demand management at two levels. The first is the integration of water supply and demand in a nonmarket setting. Allocating rights to shares of reservoir capacity, reservoir inflows, and downstream tributary flows to wetlands managers gives them control over water supplies of their own. Their demand for water stems from the desire to maintain the wetlands' historical sharing of river flows. Failure to satisfy these targets attracts penalties (or negative benefits) in the project modeling, which vary according to the magnitude and timing of the failure. The optimal dynamic balance of water supplies and demands is achieved by a simulation/dynamic programming model that maximizes the expected value of benefits (minimizes the expected value of penalties plus opportunity costs) over eighty to ninety years of historical data. This model has a similar basic structure to those previously developed and used by the author for the integrated management of irrigation farms and their capacity share of water resources (Dudley and Scott 1993).

The second level of integration is that of wetlands and other water uses via a water market. It is achieved by allowing wetlands managers to buy and sell water in an open water market, and by making them responsible for maximizing expected benefits from water sales and wetlands management. Because the market water price represents its opportunity cost per unit, maximizing the revenue from water trading, plus the dollar benefits (negative penalties) from matching wetlands supplies to their natural flows, means that water is only diverted for wetlands use when its value there exceeds its opportunity cost. If the benefits attached to water supplies to wetlands are weighted low enough in model solutions, the wetlands managers will sell all their available water to other water users. If the benefits are weighted high enough, other water users supply all the water the wetlands need to match natural flows because of the high prices paid by wetlands management, so long as its budget is sufficient.

This second level of integration has only become possible very recently with the development of models that estimate the demand and supply functions of individual cotton irrigators for all likely conditions and at different times, together with a model which aggregates the individuals' supply and demand curves and simulates regional market prices and quantities traded.

Derivation of Trade-Offs without Water Marketing

Figure 2 illustrates the nature of the problem and the objectives of this study. The location, shape, and position of the points and curves are hypothetical but the general relationships have been validated by preliminary model results. Expected net benefits from irrigation are shown on the horizontal axis. The vertical axis shows expected values of the actual flows to the wetlands in relation to the mean monthly natural or unregulated flow to them. Distance OM up the vertical axis represents the case of the wetlands always receiving their historical flows. Similarly, distance ON along the horizontal axis represents the case of nonlimiting water to irrigation.

If plenty of water existed for both the wetlands and irrigation, the zero-trade-off point, at which expected benefits to both the wetlands and cotton are maximized, would be achievable. But shortages mean trade-offs, as shown by the negatively sloped trade-off curves AC and BC. The objectives of this study include developing models to quantify trade-off points along such curves and to aid in the choice of the most preferred trade-offs by water resource managers, within a capacity sharing institutional framework.

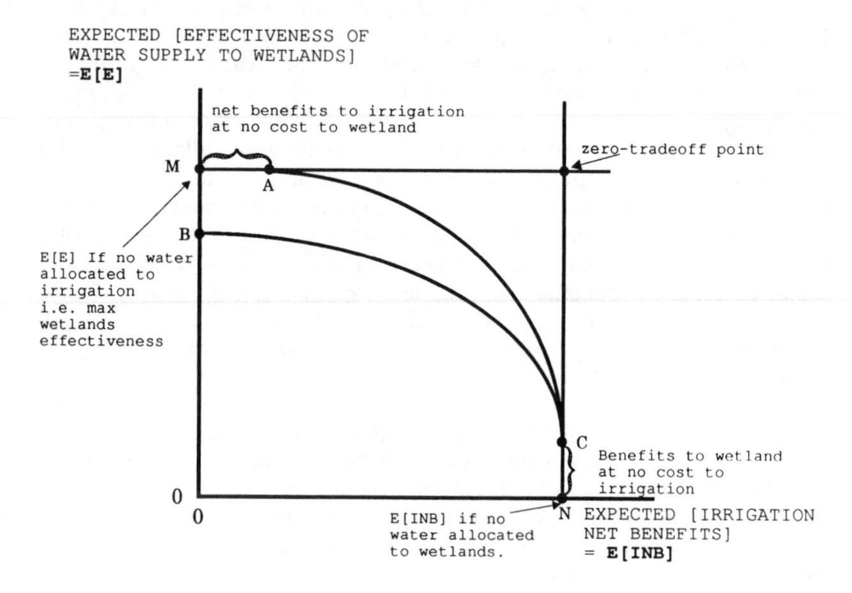

Figure 2. Illustration of Hypothetical Trade-Offs between Benefits to Wetlands and Irrigation

Point A in figure 2 sets a benchmark opportunity cost against which to judge later percentage allocations of reservoir capacity shares, reservoir inflow shares, and downstream tributary flows between wetlands management and irrigation. It is derived by allocating all tributary flows and reservoir inflows to wetlands on a top-priority basis, so that natural wetlands flows are provided before any of the river flows are allocated to other uses. This will provide a no-loss, or full-benefit, situation for the wetlands, and presumably the maximum opportunity cost or loss of benefits to cotton. Distance MA represents the expected return to irrigation from the natural overflows of the main river into the delta streams.

At the other extreme, point C is plotted on the assumption that some water is surplus to irrigation requirements in at least some years, giving an expected value of wetlands effectiveness of NC without reducing the expected value of irrigation net revenue below its maximum, ON.

Contrast the approach of this study with the granting of fixed annual volumes of water for wetlands use, which is the policy for some wetlands in Australia. It would appear that allocating fixed volumes, regardless of the seasonal supply conditions or the seasonal demands facing either wetlands or other major water users, is a very inefficient way to allocate water. It would appear to be in sharp contrast with working out how to share the water and manage reservoir releases to minimize the disbenefits and trade-offs between wetlands management and offstream diversions for irrigation. Curve BC in figure 2 depicts the results from increasing the fixed annual quantity allocated to the wetlands from zero at C to such a large quantity at B that irrigation becomes unprofitable. At point B the wetlands do not gain full historical flows, i.e., falls below M, because of the loss of water from supplying constant quantities, which often exceed the historical flows. That is, actual flows in excess of historical flows in any period do not increase effectiveness above 100 percent, so that regulating water supplies to maintain constant flows, which sometimes exceed the historical flows, has to result in a mean effectiveness of less than OM.

Maximizing effectiveness and revenue means: No water markets

Point A represents the case of 100 percent of reservoir capacity, reservoir inflows, and downstream tributary flows being allocated to the wetlands. In contrast, point C represents 100 percent of each being allocated to irrigators. Intermediate points on AC represent the maximum possible combinations of expected wetlands effectiveness and expected irrigation net revenues from allocating different combinations of reservoir capacity, reservoir inflows, and downstream tributary flows to wetlands and irrigators. For each combination considered, both the wetlands simulation/dynamic programming models and the irrigation simulation/dynamic programming models are solved to get the coordinates of the points. A curve would then be fitted through the "frontier" points closest to the zero-trade-off point. That is, some allocations of the water resources (i.e., reservoir capacity, inflows, and downstream tributary flows) result in points lying inside the frontier. However, some of these points may be preferred to frontier points when the sequence of flows to the wetlands is taken into account. That is, the frontier is in terms of means only, with no concern about risks of wetlands damage and farm financial viability due to undesirable sequences of flows to each.

Reducing risk: No water markets

Risk to irrigators can be taken into account by modifying the dynamic programming objective function. This would result in distance ON in figure 2 being reduced, and curve AC becoming steeper. However, whereas soil-water-plant growth simulation models can be used to derive quite accurate estimates of the effect of variable water supplies on crop yields and incomes, the effect of undesirable sequences of supplies to wetlands is much more subjective. The study will plot monthly flows to the wetlands for various resource combinations, giving points on and off the frontier in figure 2. These monthly plots will then be shown to ecosystems experts familiar with the Gwydir wetlands to have them choose the most desirable sequences of flows from alternatives that don't make irrigators worse off. That is, they may choose a point vertically below, or to the "southeast," of a given point on the frontier curve, rather than that frontier point. This would mean that the irrigators would be no worse off than at the frontier point, and the sequence of flows to the wetlands would be judged to be less risky than the sequence corresponding to the frontier point.

For example, figures 3 and 4 show two sets of simulated actual flows from preliminary model runs in comparison with natural flows. Figure 3 shows the case in which 100 percent of reservoir capacity and inflows are allocated to irrigation while 100 percent of downstream tributary flows are allocated to the wetlands. Because mean downstream tributary flows equal 43.9 percent of total river mean flows, this allocation meant that the wetlands received 43.9 percent of the total river flows on average. To preserve the total proportion of flows, figure 4 shows the case where wetlands have a 43.9 percent share in reservoir capacity, reservoir inflows, and downstream tributary flows.

The "actual" flows in figure 3 are the unregulated downstream tributary flows. For the flood in months 5, 6, and 7, the unregulated flows are less than half the natural flows in months 5 and 7, but over two-thirds of the natural flow in month 6. They are less than half in most of the other high-flow periods. In figure 4, the actual flows match natural flows better in low-flow events than in figure 3, but the match is usually worse in high-flow events. Month 7 is a notable exception.

The ecosystems experts comparing figures 3 and 4 may choose one in favor of the other, or they may wish to see the effects of rerunning the models with the same allocation of resources that resulted in figure 4 but with higher weights attached to the benefits (negative penalties) from matching high natural flows, at least up to some threshold level. The dynamic programming optimization model would then save more water during low flows to better match the high flows. Experimenting with different weightings may improve the results. However, there appears to be scope for both moving the frontier toward the zero-tradeoff point and reducing risks by water markets.

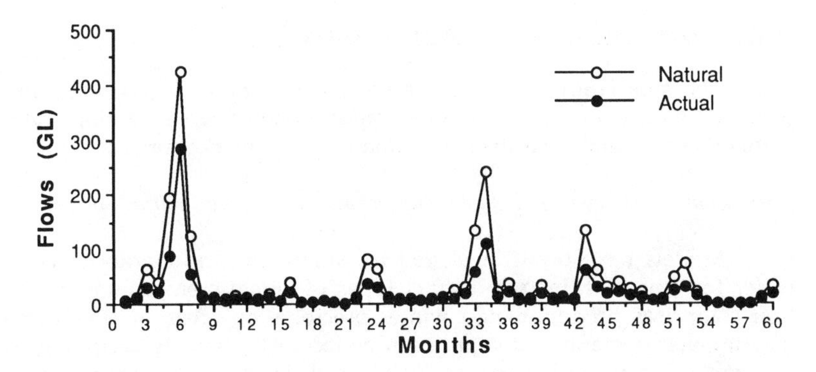

Figure 3. Natural and Actual Flows to Wetlands When Wetlands Have 100% of Unregulated Flows Only

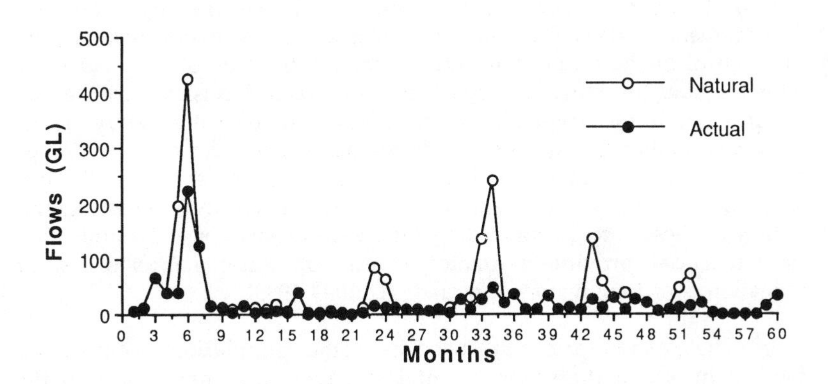

Figure 4. Natural and Actual Flows to Wetlands When Wetlands Have 43.9% of Reservoir Capacity, Inflows, and Unregulated Flows

Water Markets to Increase Means and Reduce Risk

The simulation of water market participation by irrigators and wetlands managers requires the derivation of short-run irrigation water demand curves and then the simulation of water market prices.

Derivation of short-run demand curves and simulated market prices

Models have been developed for simulating the market price of water for irrigation in two distinct types of climatic environment. The first comprises predominantly winter precipitation areas in which irrigation water demand and supply can be forecast relatively accurately at the start of an irrigation season. Traditional deterministic mathematical programming methods for deriving irrigation water demand curves are quite suitable for predominantly winter precipitation areas where water supplies and demands are rather free of climate-based uncertainty, but are quite inappropriate for the second type of climatic environment, where summer rainfall causes highly variable and uncertain irrigation water supplies and demands, as in the Gwydir system.

The author has developed an alternative, dynamic programming approach to the derivation of irrigation water demand for highly variable and uncertain climatic environments. These are farm/reservoir management models. Their state variables are (i) volume of water under the control of the individual decision maker, (ii) area of crop currently irrigated, and (iii) current price of reservoir water. Decision variables are (i) reduction in the irrigated area and (ii) quantity of water to buy or sell. This new method produces short-run demand curves for each of a range of discrete system states in each discrete stage of the season for individual irrigators in highly stochastic summer rainfall environments. Aggregation of these into regional demands, with the use of a dynamic simulation model, provides estimates of reservoir water market prices for irrigation over long historical periods (Dudley 1993).

These water prices are used in further simulation/optimization models in which the environmental managers also participate in the water market. Attaching different weights to environmental uses causes the quantities of water purchased for environmental purposes to vary with the weights. With low weights, water is only purchased for environmental purposes when it has low value to irrigators. On the other hand, weighting environmental uses highly causes water to be purchased for environmental uses even when it is very valuable to irrigators.

In this way, water allocated for environmental purposes can be costed according to the irrigation production forgone. As well as the different water purchases associated with the different weights, the different levels of environmental effectiveness are also recorded. This allows environmental cost-effectiveness relationships to be derived and plotted on figures like figure 2.

These methods have been applied in an exploratory way to environmental uses, such as instream flows for recreation and for the dilution of some pollutants, in conjunction with transferable discharge permits for which the amount of discharge allowed is conditional on stream conditions (Dudley 1991; Dudley, Coelli, and Pigram 1993). They are now being applied comprehensively to wetlands-irrigation trade-offs.

Water markets to expand effectiveness and revenue means

It is expected that participation in the market for reservoir water by wetlands managers and irrigators, who hold capacity shares in the reservoir, will lead to a movement in the trade-off curve in figure 2 toward the zero-trade-off point. If the objective of wetlands managers is to minimize the divergence between actual and natural flows to the wetlands for any resource mix, as assumed in the study, the wetlands demand for water will tend to be high in flood times and low in drought times. In contrast, the irrigation demand is typically the reverse. Hence, it would appear that market exchange of water should help each of these users increase their mean objectives. That is, it is expected that mean wetlands effectiveness can be increased without reducing mean irrigation net revenues (i.e., a vertical shift in figure 2) or that mean irrigation net revenues can be increased without decreasing mean wetlands effectiveness (i.e., a horizontal shift in figure 2) or some combination of the two (i.e., a "northwest" shift in figure 2). Determining the degree to which markets are able to shift the trade-off curves like those in figure 2 under various conditions is an important objective of the study.

However, as well as providing a way of increasing the means, markets for reservoir water would also provide a way of reducing the risk of undesirable outcomes for irrigation and wetlands, as discussed in the next section.

Water markets to reduce risk

Figures 3 and 4 indicate times and situations where it may be desirable to modify actual wetlands flows by water market trading. For example, months 33 and 34 in figure 4 show a limited response to the major flood over those two months, after a limited response to the minor flood in months 23 and 24. It may be desirable to purchase water in such cases to match historical flows better, even if it means selling water at low-flow times, and failing to match those low flows, to have funds to purchase water to match the flood flows better. This could be achieved in the marketing models by weighting higher flows more than lower flows, as introduced when discussing the nontrading models above. The dynamic programming model would "look ahead" and possibly purchase more water in low-demand periods in anticipation of the need to meet the larger flood flows. This may reduce the cost of matching these natural flood flows better compared to myopically waiting for the high natural flows to arrive before purchasing water.

Such water market trading between wetlands managers and irrigators may destabilize irrigation production, but irrigators would not sell water unless the price was favorable to them. Risk-neutral irrigators would only sell if it increased their expected net revenues, whereas those who are sufficiently risk averse would trade to balance their reduced financial risk and increased expected net revenues.

Hence, introducing markets for reservoir water would reduce the risk of undesirable flows to wetlands. It would also reduce the risk to irrigators desiring such reduction, and increase the expected net returns of those who do not.

Balanced wetlands trading budget for political sustainability

It was noted above that varying the weights attached to wetlands effectiveness changes the optimal quantities of water to be bought and sold by the wetlands managers. In this study, weights are chosen for each combination of reservoir capacity, reservoir inflows, and downstream tributary flows by experimenting with different weights and selecting those that result in expenditures on water purchases that just offset revenue from water sales over long historical data series. It is believed that this long-term balance of the wetlands' water trading budget will provide greater political sustainability in practice for such resource sharing than where wetlands managers are dependent on regular injections of funds, or where they are making profits. Of course, if the wetlands managers have control of a large share of the valley's water resources, they will be able to achieve high wetlands effectiveness compared to a situation in which they have a small share of the resources, while maintaining balanced budgets in the long term.

If a reallocation of resources of sufficient size to result in marked improvement in wetlands effectiveness is obtained by purchase of streamflow and reservoir shares from irrigators, the investment required would be considerable. To this an underwriting commitment needs to be added to ensure solvency of the wetlands management in case the sequence of streamflows is such that water purchases outweigh sales in the early years of the scheme's operation.

Conclusion

How best to allocate water resources between environmental and commercial uses over both short and long time horizons is an important problem in many parts of the globe. This paper reports on a study in progress in Australia that is innovative in a number of ways. It is developing methods to help quantify and minimize the trade-offs between these increasingly competitive users of water. Capacity sharing is used to define property rights to water resources close to their source—at the headworks reservoir for rights to reservoir empty space and reservoir inflows, and at their point of entry to the main river for downstream, unregulated tributaries. Such definitions provide a very sound base for both short-term trading of water already in the system and long-term trading of rights to shares of reservoir capacity and streamflows. Managers of both environmental and commercial uses of water participate in these markets. Whereas the environmental use may require external funding to provide its initial asset share of long-term rights to reservoir capacity and streamflows, its water trading is managed so that water sales at times when water is of limited marginal value to the environmental use will, on balance, pay for purchases of water when it has a high marginal value to the environment. Such trading is expected to result in finely tuned integrated management of water resources, and maximize the aggregate benefits, including risk reduction, to both uses for a given long-term allocation of the water resources. Changing the long-term mix of resources will make one better off at the expense of the other, but the aggregate benefit should still be the maximum possible for that mix. Later papers will report the degree to which results of the modeling validate the approach.

References

Alaouze, C. 1991. The optimality of capacity sharing in stochastic dynamic programming problems of shared reservoir operation. *Water Resources Bulletin* 27 (3): 381–6.

Dudley, N. 1990. Alternative institutional arrangements for water supply probabilities and transfers. In *Transferability of water entitlements—An international seminar and workshop,* edited by J. Pigram and B.

Hooper, 79–90. Armidale, NSW, Australia: University of New England Centre for Water Policy Research.

Dudley, N. 1991. Management models for integrating competing and conflicting demands for water. In *Water allocation for the environment—An international seminar and workshop,* edited by J. Pigram and B. Hooper, 169–82. Armidale, NSW, Australia: University of New England Centre for Water Policy Research.

Dudley, N. 1992a. An innovative institutional arrangement which incorporates the risk preferences of water users. In *Risk-based decision making in water resources V,* edited by Y. Haimes, D. Moser, and E. Stakhiv, 174–99. New York: American Society of Civil Engineering.

Dudley, N. 1992b. Water allocation by markets, common property and capacity sharing: Companions or competitors? *Natural Resources Journal* 32 (4): 757–78.

Dudley, N. 1993. Derivation of irrigation demand curves and water market simulation in highly variable environments. Submitted to *Water Resources Research.*

Dudley, N., M. Coelli, and J. Pigram. 1993. An integrated approach to tradeable discharge permits and capacity sharing under Australian conditions. Discussion paper prepared for the Environmental Research Trust and the Sydney Water Board, NSW, Centre for Water Policy Research, University of New England, Armidale.

Dudley, N., and W. Musgrave. 1988. Capacity sharing of water reservoirs. *Water Resources Research* 24 (5): 649–58.

Dudley, N., and B. Scott. 1992. Integrating irrigation water demand, supply and delivery management in a stochastic setting. *Water Resources Research* 29 (9): 3093–101.

Musgrave, W., C. Alaouze, and N. Dudley. 1989. Capacity sharing and its implications for system reliability. In *Proceedings of the national workshop on planning and management of water resource systems: Risk and reliability,* edited by G. Dandy and A. Simpson, 176–85. Adelaide, SA, Australia: Australian Government Publishing Service, Canberra.

Climate Warming and Great Lakes Management

Benjamin F. Hobbs,[1] Philip T. Chao, Mohammed Nayal, and William T. Bogart[2]

Abstract

A decision-analysis approach to determining the relevance of information on climate change for near-term Great Lakes management decisions is outlined. Simplified applications to decisions to preserve coastal wetlands and to regulate outflows from Lake Erie are presented as illustrations of the methodology.

Introduction

Greenhouse gas-induced climate change could significantly alter the availability and quality of water resources, with wide-ranging social, economic, and environmental impacts (Waggoner 1990). As an example, the following specific impacts have been studied for the Great Lakes (Smith and Tirpak 1990):

- direct physical impacts (lake levels and ice cover, and their impacts on shoreline land uses and wetlands)
- water quality (pollution concentrations, dissolved oxygen, lake thermal structure)
- aquatic ecology and fisheries (effects of changes in water levels, wetlands, and water quality on ecosystem productivity and stability, along with fishery composition and abundance)
- energy (demands and hydropower production)
- navigation (shipping costs, dredging requirements)
- socioeconomic impacts (immigration, economic growth, recreation)

---

[1]Associate Professor of Systems Engineering and Civil Engineering, Case Western Reserve University, Cleveland, OH 44106
[2]Respectively, Ph.D. Candidate, Student, and Assistant Professor of Economics, Case Western Reserve University, Cleveland, OH 44106

- policy (sharpened conflicts among different interests concerning level and flow regulation, diversions, consumptive uses, and pollution control; development of institutions for management of those conflicts; the need to consider and manage the large uncertainties that exist about potential physical and socioeconomic impacts and adjustments) (Hartmann 1990)

These studies focused on impact analysis: understanding the possible effects of climate change on the Great Lakes' socioeconomic and ecologic systems without an explicit consideration of alternative response strategies. More recently, researchers and policymakers have begun to consider the relevance of this information for management of the lakes (Cohen 1989; Donahue 1993; Hartmann 1990; Mortsch, Koshida, and Tavares 1993; Quinn 1993). There are at least three basic questions that need to be addressed:

1. How do the potential impacts of climate change affect the responsibilities of Great Lakes managers?

2. What are the uncertainties that might affect our decisions, how large are they, and how quickly will those uncertainties be resolved?

3. Is there sufficient adaptability in our management systems to cope with the impacts, or are some options so inflexible that we might regret choosing them?

In short: *Might the prospect of possible climate change alter decisions being made today concerning the management of the Great Lakes?*

The purpose of this paper is to present a decision-analysis framework for analyzing this general question, and to present prototypical applications of that framework to wetlands preservation and Lake Erie regulation. Before doing so, however, we discuss the characteristics of some Great Lakes management decisions that could make them sensitive to climate change.

Potentially Affected Great Lakes Decisions

Below we describe several Great Lakes management decisions for which information on climate change might be relevant. Each decision has the following characteristics:

1. One or more options involve a commitment of resources that has long-lived effects and is irreversible or costly to alter. Such options limit the adaptability of the system.

2. These resource commitments can be delayed, allowing managers to obtain more information on the likelihood and nature of climate change.

3. The benefits and costs, both in economic and ecologic terms, of those options would be altered by climate change. Some options might pose a risk of potentially catastrophic effects, such as species loss or severe property damage (Smith 1993).

4. Delays in commitments also mean that some interim benefits will be forgone.

Lake Levels Management. One question considered in the recent International Joint Commission (1993a) Lake Levels Reference Study was, Ought we invest a billion dollars or more in control structures for Lakes Erie, Huron, and Michigan? Climate warming would decrease lake levels, diminishing the value of regulatory structures for preventing flooding and erosion. On the other hand, a climate-change-induced drop in Lake Erie's level would exacerbate anoxia problems in its Central Basin hypolimnion; regulation could mitigate some of that change.

Water Use/Diversion. Commitments to divert water into or out of the lakes might be regretted later if water levels in the lakes dramatically change.

Shore Protection. The question here is, What type of investments ought to be made to prevent flooding and erosion? If lake levels fall because of climate warming, then the benefits of the more permanent of such investments might decrease, diminishing their attractiveness. On the other hand, the lessened ice cover that warming would bring might enhance their value. As in lake levels management, delaying decisions would allow the likelihood of climate change to be better assessed before making a commitment, but might also mean continued damages.

Navigation. A question managers might face is, If lake levels should drop precipitously, are there harbors and channels for which the increased dredging cost would not be justified by the benefits of keeping them open (Donahue 1993)? When would we know enough about the effects of climate change on the Great Lakes to decide to cease dredging some areas? These decisions have ecologic implications in terms of dredged material disposal and resuspension of toxic materials in sediments.

Public Infrastructure. Designs for new facilities might be altered at a relatively small cost now to avoid potentially huge rebuilding or relocation costs in the future, should lake levels greatly change (Smith 1993). Examples include water supply intakes, waste outfalls, public docks and other park facilities, and roads.

Wetlands. If lake levels drop, then many wetlands will be destroyed, but others will be created. At what point should we allow threatened wetlands to be converted to other valuable uses? Alternatively, when might diking and other strategies for wetland preservation be justified? Ought habitat-enhancement/wetlands-creation programs be increased, anticipating the degradation of existing wetlands (Donahue 1993)?

Water Quality. Decreases in lake levels could expose contaminated sediments to resuspension. This possibility could alter decisions being made today concerning remediation (Rhodes and Wiley 1993). Decreased flows and warmer waters might also exacerbate water quality problems (Mortsch, Koshida, and Tavares 1993). If the incremental cost of additional cleanup is in some cases much less now than it would be later, then it might be worthwhile to alter the remedial action plans now being drawn up in the Great Lakes.

Fisheries Management. If climate change leads to severe changes in the hydraulic or temperature regimes of tributaries or other critical habitat, it may not be possible to sustain populations of certain species in which we are presently investing resources or in which we plan to invest resources. At what point might it be concluded that climate change is so likely that we should shift our emphasis to species that are better adapted to a warmer climate? Should strategies for controlling nuisance species such as the zebra mussel and the ruffe, an exotic fish in Lake Superior, be redirected, enhanced, or terminated (Donahue 1993)?

These decisions differ from the type described by Rogers (1993), in which small increments of capacity are added to a system to meet growing demand. Examples of the latter type of decisions include capacity expansion for water supply systems or electric utilities that use a single type of fuel. Near-term capacity decisions are unlikely to be altered by the prospect of climate change several decades down the road, since the timing and size of later additions can be altered as the effects of climate change become clear. The management decisions we discuss below are different; what is done now can significantly affect benefits and costs later, and corrections of errors may be expensive.

Decision Analysis

We agree with other researchers (e.g., Fiering and Rogers in press; Patwardhan and Small 1992; Yohe 1991) that much insight is to be gained by analyzing management decisions under climate uncertainty using the tools of decision trees and Bayesian analysis. In the decision-tree methodology, explicit recognition is made of the range of possible outcomes and of when decisions are made and knowledge acquired. Such an approach allows for explicit quantification of the worth of information and the value of flexible strategies that leave options open.

Figure 1 presents a decision tree that represents the essential characteristics of the Great Lakes management decisions we discussed in the last section. Like any decision tree, this one has three major elements:

1. *Decision nodes,* represented as squares. Here, the options are to make an irreversible commitment of resources, or to put off the decision.

2. *Chance nodes,* represented as circles. We show three types of uncertainties. The first includes lake-related variables, such as levels or temperatures, which fluctuate randomly and have an uncertain trend due to the possibility of climate change. Socioeconomic variables, such as values of shoreline property, make up the second category. The third group includes climate variables, which provide information that can be used to update the decision makers' beliefs concerning the likelihood and magnitude of climate change. These may overlap with the lake variables. Probabilities are associated with each branch on a chance node.

3. *Outcomes,* which we show as benefits and costs (B&C) that accrue over time. In general, there may be several incommensurable types of benefits and costs, which means that the decision problem is a multicriteria one.

The solution to a decision tree consists of the optimal strategy, i.e., the option at each decision node that maximizes the expected net benefits from that point on. However, the presence of multiple criteria implies that the notion of net benefits is an ambiguous one.

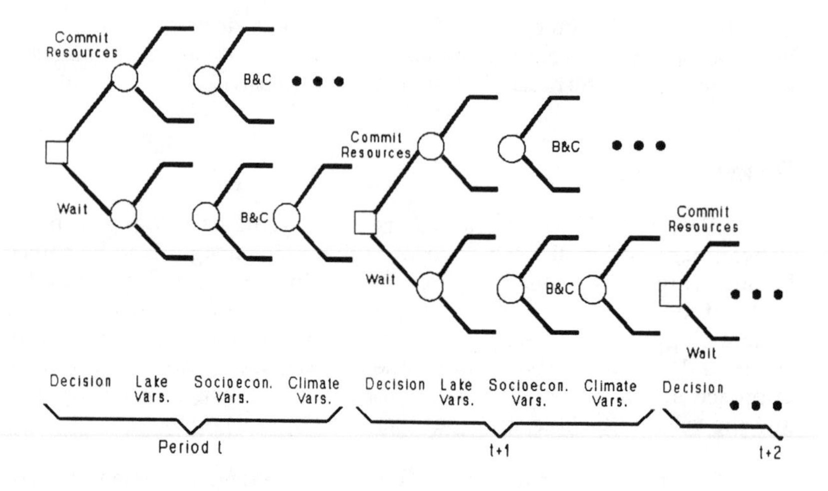

Figure 1. Decision Tree for Management Decisions under Climate Change

The four fundamental problem characteristics we described in the last section are represented in the tree as follows:

1. Resource commitments that are substantially irreversible are shown as having no subsequent decision nodes. For example, figure 1 shows that if a commitment is made in period $t = 1$, then the topmost sequence of chance nodes follows without an opportunity to reverse the decision.

2. We represent knowledge about climate change using a Bayesian framework. In this approach there are two or more possible climatic "states of nature" θ_i; for example,

   θ_1 = Future lake levels will not be altered by climate warming, and will be distributed similar to historical levels

   θ_2 = The expected value of future lake levels will decrease linearly over the next forty years to the levels projected by Croley (1990) under the GFDL doubling of CO_2 scenario

At time t, the decision maker has a prior probability, or degree of belief, $P_t(\theta_i)$, for each state of nature. In the beginning, the decision maker might be highly uncertain whether climate will change; this might be simplistically represented by giving equal prior probabilities to each state of nature. As time goes on, observations of climate variables C_t accumulate, e.g., annual mean temperatures in the southern hemisphere (Solow 1987), which allow the decision maker to modify his or her beliefs following Bayes' Law:

$$P_{t+1}(\theta_i \backslash C_t) = P(C_t \backslash \theta_i)\, P_t(\theta_i) / \Sigma_j\, P(C_t \backslash \theta_j)\, P_t(\theta_j)\,, \qquad (1)$$

where $P(C_t \backslash \theta_i)$ is the probability of a particular realization of climate variables C_t, given that θ_i is the true state of nature. With the passage of time, uncertainty may decrease significantly (i.e., one of the $P_t(\theta_i)$ will begin to dominate), which will reduce uncertainty about future benefits and costs of different actions. (Yet as Dowlatabadi and Morgan [1993] point out, uncertainty may increase over time if research reveals surprises, as it is likely to do.)

3. Figure 1 shows that the benefits and costs depend directly on lake variables, socioeconomic variables, and the decisions that have been made thus far. Benefits and costs are indirectly a function of climate in that climate change alters the probability distribution of lake variables and perhaps also socioeconomic variables. The decision tree relates lake and socioeconomic variables to climate as follows. The probability distribution of, for example, Lake Erie annual levels, L_{Et}, might be modeled as a probability distribution, $P(L_{Et} \backslash \theta_i, L_{E\tau})$, conditioned on the state of nature and previous lake levels, $L_{E\tau}$, $\tau = t - 1, t - 2, \ldots$. Because the state of nature is not known, the overall probability of L_{Et} is expressed thus:

$$P(L_{Et} \backslash L_{Et}) = \Sigma_i\, P(L_{Et} \backslash \theta_i, L_{Et})\, P_t(\theta_i) \qquad (2)$$

4. The cost of delaying a resource commitment would be reflected in differences in B&C between branches that follow a commitment and those that instead involve a delay.

Lake Erie Wetlands: Development versus Preservation

We present preliminary results of applications of figure 1's framework in this section and the next. The applications are highly simplified because our purpose is to illustrate some of the elements of the framework, rather than to make a definitive statement about the management problem in question.

An increase in greenhouse gas concentrations equivalent to a doubling of CO_2 concentration could drop levels of Lake Erie by one to two meters (Croley 1990). This decrease would have an enormous effect on coastal wetlands. Many existing wetlands would be deprived of the periodic inundation they require, while new wetlands would be created from former lake bottom.

To illustrate the application of figure 1's framework, we consider an existing wetland that would disappear if lake levels drop that far. We assume that this wetland is situated so that the land would be valuable for commercial development. We also assume that the expense of maintaining the wetland by diking and pumping could not be justified if lake levels drop, especially since diking would cut off access to the lake and would destroy the wetland's value for aquatic habitat.

The elements of the decision tree representing this problem are as follows:

1. *Decision nodes,* whether to allow commercial development or to preserve the wetland ("wait")

2. *Chance nodes,* yearly mean lake levels. Observations of these levels are used in (1) to infer whether lake levels are stationary around the historical mean (θ_1) or whether they vary around a mean trend that decreases linearly over forty years to a level 1.5 meters below the historical mean (θ_2). For simplicity, we initially assume that lake levels are a first-order autoregressive process, and our prior probabilities are $P(\theta_1) = P(\theta_2) = 0.5$.

3. *Outcomes,* benefits of development, if allowed, and permanent loss of wetlands. We assume, for illustration purposes, that the wetland has a high (but unspecified) ecological and social value if lake levels do not drop permanently (θ_1). However, because the wetland would degrade in quality and other wetlands would be created if levels fall permanently (θ_2), we assume the wetland would have relatively little value if θ_2 is realized.

This problem has the four fundamental characteristics, identified above, of management problems that might be affected by climate change:

1. Commercial development is assumed to be irreversible.

2. Delaying development would allow managers to obtain better information on whether lake levels are going to drop permanently, via (1).

3. The benefit of not developing the wetland depends on whether lake levels permanently decrease.

4. If development is delayed, economic benefits will be forgone in the meantime.

The heart of the problem is whether to develop now, and thus run the risk that a valuable wetland will be destroyed, or to wait, and thereby give up economic benefits in the interim.

One possible approach to solving this problem is to assume a monetary value for the worth of wetlands, and then combine that value with the economic value of commercial development. The resulting single performance measure can then be optimized using the usual tools of decision analysis. Monetary values for wetlands that have appeared in the literature range from a few dollars to $2000/acre/year, depending on the functions that the wetland provides (Pearce and Turner 1990). The difficulties associated with determining nonmarket values of wetlands motivate us to turn to an easier and, in some ways, more insightful approach: multiobjective analysis.

The focus of our multiobjective analysis is on generating tradeoffs among competing objectives. Here, we propose the following two objectives:

1. *Worth of commercial development,* equal to the expected present value of the land for commercial purposes at the date of development. An expected value is calculated because it is uncertain when, if ever, commercial development would occur.

2. *Loss of wetlands,* equal to the probability that a valuable wetland is developed (i.e., that both commercial development and θ_2 occur). In a sense, this is the long-run expected value of the wetland.

Different strategies will result in different levels of performance of each objective, as illustrated in figure 2. The commercial development axis has been normalized so that "1" represents the maximum possible economic value of development (resulting from development immediately at $t = 0$). The points on the lower left represent conservative strategies that prohibit development unless it is almost certain that lake levels will drop; as a result, there is little chance of developing a valuable wetland but also a relatively low present worth of development. The points on the upper right represent more aggressive strategies that allow development even if climate change is highly uncertain. Consequently, the probability of mistakenly developing a valuable wetland is higher, but so are the development benefits.

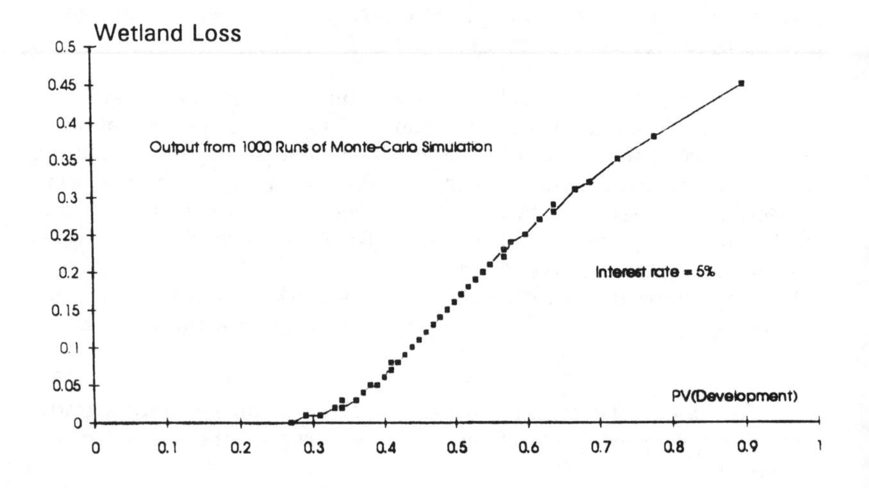

Figure 2. Trade-Off between Expected Development Benefits and Expected Loss of Valuable Wetlands

The points in figure 2 are generated as follows. Rather than discretizing probability distributions of lake levels and then solving the resulting (enormous) tree by the usual "folding back" method, we opted to define a range of simple policies and then simulate their performance using Monte Carlo simulation. (The curve is not smooth due to sample error in the simulations.) The policies are of the form:

> *If in year* t *the posterior probability of decline in lake levels,* $P_t(\theta_2)$*, exceeds a threshold* T*, then develop the wetland. Otherwise, wait.*

A low threshold T (e.g., $T = 0.7$ or less) yields the upper right points in the figure, while a high threshold ($T = 0.9$ or more) gives the lower left points. This is because higher thresholds delay development and decrease the probability that it will occur by mistake (e.g., when lake levels aren't actually dropping permanently). This lowers the present worth of development, but also the probability that a valuable wetland is lost.

The exact form of the trade-off curve depends on the assumed interest rate and the prior probabilities. There may be (indeed, are likely to be) better strategies that lie below and to the right of this curve; they might be discovered either by simulating more complex policies or by solving a simplified version of the decision tree by folding back.

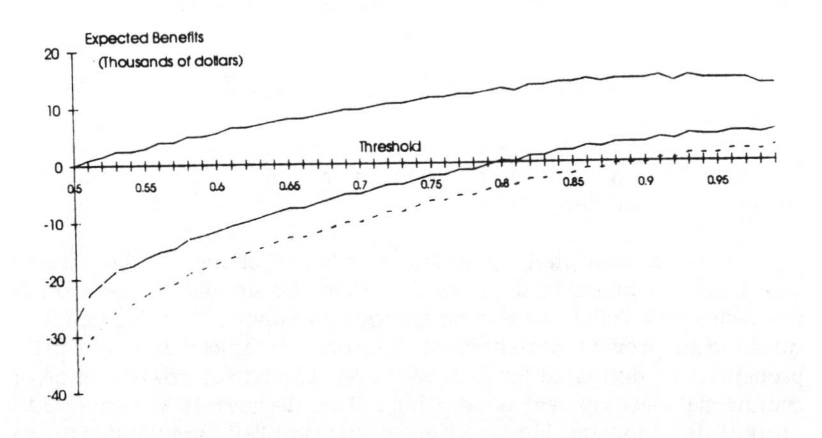

Figure 3. Net Benefits of Wetland Development Strategy as Function of Threshold T for the Probability of Climate Change

When managers are presented with a curve such as figure 2, they can decide what trade-off between development benefits and risk of wetland loss is acceptable. The decision depends on the relative weight given to wetlands and commercial development. If the manager is willing to give explicit weights, then the two objectives can be combined into a net benefit objective, as in figure 3. That figure shows the net benefits (expected present worth of development benefits minus wetland losses) of using different thresholds T under the following assumptions:

- a discount rate of 5% per year
- prior probabilities $P(\theta_1) = P(\theta_2) = 0.5$
- the social value of a valuable wetland has a present worth of $100,000
- the value of commercial development today is either $10,000/acre, $20,000/acre, or $50,000/acre (corresponding to figure 3's bottom, middle, and upper curves, respectively)

A point on figure 2 is translated into a point in figure 3 by multiplying the PV(Development) and Wetland Loss axes by the assumed dollar values, and then plotting their difference against the threshold T that yielded those values.

The resulting curves show that if commercial development is relatively valuable ($50,000/acre), then a threshold of between 0.9 and 0.95 for $P(\theta_2)$ is optimal. However, for less lucrative development, a higher threshold should be used to minimize the probability of mistakenly developing a valuable wetland.

Once a desirable strategy has been identified, then the question we posed in our introduction, does the possibility of climate change alter decisions now? can be answered. In this case, if the initial decision ("develop" or "wait" in year 0) depends on the prior probability, the answer to the question is "yes."

If we assume that wetlands are valuable enough under present lake levels to preclude development, then the immediate decision is unaffected by a belief that climate change may happen. This decision is to continue to prevent development. The only exception is if the prior probability of decreased levels is well over 0.5 and the relative value of commercial development is very high. This decision is to continue to prevent development. However, given our simplistic assumptions, two decades will suffice to tell whether lake levels will drop permanently, and by that time there is a high probability that a decision to develop will have been made. So climate change information can make an important difference in wetlands decisions, but not immediately.

Future work will emphasize the development of a more realistic wetlands decision model with the following features:

- inclusion of the options of diking and pumping as a means of preserving wetlands in the face of declining lake levels
- more complex time series models of lake levels, incorporating the possibility of nonstationarity due to causes other than climate change (Slivitsky and Mathier 1993)
- several climate change scenarios, rather than one
- separate representation of uncertainties in global climate change and the lakes' hydrologic response to it
- accounting for the near-term value of wetlands, rather than just long-term values
- testing of more sophisticated and, it is hoped, better strategies

Lake Erie Outflow Regulation

The International Joint Commission (IJC) recently finished Phase II of its Lake Levels Reference Study (IJC 1993a). The study's purpose was to "examine and report upon methods of alleviating the adverse consequences of fluctuating water levels in the Great Lakes–St. Lawrence River Basin" (U.S. Dept. of State and Canadian Dept. of External Affairs 1986). The study was motivated by record high lake levels in the mid-1980s that caused extensive property damage, and, to a lesser extent, low water levels in the early 1960s that adversely impacted hydropower production and shipping (Horvath, Jannereth, and Shafer 1989).

The study investigated a range of alternatives, or "measures," for controlling and mitigating the effects of extreme levels of the Great Lakes. The measures considered can generally be classified into the following groups:

- operation and design of structures to regulate water levels (there are existing structures at the outlets of Lakes Ontario and Superior, and regulation structures have been proposed for Lake Erie and for Lakes Michigan and Huron)
- land use regulation
- land use incentives, including taxes and subsidies
- shore protection alternatives
- adaptive practices, including emergency response measures

Many of these measures are flexible ones that can be implemented or altered as circumstances change. Examples include operating policies for existing flow regulation works that control Lakes Superior and Ontario or emergency actions to be taken during crises. Should climate change cause lake levels to fall significantly, these actions could be altered at little cost. Consequently, the merits of implementing such measures in the near future should not be affected by the prospect of climate warming in the twenty-first century.

Yet there are other, more long-lived measures whose benefits and costs could be greatly affected by climate change. These include construction of structures to regulate flows and restrictions on development in flood-prone areas. In the case of measures designed to deal with the effects of high lake levels, global warming might decrease their benefits because such levels would probably occur less frequently. Alternatively, warming might actually enhance their benefits because the lakes would be ice-free for more months of the year, rendering shorelines vulnerable during severe winter storms. In contrast, the benefits of measures that would prevent or ameliorate the effects of low levels might be increased.

The study recognized that the implications of possible climate change could be important. However, the alternatives they considered were not systematically analyzed under alternative climate scenarios, and no probabilities were assigned to the scenarios considered. Consequently, a formal decision-tree analysis, such as we propose above, could not be conducted. Instead, the information was used as a sensitivity analysis, in the spirit of Liebetrau and Scott (1991).

In this section, we outline a highly simplistic decision-tree analysis of two alternatives:

1. *Three-lake regulation,* consisting of construction of regulatory works in the Niagara River, which would permit control of Lake Erie's outflows. This is the "resource commitment" option in figure 1.

2. *Two-lake (Superior and Ontario) regulation,* as currently conducted. This is the "wait" option in that figure.

The IJC report concludes that the benefit/cost ratio of choosing the former option over the latter is less than 0.2; further, if lake levels drop, the ratio is likely to worsen. However, we arbitrarily adjust these numbers to make a point. That point is to show how analysis can answer the question we posed at the start of this paper: might the possibility of climate change alter decisions being made now?

Figure 4 shows a simplified version of figure 1's tree. Only monetized benefits and costs (B-C) are included. There are just two decision points: the years 2000 and 2030. In the year 2000, we assume for this analysis that three-lake regulation can be instituted at a capital cost of \$813 million, or we can choose to wait. If we wait, then we can consider the three-lake regulation option again in the year 2030. Between the years 2000 and 2030, we assume that we will learn whether climate warming takes place in such a way as to lead to a significant decrease in lake levels. If warming takes place, the effects on the lakes are assumed to be consistent with the $2xCO_2$ scenarios in IJC (1993a). Thus, deferring the decision until 2030 has the advantage of allowing us to obtain potentially valuable information, whereas if we decide now, we take a chance that the project will later turn out to be relatively useless.

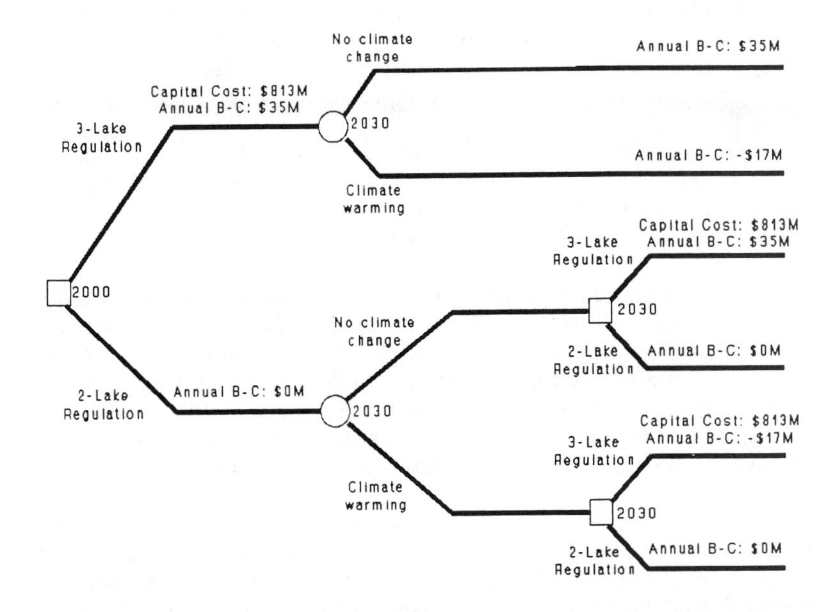

Figure 4. Simplified Decision Tree for Lake Erie Regulation Problem

If the climate does not change, we assume that three-lake regulation would yield annual benefits of \$35 million from flood- and erosion-damage reduction, net of hydropower losses, operations and maintenance costs, and increased navigation costs. If climate warming occurs, then the annual benefits are negative: -\$17 million. Both numbers represent the difference with respect to two-lake regulation, so the latter strategy's annual benefits are, by definition, zero. These costs, along with the capital cost of three-lake regulation, are shown in the appropriate locations on the decision tree.

The cost assumptions are based loosely on the IJC's (1993a) characterization of the alternatives, with the following important exceptions:

1. We include only 10% of the anticipated $2854 million capital cost of mitigating the effects of more highly variable flows in the St. Lawrence River, and a similar fraction of the associated O&M cost. We do this in part because there is a fair probability that the IJC will substantially lower its estimates of those mitigation costs.

2. We assume, consistent with IJC (1993b) and an analysis we performed of Lake Erie (Chao, Hobbs, and Stakhiv 1993), that there are no flood- or erosion-control benefits from three-lake regulation if climate warming occurs.

3. We presume, based on our previous analysis (Chao, Hobbs, and Stakhiv 1993), that the hydropower losses resulting from three-lake regulation are approximately the same under either the present climate or climate warming.

4. As a rough approximation, consistent with (Chao, Hobbs, and Stakhiv 1993), we assume that three-lake regulation increases navigation costs by $5 million annually under the present climate because regulation would then lower lake levels, forcing shippers to load their ships more lightly. However, if climate change occurs, then regulation is assumed to decrease navigation costs by $5 million for the opposite reason.

This analysis is simplified for several reasons. First, implementation of three-lake regulation could occur at any time, not only at the years 2000 and 2030. Second, the climate will not suddenly shift from the present climate to a 2xCO_2 equilibrium climate in the year 2030. The change may instead be gradual, and would likely continue beyond the year 2030. Third, we may not know for sure that the climate is changing, even as late as the year 2030. Fourth, benefits and costs will change over time due to changing prices, population levels, etc. Fifth, environmental and other nonmonetized objectives are very important in this problem.

Nonetheless, our simplifications permit some insights that illustrate the potential of the decision-analysis approach. The tree in figure 4 still preserves the four characteristics of decisions that might be affected by the prospect of climate change:

1. Three-lake management is costly and irreversible.

2. Three-lake management can be delayed, permitting managers to obtain better information on climate change's effects on the Great Lakes, if any.

3. The benefits of three-lake management are assumed to be greatly diminished if the climate warms and lake levels drop.

4. Delaying three-lake management means that the potential erosion- and flood-control benefits of the project would be forgone.

Thus, figure 4's tree can be used to demonstrate the types of results that can be obtained from a decision analysis of Great Lakes management decisions under climate change. For instance, the value of better information on climate change can be readily ascertained. Below we discuss two other types of results: the dependence of the optimal strategy on the assumed climate warming probability and interest rate example, and the effect on the initial decision of including climate change.

Figure 5 shows that one of three strategies will be optimal for figure 4's decision tree:

1. *Implement three-lake regulation immediately.* This occurs if the interest rate and probability of climate warming are both relatively small (area I in figure 5). The low interest rate makes it easier to justify the project's high capital cost, and the low probability of climate change means that the possibility of -$17 million annual benefits following year 30 is relatively insignificant.

2. *Keep two-lake regulation for all years.* This strategy is optimal if the interest rate is so high (over 4.3%/yr in real terms, area IV in the figure) that the capital investment in three-lake regulation cannot be justified even under the best possible annual benefits ($35 million/yr).

3. *Wait and see:* keep two-lake regulation for the moment, but implement three-lake regulation if it becomes apparent that climate change will not lower lake levels. This is the best strategy for relatively high probabilities of climate warming if interest rates are low (areas II and III in figure 5).

These results are based on a criterion of maximizing expected present worth of net benefits over an infinite time horizon.

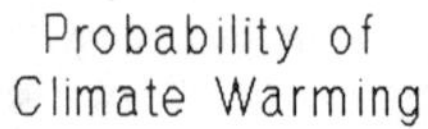

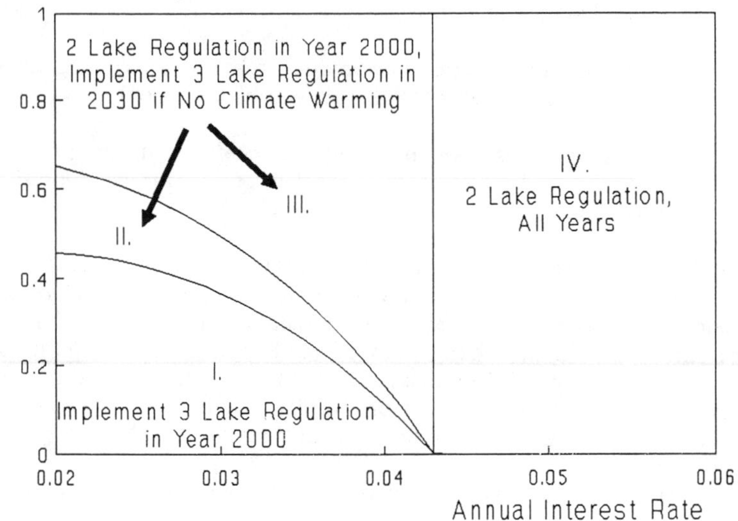

Figure 5. Dependence of the Optimal Strategy on Discount Rate and Probability of Climate Warming

We now turn to the central question of this paper: Under what circumstances might information on climate change alter our initial (year 2000) decision? There are at least two ways of viewing this question. The first is to compare decisions that assume that the probability of warming is zero (i.e., the possibility of climate change is ignored) with the optimal decision if a nonzero probability is assumed instead (i.e., the possibility is acknowledged). Figure 5 shows that for interest rates below 4.3%, including the possibility of climate change does make an important difference. Under a sufficiently high probability of warming, instead of adopting three-lake regulation immediately, the optimal strategy is "wait and see."

Another way of answering this question is to determine whether including the option of getting more information on climate change (i.e., "wait and see") could alter the initial decision. This option can be deleted from the tree of figure 4 by removing the alternative of implementing three-lake regulation in the year 2030. In that case, the resulting optimal strategies are

- *implement three-lake regulation now (areas I and II in figure 5)*
- *never implement three-lake regulation (areas III and IV)*

Then restoring "wait and see" as an option makes the following difference in the initial decision:

> *If interest rates and the probability of warming are in area II in figure 5, then "wait and see" instead of implementing three-lake regulation immediately.*

In summary, if areas II and III encompass a relatively wide range of values for climate change probabilities and discount rates, then the prospect of possible climate change is potentially important for today's management decisions.

Future analyses of lake levels management will focus on several aspects of the problem:

- appropriate characterization of alternative structural and non-structural strategies for managing extreme lake levels and ameliorating their impacts
- inclusion of additional and more realistic climate change scenarios
- integration of the decision-tree framework with a simulation model for projecting impacts on Great Lakes hydrologic, economic, and environmental systems (Chao, Hobbs, and Stakhiv 1993)
- evaluation of the decision framework by Great Lakes managers

Conclusion

Although significant climate warming may be uncertain and several decades away, we have shown that information on climate change can affect decisions being made in the near future about the management of the Great Lakes. Research is needed that not only addresses the wetlands preservation and lake levels management issues we have outlined in this paper, but also other Great Lakes problems. These may include water quality improvement, toxic sediment removal and treatment, shore protection, dredging, infrastructure design, and fisheries management.

Acknowledgment

This work was sponsored by the National Science Foundation, Grant SES 9223780, and the U.S. Army Corps of Engineers Institute for Water Resources. This paper has benefited from extensive discussions with J. A. Bloczynski, J. F. Koonce, E. Z. Stakhiv, and B. Venkatesh.

References

Chao, P. T., B. F. Hobbs, and E. Z. Stakhiv. 1993. Evaluating climate change impacts on the management of the Great Lakes of North America. Paper presented at Engineering Risk and Reliability in a Changing Physical Environment, May 24–June 4, at NATO Advanced Study Institute, Deauville, France.

Cohen, S. J. 1989. Great Lakes levels and climate change: Impacts, responses, and futures. In *Societal Responses to Regional Climate Change*, edited by M. H. Glantz. Boulder, Colorado: Westview Press.

Croley, T. E. 1990. Laurentian Great Lakes double-CO_2 climate change hydrological impacts. *Climatic Change* 17 (1): 27–47.

Donahue, M. J. 1993. *Great Lakes global climate change: Implications for water policy and management*. Ann Arbor, Michigan: Great Lakes Commission.

Dowlatabadi, H., and M. G. Morgan. 1993. A model framework for integrated studies of the climate problem. *Energy Policy* 21 (March): 209–21.

Fiering, M. B., and P. Rogers. In press. *Climate change and water resources planning under uncertainty*. Draft report, Institute for Water Resources, U.S. Army Corps of Engineers, Ft. Belvoir, Virginia.

Hartmann, H. C. 1990. *Great Lakes climate change contingency planning*. Draft. Ann Arbor, Michigan: Great Lakes Environmental Resource Laboratory.

Horvath, F. J., M. R. Jannereth, and C. A. Shafer. 1989. Impacts of water level fluctuations. In *Great Lakes water levels: Shoreline dilemmas*. Washington, D.C.: National Academy Press.

International Joint Commission. 1993a. *Levels reference study: Great Lakes–St. Lawrence River basin*. March 31.

International Joint Commission. 1993b. *Levels reference study: Great Lakes–St. Lawrence River basin: Annex 3, existing regulation, system-wide regulation and crises conditions.* March 31.

Liebetrau, A. M., and M. J. Scott. 1991. Strategies for modeling the uncertain impacts of climate change. *J. Policy Modeling* 13 (2): 184–204.

Mortsch, L., G. Koshida, and D. Tavares, eds. 1993. *Adapting to the Impacts of Climate Change and Variability. Proceedings of the Great Lakes–St. Lawrence Basin Project Workshop, Quebec City, Feb. 9–11, 1993.* Environment Canada, Canadian Climate Centre.

Patwardhan, A., and M. Small. 1992. Bayesian methods for model uncertainty analysis with application to sea-level rise. *Risk Analysis* 12 (4): 513–25.

Pearce, D. W., and R. K. Turner. 1990. *Economics of natural resources and the environment.* Baltimore: The Johns Hopkins University Press.

Quinn, F. H. 1993. The sensitivity of water resources management to climate change: Great Lakes case study. In *Proceedings of the First National Conference on Climate Change and Water Resources Management,* edited by T. Ballentine and E. Z. Stakhiv. Fort Belvoir, Virginia: U.S. Army Corps of Engineers, Institute for Water Resources.

Rhodes, S. L., and K. B. Wiley. 1993. Great Lakes toxic sediments and climate change: Implications for environmental remediation. *Global Environmental Change* September: 292–305.

Rogers, P. 1993. What water managers and planners need to know about climate change and water resources management. In *Proceedings of the First National Conference on Climate Change and Water Resources Management,* edited by T. Ballentine and E. Z. Stakhiv. Fort Belvoir, Virginia: U.S. Army Corps of Engineers, Institute for Water Resources.

Slivitsky, M., and L. Mathier. 1993. Climatic changes during the 20th century on the Laurentian Great Lakes and their impacts on hydrological regimes. Paper presented at Engineering Risk and Reliability in a Changing Physical Environment, 24 May–6 June, at NATO Advanced Study Institute, Deauville, France.

Smith, J. B. 1993. *Adaptation: A focus for Great Lakes climate change impacts research.* Boulder, Colorado: RCG/Hagler Bailly, Inc.

Smith, J. B., and D. A. Tirpak. 1990. *The potential effects of global climate change on the United States.* New York: Hemisphere Publishing.

Solow, A. R. 1987. Testing for climate change: An application of the two-phase regression model. *J. Climate and Applied Meteorology* 26 (October): 1401–5.

U.S. Dept. of State and Canadian Dept. of External Affairs. 1986. Letters to the International Joint Commission, Aug. 1.

Waggoner, P. E., ed. 1990. *Climate change and U.S. water resources.* New York: J. Wiley.

Yohe, G. W. 1991. Uncertainty, climate change, and the economic value of information: An economic methodology for evaluating the timing and relative efficacy of alternative responses to climate change with application to protecting developed property from greenhouse induced sea level rise. *Policy Sciences* 24: 245–69.

Coastal Eutrophication and Temperature Variation

Jacques Ganoulis,[1] Stilianos Rafailidis,[2] Istvan Bogardi,[3]
Lucien Duckstein,[4] and Istvan Matyasovszky[5]

Abstract

A 3-D hydroecological model has been developed to simulate the impact of climate-change-induced daily temperature variation on coastal water quality and eutrophication. Historical daily temperature time series over a thirty-year period have been used to link local meteorological variables to large-scale atmospheric circulation patterns (CPs). Then, CPs generated under a $2xCO_2$ scenario have been used to simulate climate-change-induced local daily temperature variations.

Both historical and climate-change-induced temperature time series have been introduced as inputs into the hydroecological model to simulate coastal water quality and eutrophication. Subject to model validation with available data, a case study in the bay of Thessaloniki (N. Greece) indicates a risk of increasing eutrophication and oxygen depletion in coastal areas due to possible climate change.

Introduction

Coastal water quality and eutrophication phenomena are characteristic indicators of the levels of pollution not only in coastal waters but

[1]Professor, Civil Engineering Department, Aristotle University of Thessaloniki, Thessaloniki 54006 Greece
[2]Research Associate, Civil Engineering Department, Aristotle University of Thessaloniki, Thessaloniki 54006 Greece
[3]Professor, Civil Engineering Department, University of Nebraska-Lincoln, Lincoln, NE 68588-0531, USA
[4]Professor, Systems and Industrial Engineering Department, University of Arizona, Tucson, AZ 85721, USA
[5]Assistant Professor, Department of Meteorology, Eotvos-Lorand University, Budapest, Hungary

also over an entire coastal catchment area. Water drained by rivers, small streams, and overland flows carries organics, nutrients, fertilizers, and pesticides, with ultimate deterioration of coastal waters in estuaries.

Especially in semienclosed seas, such as the Mediterranean, and near major river deltas, high concentrations of nutrients and pollutants may increase eutrophication and deteriorate marine water quality. Large amounts of algae and the related eutrophication have the following adverse environmental consequences:

- reduction of water transparency in the euphotic zone near the free surface of the sea
- decrease of solar penetration and subsequent changes in chlorophyll concentration below the euphotic depth
- uptake of nutrients and oxygen production in the euphotic zone
- settling of algae in deeper waters and oxygen reduction in the oligotrophic, deeper water layers

The complex phytoplankton kinetics (Duckstein, Casti, and Kempf 1979) are also affected by temperature fluctuations at diurnal, seasonal, or long-term scales. Depending on the specific algal type (i.e., diatoms, dinoflagellates, green or blue algae), the phytoplankton growth rate generally increases with temperature. Episodes of high temperature in spring or summer, moderate winds, and weak water circulation may produce eutrophication and algal blooms and should lead to oxygen depletion (Ganoulis et al. 1993). This could affect the biological and chemical equilibrium in coastal waters and be detrimental to marine organisms and fishes consuming oxygen (Quétin and De Rouville 1986; WHO/UNEP 1985).

This paper investigates how coastal water quality, and especially eutrophication, may be affected by daily temperature variations under conditions of climate change. Using historical temperature data over a thirty-year period with the methodology developed by Bardossy and Plate (1990, 1992), the temperature time series under a 2xCO_2 scenario is generated for the Thessaloniki bay area (N. Greece). Then, a coupled hydrodynamic-ecological 3-D model is used to evaluate coastal eutrophication under such climate-change-induced daily temperature time series. The results anticipate possible changes in water quality due to potential global climate change.

Daily Temperature Time Series: Historical and Under Climate Change

The approach used in this work is based on the methodology initially developed by Bardossy and Plate (1990, 1992) and modified by

Matyasovszky et al. (1993b). The background of the method is that local meteorological characteristics, such as temperature and precipitation, may be related to large-scale circulation patterns (CPs). Knowing the CPs over a grid covering the northern hemisphere, a downscaling to local meteorological parameters should be possible. For this operation a large sample of existing data is used.

The method takes into account data at two different scales: (a) a small scale for local meteorological data, and (b) a large scale for global circulation patterns. The linkage between the two kinds of data is made using a multivariate stochastic model with historical data. This is a rather consistent and scientifically well-founded approach, as opposed to the one that assumes different climate scenarios, such as an arbitrary increase or decrease of average temperature. To generate local meteorological time series under global climate change conditions, the methodology is applied in the following successive steps:

1. Characteristic types of daily CPs are classified for the region under study and a Markov chain modeling of the various CP types is undertaken.

2. Local meteorological variables such as temperature and precipitation are analyzed probabilistically, with probability distribution functions conditioned on a given CP type.

3. An autoregressive space-time model is developed for linking local variables, such as temperature, to various CP types.

4. CPs are reproduced by use of outputs of global circulation models (GCMs) for 1xCO_2 conditions and downscaled local time series are generated and compared with historical data.

5. 2xCO_2 CPs are simulated from GCMs and local time series of temperature are obtained, reflecting the effect of global climate change.

Results shown in figures 1, 2, and 3 refer to temperature data collected at Mikra, the location of the Makedonia airport of Thessaloniki, N. Greece. As seen in figure 1, temperature generally tends to increase under climate change due to doubling of CO_2. The increase is not equally distributed over the year: the highest increase is about 4°C in January, April, and September, while there is no significant change in February, March, June, July, August, October, and November. Research is going on (Matyasovszky et al. 1993a), taking into account data from other meteorological stations in northern Greece, to estimate more accurately the impact of climate change on temperature and on coastal water quality and eutrophication.

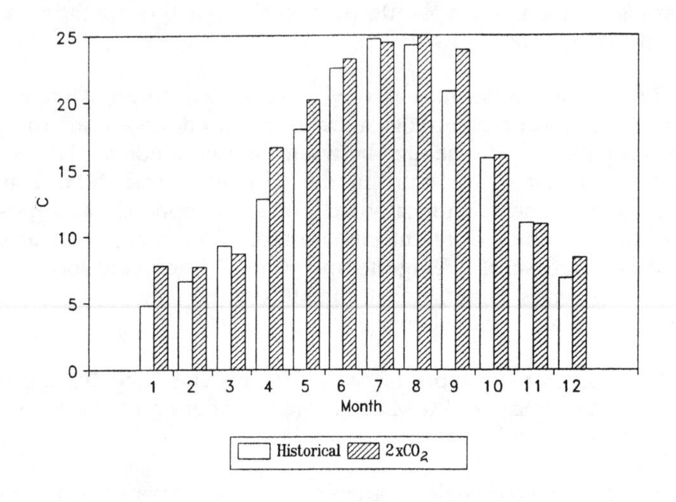

Figure 1. Monthly Means of Daily Mean Temperature at Mikra

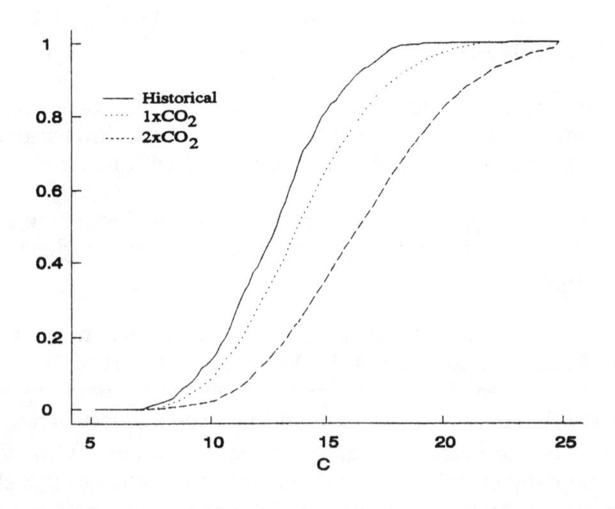

Figure 2. Probability Distribution Function of Daily Mean Temperature for April at Mikra

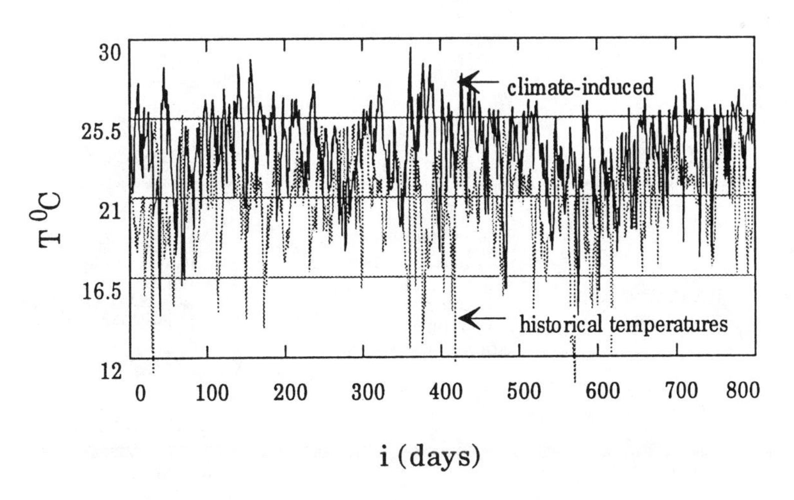

Figure 3. Historical and Climate-Change-Induced Daily Temperature Time Series for September at Mikra

Monte Carlo Simulation Using a 3-D Ecological Model

As shown in figure 4, temperature time series are introduced as input to a 3-D hydroecological model both as historical and simulated under climate change data. Results of model simulation include circulation patterns and pollutant concentrations at various positions and depths, including phytoplankton or chlorophyll-a concentrations. They are used either to validate the model by comparing simulation results with available measured data (Ganoulis and Koutitas 1981; Ganoulis and Krestenitis 1982, 1984; Ganoulis 1988a, 1992) or to predict climate-change-induced temperature impacts on water quality.

The 3-D hydroecological model describes water quality and eutrophication in coastal areas by taking into account three main processes:

- convection by currents
- dispersion due to turbulence
- variation of phytoplankton biomass due to biochemical interactions with other physical and chemical systems

Phytoplankton exists under many different forms of algae, and it is customary to consider algal concentrations in terms of chlorophyll-a concentrations. As shown in figure 5, the chlorophyll-a growth rate,

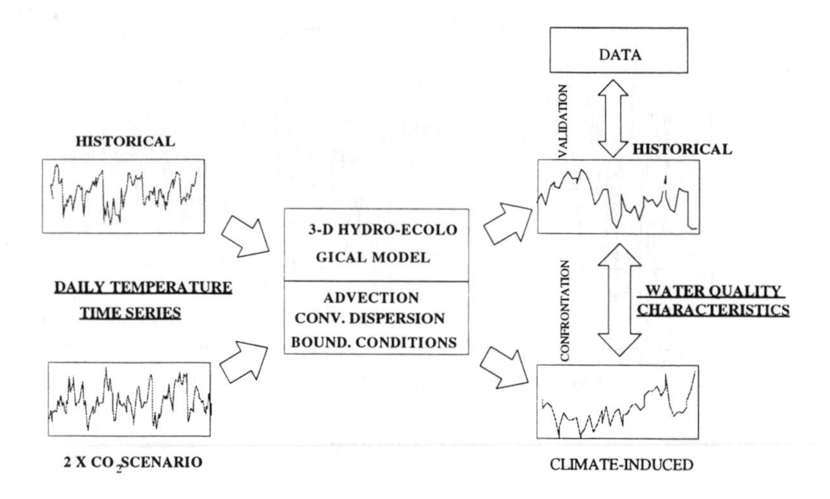

Figure 4. Operation Modes of the Hydroecological Model

$S_A = dA/dt$ reflects the uptake of nutrients such as NH_3, NO_3, and PO_4. Under the influence of solar insolation and temperature, chlorophyll is recycled by

(i) respiration
(ii) decay (nonpredatory)
(iii) settling

Inorganic nutrients are reproduced from phytoplankton biomass back into the system through respiration and nonpredatory mortality. Organic matter is converted to dissolved inorganic substances at a temperature-dependent rate.

Concerning interactions with dissolved oxygen, algae produce oxygen by photosynthesis in the euphotic zone, but this process is reversed at night due to respiration. Furthermore, algae settling at the bottom contribute to oxygen uptake by biodegradation.

In terms of chlorophyll-a concentration A, the phytoplankton kinetics may be described as

$$S_A = \frac{dA}{dt} = (\mu - r_A - e_x - S - m_A)\, A - G_A, \tag{1}$$

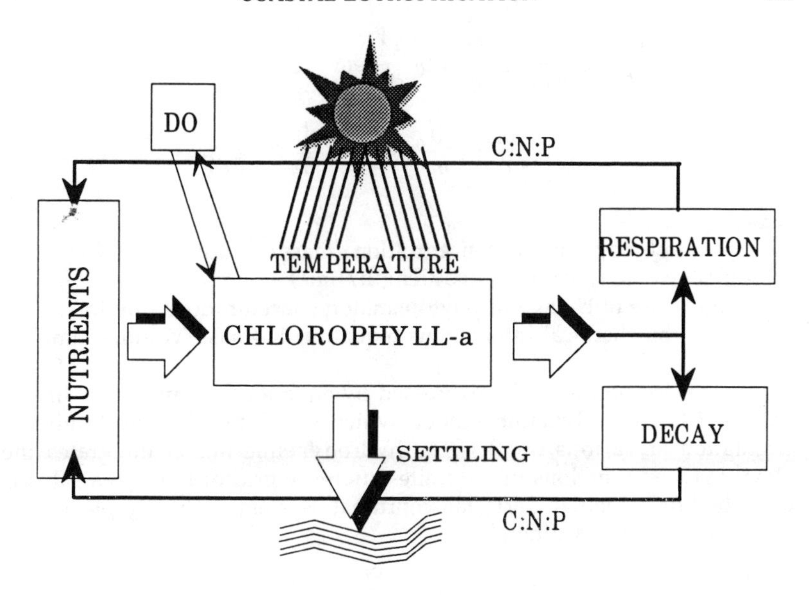

Figure 5. Phytoplankton Kinetics

where

A: chlorophyll-a concentration (mass/volume)

μ: phytoplankton growth rate (T^{-1})

r_A: respiration rate (T^{-1})

e_x: excretion rate (T^{-1})

S: settling rate (T^{-1})

m_A: nonpredatory mortality (T^{-1})

G_A: loss rate by grazing (mass/volume/time)

Temperature plays a key role in all biochemical transformations. Constants involved in usually first-order kinetic relations (i.e., oxidation, nitrification, denitrification) are related to temperature according to an Arrhenius-type relationship,

$$\mu = \mu_{20^\circ C} \cdot T^{(t-20^\circ C)} . \tag{2}$$

The growth rate $S_A = dA/dt$ in chlorophyll-a expressed in equation (1) should be incorporated into the advective-dispersion balance equation (Fischer et al. 1979), which has the form

$$\frac{\partial A}{\partial t} + u\frac{\partial A}{\partial x} + v\frac{\partial A}{\partial y} + w\frac{\partial A}{\partial z} =$$

$$\frac{\partial}{\partial x}(k_x\frac{\partial A}{\partial x}) + \frac{\partial}{\partial y}(k_y\frac{\partial A}{\partial y}) + \frac{\partial}{\partial z}(k_z\frac{\partial A}{\partial z}) + S_A, \quad (3)$$

where

A: phytoplankton concentration (mass/volume)
u,v,w: velocity components (length/time)
S_A: rate of change of phytoplankton concentrations due to biochemical interactions (equation 1) (mass/volume/time)

The ecological model expressed by equations (1) and (3) is linked with a 3-D hydrodynamic model, which simulates the wind-induced circulation at various depths. The hydrodynamic model integrates the Navier-Stokes equations in the finite-difference grid of figure 6(a); this is described on a regular Cartesian coordinate system in the x-y plane, and the transformed coordinate

$$\sigma = \frac{z - \zeta}{H}$$

along the vertical axis z. H is the water depth and ζ, the surface elevation.

A less-refined discretization is needed for the ecological model than for the hydrodynamic. The horizontal spatial relation between the two grids is shown in figure 6(b). The same segmentation is replicated along the vertical direction, following the transformed depth coordinate σ. Software effecting the automatic linkage between hydrodynamic and ecological computations has been developed specifically for this purpose.

Application to the Bay of Thessaloniki

The bay of Thessaloniki occupies the upper north part of the greater Thermaikos gulf, located at the northwest corner of the Aegean Sea (figure 7). The maximum opening between Aherada peninsula on the west and Epanomi on the east is about 15 km. From north to south the maximum "height" of the bay is about 45 km. The bay receives waters from three major rivers: the river Axios, 80% of the discharge basin of which is located in the former Yugoslavia; the river Alikmon, which is the longest Greek river; and the river Loudias, which drains the plain of Thessaloniki.

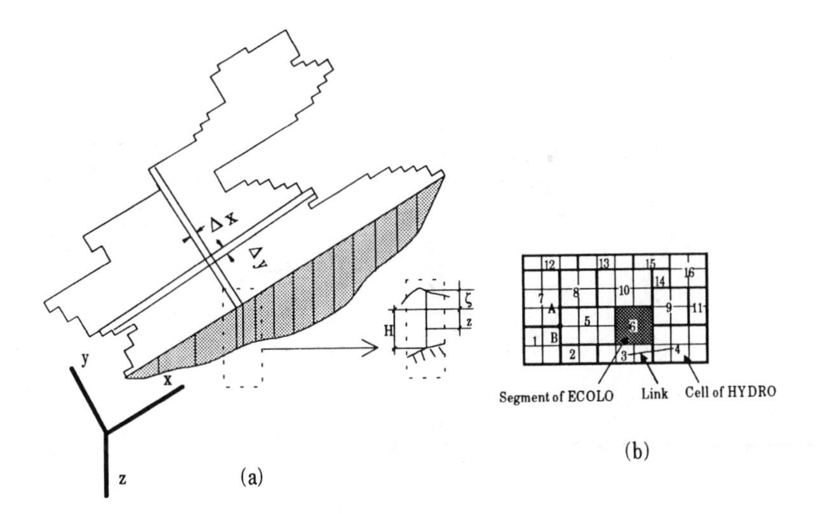

Figure 6. The 3-D Grid Used in Hydrodynamic Computations (a), and Connection with Grids of the Ecological Model (b)

Due to local climatic conditions and irregular seasonal drainage from the rivers because of agricultural irrigation, flow rates and pollutant loads flowing into the bay vary greatly. All three rivers carry water year-round, with flow rates varying from 10 m^3/s to 400 m^3/s.

The main pollutant loads of the bay are (a) partially treated sewage of the city of Thessaloniki (1,000,000 inhabitants), (b) industrial effluents, and (c) organic and pollutant loads discharged by rivers. The most polluted parts of the bay are the northern part, located close to the city; and the shallow delta, close to the river mouths. In these parts of the bay, local eutrophication has been taking place showing irregular variation in time, with algal blooms in spring and autumn.

Since 1984, the university has undertaken an extensive environmental research program in the bay, including collection of water quality data, mathematical modeling, and investigation of remedial measures for cleaning the bay.

Hydrodynamic circulation has been monitored by using floating drogues and current meters. Currents are mainly due to winds. The prevailing winds are from the north, especially in winter. During summer, sea breezes from the south create residual water circulation,

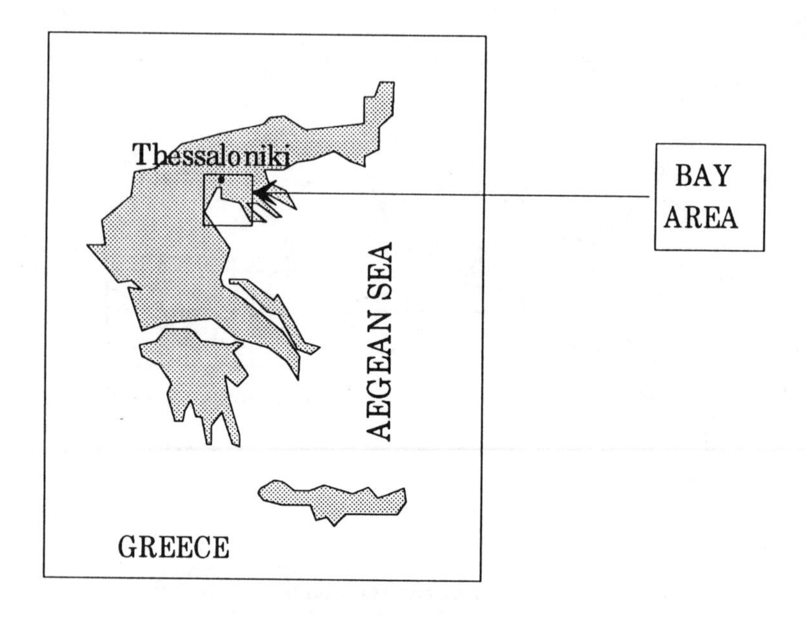

Figure 7. The Bay of Thessaloniki

which influences pollutant transport. In fact, this is the most critical circulation pattern for pollution, because currents recirculate pollution from the outer to the inner bay, where the pollutant sources are located.

Several water quality data have been collected by sampling at twelve stations, located at characteristic positions covering the whole bay area. Apart from currents and winds, the following parameters have been monitored with seasonal frequency near the surface, at mean depth, and near the bottom of the water column:

- temperature, salinity, density, dissolved oxygen, pH
- nutrients such as NO_2^-, NO_3^-, NH_4^+, PO_4^{3-}, SiO_4^{4-}
- total coliforms and E-coli
- heavy metals such as Cd, Pb, and Cu

The results obtained by simulation using the hydroecological model have been validated in two and three dimensions by comparison with available data (Ganoulis 1988b, 1988c, 1989, 1992). The capability of the model to predict trends in water quality characteristics under climate change has also been tested by taking an average increase of temperature of $4^{\circ}C$ (Ganoulis et al. 1993).

Daily time series of minimum dissolved oxygen (DO), under historical and climate-change-induced daily temperature time series, are shown in figures 8(a) and 8(b). Dissolved oxygen is a characteristic parameter, reflecting the overall influence of many pollutants and eutrophication on the water quality. Simulated daily values of DO shown in figures 8(a) and 8(b) are minimum values near the bottom. Comparing DO time series presented in figures 8(a) and 8(b), one can see that on a daily scale the decrease in DO due to climate-change-induced increase of

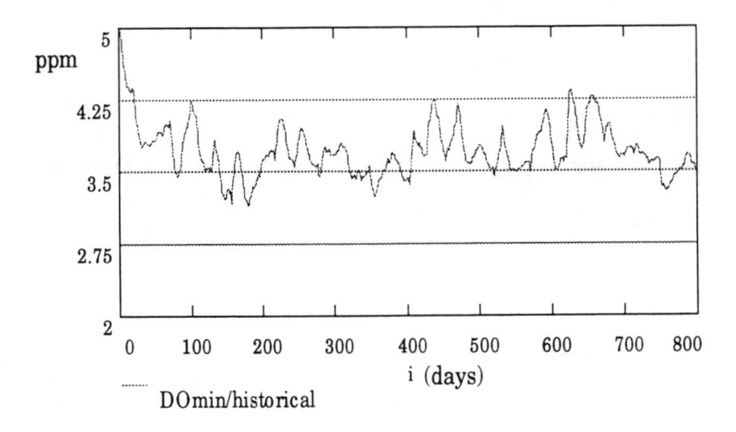

Figure 8(a). DOmin Concentrations (ppm): Historical

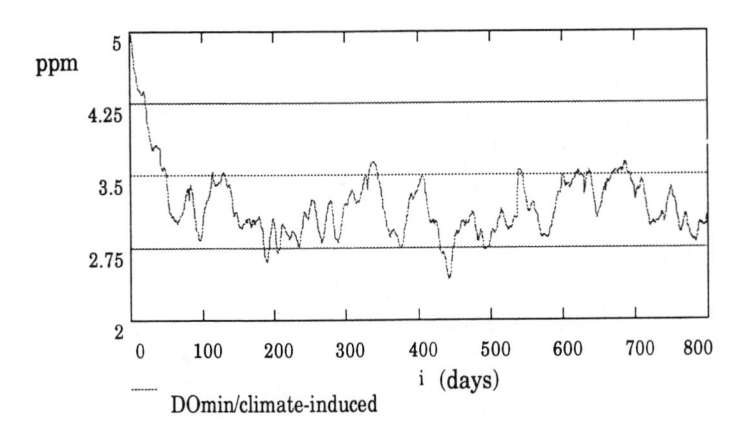

Figure 8(b). DOmin Concentrations (ppm): Under Climate Change

temperature may be much larger than the average decrease over a longer time scale. There is an average decrease of oxygen, which could threaten species living near the bottom.

The climate-change-induced change in phytoplankton is related to total nitrogen concentrations, minimum values of which over the total area of the bay are shown in figures 9(a) and 9(b). An average increase of temperature of 4°C also causes an increase in total nitrogen that is rather small on the average, but more pronounced on a daily scale.

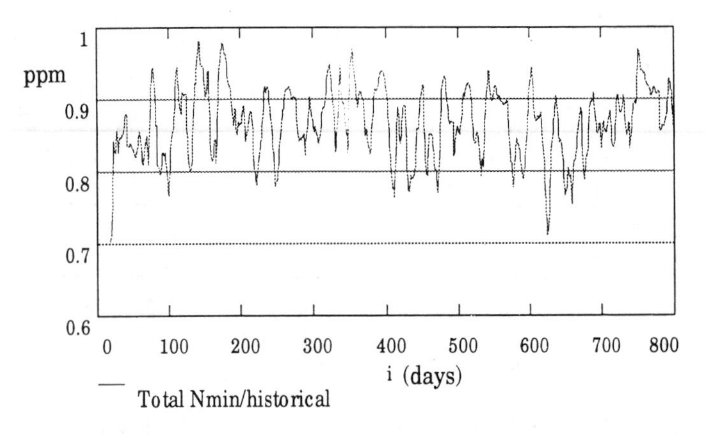

Figure 9(a). Total Nmin Concentrations (ppm): Historical

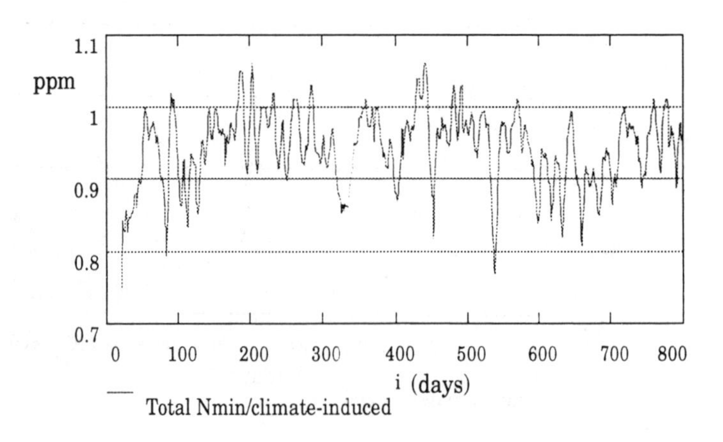

Figure 9(b). Total Nmin Concentrations (ppm): Under Climate Change

Conclusions

A 3-D hydroecological model has been coupled with a meteorological multivariate stochastic model to analyze the impacts of a doubling of atmospheric CO_2 on coastal water quality. The study has proceeded in two parts:

1. Historical daily temperature time series have been used to link atmospheric circulation patterns to local meteorological data.

2. Daily temperature time series, generated under a doubling of CO_2 in the atmosphere, have been introduced into the hydroecological model to analyze impacts on coastal water quality.

For the case study investigated in the bay of Thessaloniki (N. Greece), although no significant change has been found in the average temperature over a year, there is a risk of oxygen depletion on a daily scale. Impacts on phytoplankton concentration and eutrophication change are currently under investigation.

Acknowledgment

Research leading to this paper has been supported in part by a NATO collaborative research grant (CRG 920814) and funds from the U.S. National Institute for Global Change (DOE).

References

Bardossy, A., and E. Plate. 1990. Modeling daily rainfall using a semi-Markov representation of circulation pattern occurrence. *J. Hydrol.* 66: 33–47.

Bardossy, A., and E. Plate. 1992. Space-time model for daily rainfall using atmospheric circulation patterns. *Water Resour. Res.* 28: 1247–59.

Duckstein, L., J. Casti, and J. Kempf. 1979. Modeling phytoplankton dynamics using catastrophe theory. *Water Resour. Res.* 15 (5): 1189–94.

Fischer, H., J. List, R. Koh, J. Imberger, and N. Brooks. 1979. Mixing in inland and coastal waters. New York: Academic Press.

Ganoulis, J. 1988a. Environmental impacts from the sewage works to the bay of Thessaloniki. Phase B': 1988–91. Report submitted to the Ministry of Environment and Public Works, (in Greek).

Ganoulis, J. 1988b. Pollutant dispersion in oceans. In *Disorder and mixing,* edited by E. Guyon et al. The Netherlands: E. M. Nijhoff.

Ganoulis, J. 1988c. Modelling the sewage assimilative capacity of coastal waters in the Mediterranean. *Rapp. Comm. int. Mer Medit.* 31 (2): 321.

Ganoulis, J. 1989. Marine pollution assessment from wastewater discharges. *Proc. IAHR Congress, D. 123–9.* Ottawa.

Ganoulis, J. 1992. Dispersion et disparition des bactéries coliformes dans la baie de Thessaloniki. *Revue des Sciences de l'Eau.* 5: 541–54.

Ganoulis, J., L. Duckstein, I. Bogardi, and I. Matyasovszky. 1993. Water quality in coastal zones under variable climatic conditions. To appear in *Engineering Risk and Reliability in a Changing Physical Environment.* NATO ASI Series, Amsterdam.

Ganoulis, J., and C. Koutitas. 1981. Utilisation de données hydrographiques et de modeles mathematiques pour l'étude hydrodynamique du Golfe de Thessaloniki. *Rapp. Comm. int. Mer. Medit.* 27: 41.

Ganoulis, J., and J. Krestenitis. 1982. Choix du site optimum d'évacuation des eaux usées dans une région côtiére. *Proc. XVIIèmes Journées de l'Hydraulique.* SHF, Nantes.

Ganoulis, J., and J. Krestenitis. 1984. Modélisation numérique du transport côtier des pollutants en Méditerranée. *Proc. XVIIIèmes Journées de l'Hydraulique.* SHF, Marseille.

Matyasovszky, I., I. Bogardi, A. Bardossy, and L. Duckstein. 1993a. Estimation of local precipitation statistics reflecting climatic change. *Water Resour. Res.* 29 (12): 3955–68.

Matyasovszky, I., I. Bogardi, J. Ganoulis, and L. Duckstein. 1993b. Local climatic factors under global climatic change in a Greek watershed. Working paper, Dept. of Civil Eng., Univ. of Nebraska, Lincoln, 68588, USA.

Quétin, B., and M. De Rouville. 1986. Submarine sewer outfalls: A design manual. *Marine Pollution Bulletin* 17 (4): 132–83.

WHO/UNEP. 1985. Guidelines for the computations concerning marine outfall systems for liquid effluents. Athens.

SESSION I

Federal Risk Management Policy: Where Is the Federal Government Heading? Where Are the Problems?

Jonathan W. Bulkley, F. ASCE[1]
Rapporteur

Yacov Haimes welcomed everyone to the conference. After introductions of all the participants, he and Dave Moser presented the mission statement for the conference. The mission statement is as follows:

1. We need to listen to each other and then exchange information and knowledge on risk-based decision making.

2. We need to listen to each other to facilitate getting to know each other on a personal basis.

3. We want to create an environment that encourages dialogue and communication—for example, fewer lectures and more discussion.

4. We must open and maintain new lines of communication.

5. We must be able to challenge ourselves and challenge our old assumptions.

6. We must free ourselves of prejudices.

7. We must be ready to reexamine our biases—both professional biases and personal biases.

---

[1]Professor of Natural Resources and of Civil and Environmental Engineering; Director of the National Pollution Prevention Center for Higher Education, University of Michigan, 2506 B Dana, Ann Arbor, MI 48109-1115

8. We should be open to discovery—new theories and new methodologies.

9. We should be ready to accept the premise of risk assessment as part of management.

The two co-chairs, Yacov Haimes and Dave Moser, posed four questions for the participants to keep in mind during this first session. These questions are as follows:

1. What is the efficacy of models on risk analysis and decision making?

2. To what extent are the models and methods of risk analysis credible?

3. How important are the methodologies for the quantification of risk as part of the risk-assessment process and how critical is the understanding of the process itself?

4. Who should decide what risks are acceptable for whom, in what terms, and why?

Dave Moser observed that many of us are being asked to do more with less fiscal resources. This is one reason why the concept of risk and risk analysis is being encouraged. Resistance to this approach is observed.

Federal Risk Management Policy: Where Is the Federal Government Heading? Where Are the Problems?

Anthony Thompson, Esq.

Anthony Thompson, at the time of the conference an attorney with Perkins Coie in Washington, D.C., but by press time for these proceedings a partner in the Environmental Law Group at Shaw, Pittman, Potts, and Trowbridge, also of Washington, D.C., gave the keynote presentation for the session. He focused on human health risk and outlined the elements of risk assessment and risk management. He urged the participants to become involved in risk assessment activities. The risk assessment process is science-driven; risk management is informed by science but not determined by science.

The speaker emphasized that the field of risk assessment is moving away from providing the single, and often worst-case, analysis of a particular situation. Now the emphasis is much more on the uncertainties associated with both the exposure and the effects. These uncertainties need to be accounted for in risk management decisions and in communication regarding such decisions.

The speaker provided a number of interesting case examples regarding the historical approach taken by U.S. EPA to regulate and control toxics. What emerged from this presentation was the strong suggestion that the procedures of the past—especially the worst-case analysis—are not able to provide an adequate framework for evaluating the wide range of biological information needed. He called for avoiding the deterministic approach to risk assessment—i.e., the use of single numbers or point estimates as input to risk assessments—and for moving toward probability distribution analysis.

Following Tony's presentation, a spirited discussion took place. The first question asked for a clarification of what is meant by "real risk." Tony responded that it is important to get away from a person fixed in a single place for the worst-case scenario over a seventy-year period. Rather, it is important to obtain a best estimate of real-life exposure.

A second question noted that, clearly, certain EPA analyses had been very conservative. However, it was observed that, at present, no laws consider the holistic exposure of any individual to all the environmental insults that are present at any moment. Tony responded that this position is often articulated by environmental groups such as the NRDC—namely, that risk assessment is limited since it considers one factor at a time and subjects the individual to multiple insults.

A third question raised the issue of how one communicates the results of probabilistic approaches so that they can be understood in simple terms. Tony responded that it is necessary to explain complex relationships term by term. For example, demonstrate to the audience the range of parameter values and indicate how these ranges affect the results obtained. Tony emphasized that it is important to take sufficient time to explain the process of the findings to the audience.

The next comment pointed out that the initial, conservative assumption utilized by the EPA balanced the more crude risk models. As the models become more accurate, we should expect to move away from the very conservative assumptions. Tony observed that risk assessment/analysis needs uncertainty. The analyst needs to provide more information in the range, i.e., the confidence limits associated with the results.

Tony agreed with the next observation—namely, that regarding ecology we really are back to square one. If ecological risks are reevaluated in light of more accurate information now available, the results may indicate that certain environmental risks are not as severe as initially believed.

The next comment focused on the requirement of cost-benefit analysis and what implications, if any, this would have for risk management, especially as it relates to environmental equity and environmental justice. Tony responded that this is a very sensitive area and concerns exist that the more liberal application of risk assessment may lead to charges of environmental racism.

The final questioner asked Tony to respond to the idea that his presentation appears to move toward less rigorous environmental standards. If so, is this a surrogate for less regulation? Tony responded that some critics of risk assessment say that it is a way to return to the old idea that dilution is the solution to pollution. However, he further stated that we need to have a clearer understanding of where the federal government is going with risk assessment. If this is known, then it may be possible to address problems that are currently not on the table, such as alcohol and smoking as two activities that pose very significant risks in our society.

Use of Risk in the Corps of Engineers
Leonard A. Shabman, Virginia Polytechnic Institute

As the second speaker, Len made the point that the use of risk analysis in the Corps dates from 1984–85. It was implemented to assist in budget decisions where fewer dollars were available and the Corps needed to make the best use of limited resources. A number of issues need to be addressed when any group, including the Corps, considers the use of risk analysis. These issues include the following:

- Need to consider where risk analysis may be most usefully applied
- Need to understand the groups and interests that fight against and oppose risk analysis
- Need to agree early who bears the cost if the project fails
- Who makes these decisions?

As a general rule, the trade-off is a reduced likelihood of failure versus increased cost—if you increase the investment, the likelihood of failure is generally reduced.

The use of risk analysis brings into question certain design practices that have not been questioned in the past. As a consequence, traditional standards and rules of thumb are being called into question because of severe cost pressures. The current design practice tends to be a mix of formal standards, rule of thumb, and tradition that results in a risk-averse design procedure. The pressure to use risk analysis, then, is in conflict with long-held values of the Corps of Engineers. For example, the history of civil works by the Corps includes elements of professional reputation and conservative design. The Corps is risk averse. There is also the need to maintain the tradition of the agency—the Corps doesn't want its projects failing. In addition, there is the design system within the Corps that relies on standards. This provides a relatively effective way to maintain organizational control. Finally, there is the issue of liability in case of project failure.

Following Len's presentation, the discussion opened with the observation that engineers are responsible and that the decision process that leads to certain project designs should be open. A second individual observed that what Len presented indicates a difference between a random failure and a competent failure, i.e., is the failure random or is it because of a basic design fault? Len observed that the first time a project that has been designed/built under this new approach fails, it may result in the Corps returning to very conservative design practice.

A third individual raised the question of whether the citizens of Sacramento really understand the flood risk presently before them. It was pointed out that property with an estimated value of $37 billion is behind Sacramento's levee system. It has about 65% protection at the moment and the choices to be made will either enhance or reduce this protection for thirty years. Len observed that most people in Sacramento are not aware of the flood risk problem.

Another person then asked if the people are ready to accept the responsibility to take on costs if a major flood occurs. Len responded that it may be necessary to bring the decisions about risk and about safety back into the technical area.

It was observed from the floor that safety reflects a level of risk we are comfortable with and risk is a quantitative measure of damages expected with specified flood magnitude. The question was then placed as to what it would take to achieve a level of safety that could be uniformly imposed. Len responded that at present we do not have a good feel for what we are spending on dam safety. Dialogue is in process to address this issue.

Len pointed out that the people of Sacramento understand the National Flood Insurance Program. Since they have less than the hundred-year flood protection, the community can't grow. The tragedy will occur if people build beyond the protection afforded by the levees; if they do, the damage will be much greater than it would have been otherwise.

Len noted that the challenge is, if all of this information is placed out on the table, how will it be resolved? There is much information in public view that politicians do not want to receive.

A final question/observation related to a process to encourage the design engineer to use initiative and alter the design standards—either more conservative or less conservative. This would be done in place of risk assessment. Len responded that while any organization may talk about flexibility, the established institutional hierarchy acts to constrain the actual flexibility in the organization.

Review and Summary of the EPA/U.Va. Workshop,
"When and How You Can Specify a Probability Distribution When You Don't Know Much"
Yacov Y. Haimes, University of Virginia

This third speaker provided a review and summary of the U.S. EPA/University of Virginia workshop titled "When and How Can You Specify a Probability Distribution When You Don't Know Much?"

Yacov noted that this workshop had two primary objectives. First, to assess the state of the art in selecting input distributions for environmental risk. Second, to establish a theoretically sound and defensible foundation for specifying probability distributions under such conditions. To accomplish these objectives, five white papers were prepared. These included the following topics:

- Quality assurance
- Tails and extreme events
- Uncertainty versus variability
- What are the bases for our choices and when should we use probabilities?
- Uncertainty analysis for decision makers

The workshop proceeded in two distinct parts. Part I included (1) the principle for selecting probability distributions for uncertainty analysis, (2) uncertainty and variability in assessing the quality of uncertainty analysis, and (3) the value of uncertainty analysis to decision makers.

Part II of the workshop took three different perspectives: (1) a top-down view: Monte Carlo analysis for needs of decision maker; (2) a bottom-up view: recommended tools for selecting probability distributions and Monte Carlo techniques; and (3) a forward view.

Yacov then reported on output from the workshop. The results were presented as a series of issues under each of the six major elements considered in the workshop as a whole. It should be noted that a special issue of the journal *Risk Analysis* will publish the proceedings of this workshop.

(Note: Because of time limitations, the discussion period for this third paper was very limited.)

A number of people who were part of the workshop and who were present at the conference added key points. These included the following:

1. It was good to have representatives of the professional community coming together and bringing so many different perspectives. It was an incredible achievement.

2. It was useful to observe regulatory people move toward Monte Carlo processes.

3. A lot of ideas were put forth. Good criticism was generated. Some regulatory people were willing to alter positions; others were not willing to change.

4. There was not uniform agreement at the end of the workshop. However, it certainly opened people's minds.

SESSION II

Risk and Reliability of Water Resources Infrastructure and Engineering Approaches to Risk and Reliability Analysis

Richard Males[1]
Rapporteur

Four papers were presented in this session, dealing primarily with tools and techniques to perform risk analysis. Three of the papers examined water distribution systems, while the fourth addressed navigation and flood control structures.

Larry Mays presented a review of risk reliability procedures for water distribution systems. He pointed out that although the technology to do risk-reliability analysis is well reported in the literature, municipalities are not using these techniques. Two types of failures can be distinguished for a water distribution system—mechanical and hydraulic. Hydraulic failure implies not meeting a pressure at a certain location for a time period. Procedures for defining overall system reliability based on failure probability of each link and node were outlined. The general problem is to minimize cost, subject to flow, head, and reliability constraints; the solution is to use optimization techniques interfaced with hydraulic simulators.

Discussion centered on why this type of analysis is not more widely used if the necessary tools are available. The general lack of attention to maintenance was cited. Yacov Haimes noted that money for capital investment is different from money available for maintenance. Larry Mays distinguished between preventive and corrective maintenance. Steve Cone discussed the problems associated with projecting and budgeting maintenance costs. Dave Moser suggested that it is more effective to do maintenance when something fails, unless maintenance cost is small. Jim Heaney noted, on the positive side, that there is a growing

[1]RMM Technical Services, Inc.; 3319 Eastside Avenue; Cincinnati, OH 45208

resistance to capital investment, with the electric utility industry taking the lead in looking at maintenance approaches. George Apostolakis suggested that the training and background of engineers plays a role—water utility engineers were not trained in probabilistic approaches.

Rafael Quimpo's presentation described a technique to define a "reliability surface" for a water distribution system, based on the probability of connecting a source of supply to any given demand. The problem can be computationally demanding for very large systems. For water distribution systems, in addition to source-demand connectivity, requirements exist as to pressure, quantity, and quality. Attempts continue at developing a capacity-weighted source to demand reliability, in order to be able to present decision makers with a reliability surface. Again, discussion centered on the water industry's apparent lack of interest in these tools.

Duan Li presented the water infrastructure risk ranking and filtering (WIRRF) method, describing a methodology used to rank water distribution subsystems on the basis of risk. The method is derived from risk ranking and filtering methods developed for NASA's space shuttle. Subnetworks are analyzed individually, and the optimal resource allocation is determined for the network as a whole, based on the risk-improvement/cost ratio.

Mary Ann Leggett presented a discussion of reliability-based engineering assessment of U.S. Army Corps of Engineers' structures, oriented toward analysis of navigational and flood control structures. Specific estimations of component probabilities of failure are developed. A reliability index is developed, as the mean of the safety margin/standard deviation of the safety margin. The overall goal is to develop a condition-based prioritization for rehabilitation work. Discussion centered on the issue of simplifying assumptions made, on the difficulty of obtaining data, and again on the need for training engineers in this form of analysis.

SESSION III

Behavioral, Social, and Institutional Aspects of Risk Analysis

Richard C. Schwing[1]
Rapporteur

Decision making is relatively straightforward when the scope of a problem is narrow and everything is known. However, the world is a complicated place! In Session III, the papers presented provided a list of four items that become pitfalls for decision makers when neglected. They are as follows:

- parameter uncertainties as distinct from model uncertainties
- differing priorities and agenda among different public bodies
- resource allocation too narrowly conceived
- caution in defining the "correct" estimate of benefits

George Apostolakis provided the group with a theoretical framework for risk assessment that distinguished between model uncertainty and parameter uncertainty. By considering a number of models, Apostolakis was able to provide a general case in which the uncertainty about which model is applicable to a physical situation is reflected in a form that makes it explicitly dependent on the objectives of the analysis.

Jonathan Bulkley described two activities in the state of Michigan that involve risk analysis. The first activity, which ranked environmental issues facing the state, was patterned after the U.S. EPA's risk reduction report. Of the twenty-four environmental issues in the state, 70% are directly related to water resources. The issue "Absence of land-use planning that considers resources and integrity of ecosystems" was ranked the single most important issue by three separate committees. The second activity attempted to establish the trade-off between increased cost and risk reduction by combining sewer overflows in the

[1]Principal Research Engineer, Operating Sciences Department, Research & Development, General Motors Corporation, 30500 Mound Road, Warren, MI 48090-9055

Rouge River basin. Jurisdictional and regulatory agency conflicts required conflict resolution. A negotiated nine-element agreement was eventually achieved. Because of a lack of information on risk reduction, it was clear that the trade-off between cost and the risks of overflow from combined sewage awaits the completion of this demonstration project.

In my paper I attempted to provide sufficient data to show that measures to anticipate narrowly defined risks are usually inferior to a strategy of resiliency. This thesis of "richer is safer," coined by Aaron Wildavsky, requires that we not spend large sums of money on small risk reductions when efficient risk reduction opportunities remain unfunded.

Leonard Shabman examined three methods to estimate the benefits of flood risk reduction. The methods include property damages avoided (PDA), land price analysis or hedonic price analysis, and contingent valuation (CVM). The values obtained by these methods were compared across each of the five flood zones. Shabman concluded that the best that can be said of the PDA and land price methods is that they follow a credible logic, as benefits fell with distance from the river. However, he concludes that no estimate can be taken as the "correct" measure of benefits. Benefits cannot be the final factor determining project choice.

SESSION IV

Risk Management Strategies for Natural, Manmade, and Technological Hazards

Steven R. Cone[1]
Rapporteur

The speakers in this session presented a varied series of topics, including a general overview of risk assessment approaches in bio/chemical hazards, specific studies on people's behavior and views of natural hazards and perceptions regarding fresh water, and a prototype model for decision making for maintenance dredging. Elizabeth Anderson of Sciences International chaired the session and introduced each of the speakers.

Ms. Anderson presented the first paper, titled "Risk and Reliability: Health and Ecological Risk Assessment." The paper focused predominantly on how past practices of dealing with health risks have taken a conservation approach by representing the most plausible upper limit of potential exposure to the current, but still evolving, more site-specific risk-based approaches. The more modern risk-based approaches deal with toxicity assessment, exposure assessment, risk characterization and uncertainty analysis, and ecological risk assessment. Ms. Anderson showed examples of how better sampling can result in fewer samples, lower costs, and greater accuracy of results. She also presented a relative comparison of perceived problems of ground water contamination by showing alternative risks for similar but common exposures to other potential health hazards.

Questions and discussions on Ms. Anderson's paper included "Where are we with ecological risk assessment and where are we going?" In response, Ms. Anderson stated "very slowly coming along," the measure is survivability. EPA is endorsing newer concepts, there are

[1]Economist, Office of the Chief of Engineers, U.S. Army Corps of Engineers, 20 Massachusetts Avenue, Washington, DC 20050

many publications, but work is still developmental. Not specifically discussed, but observed by this participant, was the commonality of the use of the traditional approach of "most plausible upper limit" for exposure and toxicity in health and ecological risk issues and the traditional civil engineering approach of relying on conservative "design standards."

The second paper was presented by Mr. Kevin O'Grady of Planning and Management Consultants. His research and paper were linked to Dr. Shabman's paper presented earlier in the conference as both dealt with natural hazards and people's perceptions and preferences regarding risk. Mr. O'Grady began his presentation by discussing the increasing interest in natural hazard reduction as a result, in part, of improved media attention, which increases awareness and sympathy, and as a result of basic intolerance of the vagaries of nature. However, there are decreasing public funds available for hazard reduction. This results in pressures to reexamine public decision making. The economic model of choice is one-dimensional in dealing only with maximizing utility and is subject to increasing criticism.

The objective of Mr. O'Grady's research was to estimate time preference rates for protection from shore erosion along the Great Lakes shoreline. A contingent valuation method was employed and data was obtained from mail surveys. The findings were that people on average were risk seeking, choosing and remaining in shoreline locations, yet maintained a strong time preference for protection now through public resources. The risk-seeking attitude was likely a result of truncating probability and consequences of extreme events. Policy implications of the results, according to Mr. O'Grady include the assessment of benefits, the behavioral implications of those affected by hazards, and the level of hazard awareness.

The discussion that followed was almost rhetorical in nature rather than specific questions regarding the research and findings. One participant offered that perhaps the concept of present value, that is discounting, does not mean much to the general public. Another participant expressed some surprise at the results of discounting and time preference and suggested that the mean value is meaningless because many persons focus on the extremes. To close, the discussions, one participant remarked, in reference to the International Joint Commission Study on the Great Lakes, that some individuals and institutions suggest that more land-use controls are the answer. The idea being that people need to be saved from themselves because they are not rational. He further asked, rhetorically, does the result of Mr. O'Grady's study suggest that people are rational and need no protection?

Mr. Robert O'Connor of Pennsylvania University was the third speaker, with a paper focused on how survey questionnaires were conducted and the need to ask questions in several different ways to obtain good results. One question from the participants was "What are three do's and don'ts for surveys?" Mr. O'Connor responded with the following: 1) pretest the survey instrument, 2) hire those with experience, and 3) focus on the group you are after, don't worry about randomness. Another question pertained to public involvement with regard to hazardous and toxic substances in light of the upcoming Superfund reauthorization. This question asked what Dr. O'Connor's results tell us about how to get the public more involved in these matters. Dr. O'Connor replied that citizens are already involved, there is no reason not to bring them into the decision process more, and that greater public involvement should not slow down environmental cleanup decisions.

Mr. Sam Rattick of Clark University presented a paper on a location allocation model based on past navigation maintenance dredging practices. The model is still in developmental stages. His work is closely tied to the work Mr. Mike Walsh presented in a subsequent session. Future model extensions include incorporation of variability and uncertainty in dredge operating characteristics and costs, navigation user benefits by level of reliability of channel conditions, assessment of mobilization and demobilization costs for alternative practices, identification of nonindependent reaches of channels, and effects of dredging on shoaling rates. There were numerous questions on the specific workings of the models and the ability to deal with the variabilities associated with sediment transport and shoaling.

In summary, the first and last papers dealt largely with quantitative and qualitative risk-based analyses in decision making using models and observable and scientific data. The environmental analyses, however, having greater adverse consequences than the dredging issue, are still in their infancy. Both concerns must cope with high uncertainty of data. The middle papers dealt more with people's attitudes and perceptions of risk, and the implication of choices in decision making based on these revealed preferences. Clearly the environmental issues likewise must cope with people's perceptions in decision making involving risk. All the papers included issues pertaining to public policies regarding risk reduction.

SESSION V

Risk of Extreme and Catastrophic Events

Jim Lambert[1]
Rapporteur

Assessing the extreme tails of a probability distribution, where events of very low probability can nevertheless lead to catastrophic outcomes, is one of the most widely discussed and challenging problems of risk-based approaches to management. From minimizing performance errors in hospital surgery to planning for flood protection, this session focused on the assessment and management of the risks of worst-case scenarios.

Elisabeth Paté-Cornell observed the following in situations ranging from off-shore oil platforms to surgical anesthesia: quite often the blame for catastrophes can be attributed to the organization. Elisabeth is trying to identify and assess the risks of some negative practices and organizational features that affect surgical anesthetists in Australia. In a five-year program, surgeons in Australia and New Zealand have reported 2,000 incidents, and their distinguishing characteristics, where an error by the anesthetist led to a patient injury or fatality. There is considerable bias in the database—that is, certain types of events are reported more regularly than others. In evaluating the benefits (reduced fatalities, injuries, etc.) of changing standard procedures in the operating room, a few major challenges are to account for the irregular reporting, to "clean" the database, and to produce a credible analysis using all the best available information. An incident database in need of de-biasing is common to many situations, including the reporting of accidents leading to pipeline breakages in the Gulf of Mexico.

The approach to de-biasing the anesthesiology database is to partition the incidents into classes. Within each class one assesses the *condi-*

[1]Center for Risk Management, University of Virginia, Charlottesville, VA 22903

tional probability that an accident is reported, given that the accident in fact occurred. For example, one finds that a given type of initiating event or consequence is more often reported, or that the reporting frequency depends on hospital size, among other factors. From a model tracing the incident from its initiating event to its consequences, one estimates the reduction in risk per operation of, for example, changing the duty cycle, increasing training on a simulator, requiring recertification, or improving supervision of residents.

Stan Kaplan insisted that Bayes' theorem should be the definition of logical inference for the risk analysis professional. Probabilities are quantifications of our state of confidence about entities in the real world; it is wrong to think of the probabilities themselves as real—they don't exist in a physical sense. Moreover, probabilities are trustworthy when they are based on all the evidence and are independent of personality. They are objectively related to the evidence and there is an established science of how to develop probability curves. In contrast, the choice of a preferred curve in risk management is a subjective value judgment based on all the risks, costs, benefits, and available options. Similar to the choice of a preferred option, the generation of options is a creative endeavor. Stan pointed to the seriousness of miscommunicating probability statements. When a military analyst forecast the success of the Bay of Pigs invasion to be of the probability "fair, or about 30%," the report filtered up through the command hierarchy as simply "fair probability of success." Unlike many cookbook approaches to quantifying uncertainty, the underlying theory of probabilities is complete and consistent and has the most solid link to decision problems.

Mike Burnham spoke about the state of the practice in risk-based flood-damage reduction across the Corps of Engineers. The Corps' planning studies are evolving rapidly to incorporate risks involving hydrology, hydraulics, and economics, while the use of safety factors such as design freeboard are being phased out. Planners have the following as objectives for risk-based approaches: to be more quantitative, to have well-defined processes for sizing structures, to produce comparable analyses across projects, and to be able to interpret their studies to local sponsors. Much of the analysis supports the calculation of the expected annual damages and the expected benefits of alternative projects. There is considerable effort to assess probability distributions of discharge and stage-discharge and stage-damage relationships; Monte Carlo simulations are being tested as a means to account for uncertainty in these modeling elements. Issues include whether FEMA and other agencies with which the Corps deals will accept the risk-based approach and integration of the new results with nonmonetary benefits analyses. Economic analyses are typically second-order compared to broader congressional mandates, such as providing for a socially equitable distribution of benefits.

David Moser contended that since the basic premise of the Corps' flood program is infrastructure investment, it is natural to emphasize improving estimates of damage curves and other economic parameters over flood frequency studies. A depth-damage study involves collecting data on the locations, numbers, and types of structures, the values of the structures, the "effective age" of the structures, and the distribution of value by floor. There is considerable uncertainty about market values, potential damages, and the influence of location on value. Compounding the difficulties is that there is diminishing interest in post-flood surveys. Tools involved in improving the damage curves include calibration of different appraisers, ratios of the value of the contents to the value of the structure, probabilistic modeling of uncertainty, and moving from a 100% inventory of structures to an inference about the community from a sample. Flood warning and preparedness issues also influence the depth-damage relationship. Additional concerns include evaluating the cost of doing a study at all, relative to the value of the study, and the identification of uncertainties that matter (e.g., a two-foot versus a ten-foot resolution on contour maps). Indirect benefits of flood-damage studies can include increases in community awareness and preparedness from participation in the process.

SESSION VI

Risk Management Strategies for Natural, Manmade, and Technological Hazards

Benjamin F. Hobbs[1]
Rapporteur

This session featured three important applications of risk and reliability analysis in water resources planning. Papers were presented on incentives to use reliability and risk analysis in water supply and distribution (J. Heaney), integration of structural flood control measures and warning systems (D. Li and Y. Haimes), and risk-based budgeting for maintenance dredging (M. Walsh and D. Moser).

Questions following the water supply and distribution paper focused on whether reliability analysis will become widely adopted by water utilities. Some engineers are accustomed to performing reliability analyses of reservoirs, but very few have looked at pipelines, pumps, treatment plants, valves, etc., in the same way. Is the water utility industry where the electric utility industry was in the 1950s and 1960s, when the latter industry began to perceive a need for probability-based reliability analysis? Since then, reliability analysis has become ubiquitous for electricity generation and, to a lesser extent, transmission. Is it inevitable that reliability analysis will also become widely used by water utilities? Is liability a potential "show-stopper," in that the abandonment of standards for risk analysis will expose planners and engineers to lawsuits when (not if) failures occur? Yet risk analysis is used now for reservoir design; there is no widely adhered-to standard (e.g., meet a fifty-year drought) in that case.

The flood warning paper stimulated extensive discussion of the potential of such systems. One person asked where real data could be

---

[1]Associate Professor of Systems Engineering and Civil Engineering, Case Western Reserve University, Cleveland, OH 44106

obtained for evacuation costs, response functions, and resulting damage reductions. There are unfortunately few examples of analyses that have produced such information. Dr. Li's reply was that we have a chicken-and-egg problem. Once studies demonstrate the potential benefits of coupled analysis of flood warning and structural measures, then agencies will be motivated to obtain the needed data. Indeed, the U.S. Army Corps of Engineers now has a policy that flood warning systems must be examined in flood control studies. Other members of the audience found linkages between this paper and other work on warning systems by D. Milleti at the University of Colorado and the Corps' Hydrologic Engineering Center (see *Risk-Based Decision Making in Water Resources*, 1986, edited by Y. Y. Haimes and E. Z. Stakhiv, published by the ASCE). Some of the lessons from these sources include the following:

1. The warning system needs to be used more often than once every 10–15 years to make sure a response occurs.

2. The Corps believes that warning systems are worth building at the reconnaissance stage of a flood study, even before a structure is built.

3. A Markov model has been developed to calculate the steady-state probability of warning errors (Type I, Type II); this type of approach might be used to study combination warning/structural systems.

The maintenance dredging paper addressed the question of forecasting budgets for Corps dredging activities, given that sedimentation and dredging costs are random. This is basically a problem in cost accounting: understanding what fixed and variable costs are associated with different activities. Some questions concerned the use of the information that such an analysis generates, and what information is still missing. It was pointed out that dredged material disposal is excluded, even though it is getting more costly than dredging itself, and that no measures of the benefits of dredging have been obtained. Mr. Walsh pointed out that good estimates of dredging costs by reach or river system naturally lead to a review of whether the benefits of maintaining particular channels and harbors justify that expense. Presently, Congress just authorizes particular projects, so the Corps does not need to consider benefits. In the present budget environment, however, benefit estimates could help guide judgments about which activities to maintain. For instance, a decision has been made to cease maintenance of sixteen locks on the Fox River.

SESSION VII

Computer Software for Risk Analysis

Walter M. Grayman[1]
Rapporteur

This session was directed toward computer software that is designed for use in various aspects of risk analysis. It comprised three papers and a formal and informal demonstration of software.

In the past, risk analysis was generally performed using custom-designed computer software specifically built to address a specific problem in risk analysis. Among engineers, such software was generally written in languages native to the engineer, such as FORTRAN and BASIC, and adapted, when needed, for use with problems of similar substance.

The three papers and supporting demonstrations in this session demonstrated some recent trends in computer software that are being applied in the risk analysis field:

1) Application of user-friendly general software that may be applied to a wide range of technical problems with no need for custom programming skills;

2) Development of problem-specific, data-driven software that may be easily applied to a wide class of problems with or without some customization.

3) Development of problem-specific software designed to address a particular problem in risk analysis with emphasis on ease of use and program speed in order to support "what-if" scenarios posed by decision makers or the public.

---

[1]Principal, Walter M. Grayman Consulting Engineer, 730 Avon Fields Lane, Cincinnati, OH 45229

In the presentation titled "Commercial Software for Risk Analysis" by Charlie Yoe, fifteen software packages were reviewed. Many of these packages were designed for use with commercial spreadsheet programs to support Monte Carlo simulations incorporating a stochastic element. Other packages provided capabilities to fit distributions to data, perform statistical forecasting, utilize mathematical programming techniques, build and analyze decision trees and influence diagrams, and apply techniques in the area of neural networks. Most of the software was designed using a graphical user interface (GUI) and operated under Microsoft's "Windows" or on a Macintosh computer.

In the presentation on "Risk Simulation Model for Rehabilitation Studies," Males et al. discussed a custom-written prototype program for use in evaluating major rehabilitation alternatives. General purpose spreadsheet-based software, used in a pilot implementation in this project, proved to be too slow and inflexible. The software prototype that was developed incorporated modern object-oriented techniques under the C++ language in conjunction with data base input-output to make the process easily extensible and data-driven.

William Rowe's paper on "Uncertainty versus Computer Response Time" relied on a custom-written program. In this case, the primary objectives of the program included explicit display of different risk perspectives in a package that could be applied in "real-time" to test alternative "what-if" scenarios posed by decision makers and public participation groups.

The consensus of the conference participants was that the availability of software is progressing at a very fast rate and that the combination of general purpose software and robust custom-written programs will greatly expand the tools available in the future.

SESSION VIII

Climate Change

Jim Lambert[1]
Rapporteur

Forecasts of climate-change impacts on water resources alternately reinforce our complacencies or demand immediate shifts in policy, frustrating efforts at meaningful analyses. The wide range of predictions from global climate models most often encourages a wait-and-see approach. These uncertainties and a trust that conventional policy is sufficient to accommodate future changes are paralyzing managers into inaction—perhaps for the better, perhaps not. Regardless, risk-based approaches must be improved to (1) synthesize all the existing knowledge and evidence, and (2) reflect the credence of all underlying assumptions. This session captured some diverse thinking on how risk-based engineering tools cope with variable and uncertain climatic conditions.

Norm Dudley related that within the Murray-Darling basin, the largest in Australia and the basin with the most irrigation, the temporal variability is so great that the ratio of maximum instream discharge to the mean discharge can be on the order of a thousand. Here irrigation interests are often in direct competition with wetlands, which experienced natural inflows, including extended periods of flood and drought, until the 1940s, for the benefits of reservoir operation. Norm suggested a banking model to balance the benefits to wetlands with those to irrigators for three reservoir commodities: unused capacity, reservoir inflows, and downstream regulated flows. Trade-off curves represent different allocations of benefits to the two interests and allow comparison of the wetlands benefits with the historical conditions. In quantifying the priorities to wetlands and irrigation, Norm proposed varying the weights over a period of time to look for a sustainable situation. As a first cut, the

---

[1]Center for Risk Management, University of Virginia, Charlottesville, VA 22903

historical inflows to the wetlands provide the ideal operating scenario, but there is more work to be done to assess the wetlands benefits more effectively.

The public and political alarm on climate change must be addressed by the Corps of Engineers. Gene Stakhiv asked what the Corps, as a pragmatic and conservative agency, can say sensibly on the issue to water managers. Gene posited that the debate is driven by academicians and environmental agencies, not by water managers. Moreover, existing methodologies for handling risk and uncertainty are too specific and precise for the climate-change problem. "Speculation and ignorance" characterizes the true state of knowledge of the future of climate. The bottom line is that water managers don't have to do too much differently to deal incrementally with potential signals of climate change. It is a business-as-usual policy for water managers to make continuous adjustment for environmental, physical, and public pressures. For example, it is likely that global population will double before CO_2 will; changes in water demand and values are bigger drivers of water policy than potential impacts of climate. Gene's recommendation was for a parallel menu of research into climate change impacts, meanwhile maintaining and enhancing the Corps' current approaches for strategic, tactical, and contingency planning. He identified some important questions: How useful are the global climate models to water managers? Should the water manager adapt to or anticipate climate change? Would one do things differently if a climate change were certain to occur? What should researchers do differently to address this highly speculative problem? What kinds of impacts, including water-quality effects, can be expected for ecological and other unmanaged systems?

Ben Hobbs suggested a decision-tree approach to address possible climate change in the Great Lakes. He asked (1) How could climate change affect the Great Lakes? (2) How would management decisions be affected? and (3) Should we worry now or postpone action? The range of potential areas of impact addressed by Ben included ice cover, water level, hydropower loss, shipping costs, erosion, fishery production, and wetlands. Ben argued that many water resource projects are not for incremental adaptation, but rather are large "do it or don't" decisions about major infrastructure, such as abandonment of a harbor or preservation of a wetland. Assumptions about climate change might impact today's net benefits calculations for such projects. It is important to distinguish commitments that are irreversible from those that can be later adapted. Ben's model involves global climate forecasts under a $2xCO_2$ scenario, hydrologic, other environmental, economic elements in a multiobjective framework. The use of a multistage decision model that accounts for the evolution of uncertainties in time is essential for looking at the "wait or not to wait" dilemma.

Jacques Ganoulis asked whether we are more optimistic than we should be about a potential climate change. Climate change is not a scientific myth—indeed, climate change is a fact of history. How fast will some change occur and how ready should we be for the future in what we know is an unsteady physical environment? Specifically, Jacques is studying the influence of climate change on the determinants of coastal water quality in the Thermaikos Gulf in the Mediterranean. The importance of coastal water quality for humans, biodiversity, fishing, and other uses is well known in Greece, with its 12,000 kilometers of coastline. Climate determines wind patterns and temperature variations—each impactive on dissolved oxygen, the decay of pollutants and dead organisms, the growth of phytoplankton, rates of chlorophyllic photosynthesis, and other parameters. Jacques's models utilize the downscaling of global climate models, the time series of historic and predicted temperatures, and three-dimensional circulation and ecological models with eutrophication processes. That effort makes it possible to investigate the temporal variation of CBOD and dissolved oxygen under the $1xCO_2$ and $2xCO_2$ scenarios. Preliminary results indicate a potential reduction in both the maximum and minimum dissolved oxygen levels under the CO_2 doubling.

SUMMARY OF RESPONSES TO PARTICIPANT QUESTIONNAIRE

A questionnaire was distributed to all conference participants, requesting answers to each of the following questions. A listing of the participants' responses follows each question.

1. List the three most important issues/aspects/elements related to risk-based decision making that were raised during this conference.

- Apostolakis's formulation of model uncertainty using subjective likelihood functions: P_r[satisfactory assumptions/evidence and objective]
- Semantic disagreements/misunderstandings of the meaning of risk and probability—still unresolved
- Risk-based formulation of O and M problems (dredging, repair policy); what should be the decision criterion
- Need to start thinking about a new "social risk compact" between the public and engineers to clarify the limits of social risk acceptance
- Risk analysis now becoming a part of agency analytical procedures
- Still a controversy between statisticians: Bayesians and fuzzy set theorists
- De-biasing incident database, introduced by Paté-Cornell
- Concept of "poorer is riskier" advanced by Dick Schwing
- Talk by Tony Thompson (a lawyer) on the federal risk management policy or the lack thereof

- Communication to participants in decision-making processes
- Structure of decision-making process needs to be developed to use risk infrastructure as a consensus-building process
- Still need to gain acceptance of risk analysis from professionals and managers who think deterministically
- How to get probability distributions when you don't know much
- Comparison between present worth models and other models
- Lack of criteria for selecting acceptable risk for decisions in public policy
- Incorporate risk-cost analysis in engineering design
- Difficulty in realistic estimation of probability distributions, e.g., cancer risk rates
- Directions that the federal government is heading; highly diverse styles of EPA versus Corps in risk area
- Methods of communicating uncertainty on decision making
- Impact of model uncertainty on decision making
- Need for the ability to demonstrate that risk and reliability can be applied to other fields
- Methods for handling time preference
- Questioning "standards"
- Model uncertainty and impact on decision making
- Risk models and case studies
- Implementation of these ideas to a more general audience
- Significant progress in risk balancing across specialty areas
- Need to continue to link water resource research to other areas of risk analysis

- Relating regulatory policy to risk-based concepts
- Advances in including uncertainty
- Organizational/institutional/public constraints in using risk-based concepts
- Technology transfer: A great deal of effort is needed to transfer the knowledge about risk-based procedures to the engineering community
- Adoption of risk-based design: Methodologies have been developed that need to be used in practice
- Engineering design standards: Future engineering design standards need to be based on risk-based standards
- Use of risk analysis in engineering design will be hampered by lack of data in uncertainties and appropriate distribution to describe the uncertainties
- Public's different evaluation of "everyday" risks to which they choose to expose themselves and risks that they feel are imposed on them
- Environmental impact risk analysis
- Commercial software for risk analysis
- Rapid expansion in the federal government of risk-based decision making that is relevant to a variety of water resources applications
- Importance of training students in these areas of expertise
- Monte Carlo methods for risk analysis are software driven and often have nothing to do with reality
- Focus on uncertainty rather than risk is appropriate and healthy
- Understanding uncertainty and how to harness it is an important new direction in the field
- "Ecology risk assessment is back to square one."

- Implication for social decision making of increased use of risk analysis (changing role of public and expert)
- Need to phrase risk in terms of "lives versus lives" instead of "lives versus dollars"
- Changing understanding of risk by our legal institutions
- Public perceptions versus expert information
- Tools are there; how do they get factored into the decision process?
- Costs of complying with "irrational" regulations
- We still have a long way to go regarding state and local agencies and citizen acceptance
- General movement toward requiring risk-based decisions in federal agencies
- Confusion still exists as to the difference between parameters and model uncertainty
- Concern that the legal processes are so strong in the U.S. that collection of anonymous data may be infeasible
- Versatility of the risk analysis methods as developed for water resources applications, such as the application to anesthesia
- Consumer/customer acceptance and understanding of models
- Capability of analysts has advanced faster than capability of decision makers to process risk information
- Risk communication: How can it be done effectively?
- Breadth of applications range for risk-based methods
- Regulation-based risk control appears to be much less cost-effective than risk control aimed at individuals or specific issues that directly affect people

- Limitations of GCMs (Global Circulation Models) and the limited (or lack of) progress that has been made so far in the area of global climate change

- Use of multiobjective decision trees in studying global climate change

- Global climate change needs to be translated to regional impacts

- Risk/decision analysis may not be helpful in climate change impact studies at this time; needs a lot more thought and research

- Extent to which current water management practices may be adaptive to climate change challenges

- Uncertainty in climate change models casts serious doubt on their utility for forecasting future climate change

- Banking approaches to balancing wetland and irrigation needs/objectives

- Distinguish incremental and irreversible decisions in water resources

- Relevance of possible climate change to water management (Stakhiv says "Nyet")

2. List the three most important new ideas/concepts you have learned during this conference that would be helpful to you in your job.

- De-biasing incident databases

- Advances in decisional-risk software

- Different methods of ascertaining risk attitudes

- Four types of uncertainty described by George Apostokalis are a good separation

- Correcting data biases using expert judgment to eliminate nonsense and capture behavior patterns

- Identification of critical uncertainties or critical parameters in the myriad of uncertainty sources is a useful shortcut to extensive structuring and computations

- Comparison of valuations of risk reduction alternative methods

- Do not ask expert for his/her opinions—ask for evidence and confidence

- Combining risks and confidences in risk-cost analysis

- Concept of "de-biasing" data sets may prove useful in effective use of "all" information

- Use of commercial software packages for decision analysis with some better sense for applications that are appropriate

- New ideas for presentation strategies

- Software availability

- Advances outside of water resources

- Can point to COE activities as good precedent for other applications

- Software is very limited and inefficient; only stand-alone systems are available

- Many applications can be performed in engineering applications that have not been touched; applications are only in their infancy

- Application of software tools

- De-biasing of databases

- Use of risk/reliability in other fields

- Len Shabman's work on public perceptions of risk and of avoidance, and comparisons of CV and other methods

- Liability problems with moving from "accepted" standards with precedents to risk analysis

- Software displayed
- Importance of presentation results
- Data de-biasing
- Availability of commercial software
- Increased emphasis on statistics/probability in engineering
- Description of risk analysis in the Corps' flood damage reduction guidelines (Burnham and Moser)
- Risk analysis in major rehab projects (Leggett)
- Commercial software for risk analysis (Yoe)
- Proliferation of handy risk analysis software
- Innovative ideas and insights on structuring problems within risk context
- Look for precedent analyses in electric and gas utility problems
- Availability of software for decision trees, influence diagrams, and probabilistic project management
- Bayesian approach to de-biasing an incident database
- De-biasing biased databases
- Concepts of risk decision procedures that are being used in making public policy
- Procedures for de-biasing a database
- Meaning of probability distributions in risk decision making
- Reliability versus economic considerations for decision making
- Objective versus subjective de-biasing of databases
- Available software for risk analysis
- De-biasing incident database

- Discussion on system reliability

- Knowing who is doing what; broadening the communication network

- Significant advances are possible in risk assessment applied to a variety of topics

- Difference between risk analysis and risk management

- Rowe's range estimates for decision makers

- Use of experts to identify adjustment factors, i.e., Paté-Cornell's method with de-biasing

- Wide range of off-the-shelf computer packages

- Software risk session and the information on the availability of software for risk and decision making

- Availability and extent of software that has become available and should be tried/evaluated

- Importance of Bayes' theorem for risk-based decision making

- Techniques being developed for representing confidence limits—in combined factors

- Uncertainty in sedimentation rates

- U.S. Army Corps of Engineers' sphere of influence/system boundary—be careful of the assumption of it

- Eliminate change interesting and significant topic—one of the most important presented

- Bill Rowe's presentation/topic—perhaps the most important of the conference

- Water resources planning/management practices and contemporary climate change are adequate to deal with climate uncertainty

3. List the three most important issues that need further study in risk-based decision making in water resources.

- Uncertainty in the states of nature, i.e., how to capture uncertainty in states of nature in parameter estimates used in models of natural systems
- Specific pdf's with limited or no data
- Relationship between model and parameter uncertainty and risk-based decisions
- Cross-agency uniformity—FCCSET
- Legal constraints
- Training the public to think in terms of probability
- Training the public to understand there is no certainty
- Training the public that this has tremendous implications for expenditures on hazard mitigation and government aid and will have implications regarding our legal process (litigation)
- Communication techniques for risk
- Incorporation of risk-based techniques in decision processes
- Data needs and data availability to support techniques (plus cost of getting data)
- Examination of the potential role of markets for rights to water resources, where risk is important, i.e., do we want to move from "here" (very limited reliance on markets) to "there" (much greater reliance on markets for water resource allocation)
- Examination of the practical difficulties of moving from "here" to "there," especially legal constraints and how they may be overcome
- Examination of the lack of social costs and benefits of moving from "here" to "there"

- Still no accepted evaluation principles for risk analysis and its role in decision making

- Decision rules and modeling structures are still ad hoc—no agreement among practitioners and agencies

- Risk communication and public decision making still a major problem—experts versus public preferences

- Are our "risks" (probabilities) indifferent between epistemic and aleatory (or uncertainty and variability) sources of variance?

- What are the critical uncertainties to Corps decision making? or What organizational levels and functions does risk best address?

- "How, where, what?" of risk-based decision making adding value to our society? Explicitly—today's reality and the ideal

- Methods for incorporating risk-based decision making in sophisticated computer models

- Transfer of risk techniques used in other fields to problems in water resources

- Standardization of terms and concepts in risk analysis to improve interchange of information

- Risk management: much information on risk analysis, but need more ideas on decision making with risk and uncertainty information

- Defining and characterizing residual risk, i.e., "with project" condition

- Value of risk reduction in nonmonetary terms, i.e., ecosystems, safety, and health

- Social impacts/response

- Derivation of pdf's

- Alternatives to Monte Carlo analysis

- Risk/reliability aspects of ground water modeling
- Training on the use of pdf's
- Methods for handling time preference
- Better developed methods to incorporate uncertainty into risk-based design procedure
- Development of design standards
- Develop methodologies to interlace hydraulic, hydrologic, structural, and economic uncertainties into the risk-based procedures
- Impact of public attitudes on policy options and choices
- Overcoming organizational/institutional constraints to using risk-based methods
- Improving data validity and reliability for risk-related phenomena
- Means to identify rapidly critical decision parameters and uncertainty contributors
- Means to validate expert judgment used to capture behavior patterns
- Use of Monte Carlo simulations to explore uncertainty may be appropriate, but has little to do with risk analysis itself
- Methods for sorting through uncertainties to provide focus for decision making
- Need to develop risk communication approaches
- Issue of economic inequalities; finding ways to expand decision making beyond purely economic constraints
- Development of tools and methods for effective communication of risk-based decisions to decision makers and public
- How to use risk analysis to build consensus around decisions on risk management

- How to share liability for adverse (ex-post) outcomes that occur when sound risk management decisions were made in advance (ex-aute)

- Accounting for human factors in making risk decisions, i.e., survey data, etc.

- Verification of models. Can we or can we not validate a model?

- Expertise input to risk decisions: How much influence does it have on the outcome?

- More complete integration of probabilistic approaches in all aspects of engineering

- Continue unified approaches to the problem

- More systematic development of databases, e.g., national database on dams

- Need to understand the dominant impact of organizational behavior on risk-based decision making

- More research on the reliability of water distribution systems

- Develop theoretical foundations that relate Bayesian and fuzzy set thinking

- Combine existing knowledge (data) to probabilistic modeling for risk analysis

- Risk analysis of extremes

- Environmental risk analysis

- Hands-on workshop sessions to go through case study applications of risk

- Continue theory presentations, but mix in more application

- Look at health/ecological risk in both surface and ground water issues

- What role does pollution prevention play?

- There should be a more subjective analysis of global climate change; we obviously can disagree on the validity of the models and resulting probabilities, but we should be able to agree on a methodology and let the alternative data sets drive the results

- How to go from global climate change to regional foci

- Need to spend much more effort in looking at ecosystem/ water quality impacts of climate change

- Need for scenario modeling to test whether and under what conditions climate change will lead to a different decision

- What are incremental and irreversible decisions in long-range water planning?

- Is water distribution system reliability an important issue or not? (Males "nay," Heaney "yea")

PARTICIPANTS' LIST

ENGINEERING FOUNDATION CONFERENCE

RISK-BASED DECISION MAKING IN WATER RESOURCES VI

October 31–November 5, 1993

Santa Barbara, California

Anderson, Elizabeth
Sciences International, Inc.

Bulkley, Jonathan W.
University of Michigan

Buras, Nathan
University of Arizona

Burnham, Michael W.
U.S. Army Corps of Engineers

Chartier, Antoinette
Engineering Foundation

Cone, Steve
U.S. Army Corps of Engineers

Dudley, Norman J.
University of New England

French, Richard H.
Desert Research Institute

Ganoulis, Jacques G.
Aristotle University

Grayman, Walter M.
W. M. Grayman Cons. Eng.

Haan, Charles T.
Oklahoma State University

Haimes, Yacov Y.
University of Virginia

Huff, Dale D.
Oak Ridge Reservation DOE

Lambert, James H.
University of Virginia

Leggett, Mary Ann
U.S. Army Corps of Engineers

Li, Duan
University of Virginia

Males, Richard M.
RMM Technical Services

Mays, Larry
Arizona State University

Moser, David A.
U.S. Army Corps of Engineers

Mosher, Reed
U.S. Army Corps of Engineers

O'Connor, Robert E.
Pennsylvania State University

O'Grady, Kevin
Planning & Management Consultants

Paté-Cornell, Elisabeth
Stanford University

Quimpo, Rafael
University of Pittsburgh

Ratick, Samuel J.
Clark University

Rowe, William
Automation Research Systems

Schwing, Richard C.
General Motors Research Labs

Shabman, Leonard A.
Virginia Polytechnic Institute

Stakhiv, Eugene Z.
U.S. Army Corps of Engineers

Thompson, Anthony
Perkins Coie

Walsh, Michael
Institute for Water Resources

Yoe, Charles E.
College of Notre Dame

ENGINEERING FOUNDATION CONFERENCE
on
RISK-BASED DECISION MAKING VI

Santa Barbara, California

October 31–November 5, 1993

Co-sponsor:

The Universities Council on Water Resources (UCOWR)

Supported by:
National Science Foundation
U.S. Army Corps of Engineers

Conference Chairman:	Yacov Y. Haimes, University of Virginia
Conference Co-Chairman:	David Moser, U.S. Army Corps of Engineers, Institute for Water Resources

Proceeedings Editors
Yacov Y. Haimes, David Moser, and Eugene Stakhiv

Sunday, October 31, 1993

3:00 pm–9:00 pm	REGISTRATION
6:00 pm	DINNER
7:00 pm	INTRODUCTION TO THE CONFERENCE
8:00 pm–10:00 pm	SOCIAL HOUR

Monday, November 1, 1993

7:30 am BREAKFAST

9:00 am–Noon **Session I**

Federal Risk Management Policy: Where is the Federal Government Heading? Where are the Problems?

Opening Remarks
Yacov Y. Haimes, University of Virginia
David Moser, U.S. Army Corps of Engineers

Chair: David Moser, U.S. Army Corps of Engineers
Rapporteur: Jonathan Bulkley, University of Michigan

Anthony Thompson, Perkins Coie
"Federal Risk Management Policy: Where are the Problems?"

Yacov Y. Haimes, University of Virginia
"When and How Can You Specify a Probability Distribution When You Don't Know Much?"

Noon	LUNCH
2:00 pm–5:00 pm	AD HOC SESSION
5:00 pm–6:00 pm	SOCIAL HOUR
6:00 pm	DINNER

<u>Monday, November 1, 1993</u>

7:30 pm–10 pm	<u>Session II</u> Risk and Reliability of Water Resources Infrastructure and Engineering Approaches to Risk and Reliability Analysis Chair: George Apostolakis, UCLA Rapporteur: Richard Males Larry Mays, Arizona State University **"Methods for Risk and Reliability Analysis"** Raphael Quimpo, University of Pittsburgh **"Reliability Analysis of Water Distribution Systems"** Tom Dolezal, Yacov Y. Haimes, and Duan Li, University of Virginia **"Water Infrastructure Risk Ranking and Filtering Method"** Mary Ann Leggett, US Army Corps of Engineers **"Reliability-Based Assessment of Corps Structures"**
10 pm–11 pm	SOCIAL HOUR

<u>Tuesday, November 2, 1993</u>

7:30 am	BREAKFAST
9:00 am–Noon	<u>Session III</u> Behavioral, Social, and Institutional Aspects of Risk Analysis Chair: David Moser, U.S. Army Corps of Engineers Rapporteur: Dick Schwing, General Motors George Apostolakis, University of California **"A Theoretical Framework for Risk Assessment"** Jonathan Bulkley, University of Michigan **"Risk Analysis: Wet Weather Flows in S.E. Michigan"** Richard Schwing, General Motors **"Poorer is Riskier: Opportunity for Change"**

Len Shabman, Virginia Polytechnic Institute
"Measuring the Benefits of Flood Risk Reduction"

Noon LUNCH

2:00 pm–5:00 pm AD HOC SESSION

5:00 pm–6:00 pm SOCIAL HOUR

6:00 pm DINNER

Tuesday, November 2, 1993

7:30 pm–10 pm **Session IV**
Risk Management Strategies for Natural, Manmade, and Technological Hazards
Chair: Elizabeth Anderson, Sciences International, Inc.
Rapporteur: Steven R. Cone, U.S. Army Corps of Engineers

Elizabeth Anderson, Sciences International, Inc.
"Risk and Reliability: Health and Ecological Risk Assessment"

Kevin O'Grady, Planning & Mgmt. Consultants, Ltd.
"Uncertainty and Time Preference in Shore Protection"

Robert O'Connor, Pennsylvania State University
"Public Perceptions of Fresh Water Issues"

Sam Ratick, Clark University
"Reliability-Based Dynamic Dredging Decision Model"

10 pm–11 pm SOCIAL HOUR

Wednesday, November 3, 1993

7:30 am BREAKFAST

9 am–Noon

Session V
Risk of Extreme and Catastrophic Events
Chair: Elisabeth Paté-Cornell, Stanford University
Rapporteur: Jim Lambert, University of Virginia

Elisabeth Paté-Cornell, Stanford University
"Subjective De-Biasing of Data Sets: A Bayesian Approach"

Stan Kaplan, PLG, Inc.
"Bayes' Theorem and Quantitative Risk Assessment"

Mike Burnham, U.S. Army Corps of Engineers
"Risk-Based Analysis for Flood Damage Reduction"

David Moser, U.S. Army Corps of Engineers
"Quantifying Flood Damage Uncertainty"

Noon — LUNCH

2 pm–5 pm — AD HOC SESSION

5 pm–6 pm — SOCIAL HOUR

6 pm — DINNER

Wednesday, November 3, 1993

7:30 pm–10 pm

Session VI
Uncertainties in Data, Models, Forecasts and Their Influence on Risk Analysis
Chair: Yacov Y. Haimes, University of Virginia
Rapporteur: Ben Hobbs, Case Western Reserve University

Jim Heaney, University of Colorado
"Reliability Considerations in Water Supply Planning, Design, and Operation"

Duan Li, University of Virginia
"Flood Loss Assessment with Integrated Measures"

	Michael R. Walsh and David A. Moser, U.S. Army Corps of Engineers **"Risk-Based Budgeting for Maintenance Dredging"**
10 pm–11 pm	SOCIAL HOUR

<u>Thursday, November 4, 1993</u>

7:30 am	BREAKFAST
9:00 am–Noon	<u>Session VII</u> Computer Software Risk Chair, Jim Heaney, University of Colorado Rapporteur: Mike Walsh, IWR
	Charlie Yoe, College of Notre Dame of Maryland **"Developments in Risk Analysis Commercial Software"**
	Bill Rowe, Rowe Research & Eng. Associates **"Uncertainty versus Computer Response Time"**
	Workshop on Computer Software Risk
	Richard Males, David Moser, Mike Walsh, Walter Grayman and Craig Strus, U.S. Army Corps of Engineers **"Risk Simulation Model for Rehabilitation Studies"**
Noon	LUNCH
2:00 pm–5:00 pm	AD HOC SESSION
5:00 pm–6:00 pm	SOCIAL HOUR
6:00 pm	DINNER

Thursday, November 4, 1993

7:30 pm–10 pm

Session VIII
Climate Change
Chair: Eugene Stakhiv, U.S. Army Corps of Engineers
Rapporteur: Jim Lambert, University of Virginia

Norman Dudley, The University of New England (Australia)
"Environment-Irrigation Trade-Offs and Risks"

Eugene Stakhiv, U.S. Army Corps of Engineers
"Climate Change: Issues, Opportunities and Challenges"

Ben Hobbs, Case Western Reserve University
"Climate Warming and Great Lakes Management"

Jacques Ganoulis, Aristotle University of Thessaloniki (Greece)
"Coastal Eutrophication and Temperature Variation"

10 pm–11 pm SOCIAL HOUR

Friday, November 5, 1993

7:00 am BREAKFAST

8:00 am

Session IX
General Discussion of Themes
Chair: Yacov Y. Haimes, University of Virginia
Co-chair: David Moser, U.S. Army Corps of Engineers

11:00 am LUNCH AND DEPARTURE

Subject Index

Page number refers to first page of paper

Author Index

Page number refers to the first page of paper